FLUIDS & ELECTROLYTES

A CONCEPTUAL APPROACH

KINSEY SMITH, M.D. (Lond.), F.R.C.P., F.R.C.P.(C).
Associate Professor, Department of Medicine,
McMaster University, Faculty of Health Sciences,
Head of Nephrology Service, St. Joseph's Hospital,
Hamilton, Ontario, Canada.

Edited by:
ELIZABETH BRAIN, M.A., B.M., B.Ch.
Learning Resources Editor,
McMaster University, Faculty of Health Sciences,
Hamilton, Ontario, Canada.

Illustrated by:
JANE WILLICH
MOLLY PULVER
RICHARD ZAZULAK
McMaster University, Audio Visual Services,
Hamilton, Ontario, Canada.

Typesetting by:
VICKY BACH,
Hamilton, Ontario, Canada.

Churchill Livingstone
New York, Edinburgh and London 1980

Distributed in the United Kingdom by Churchill Livingstone, Robert Stevenson House, 1–3 Baxter's Place, Leith Walk, Edinburgh EH1 3AF and by associated companies, branches and representatives throughout the world.

First published 1980

Printed in USA

ISBN 0-443-08101-8

7 6 5 4 3

Library of Congress Cataloging in Publication Data

Smith, Kinsey.
Fluids and electrolytes, a conceptual approach.

Bibliography: p.
1. Body Fluids. 2. Water-electrolyte balance (Physiology) I. Brain, Elizabeth.
II. Title. [DNLM: 1. Body fluids. 2. Water—Electrolyte balance. QU105 S435f]
QP90.5.S63 612'.01522 80-11458
ISBN 0-443-08101-8

Preface

This illustrated text was developed primarily for undergraduate medical students at McMaster University. Some of these students do not have any background in the physical or biological sciences, and many of the basic concepts they need for the understanding of body fluid regulation are foreign to them. Moreover, the constraints of a three year undergraduate course emphasize the value of clear introductory resources which link basic concepts to relevant clinical problems.

This book attempts to introduce the idea of a regulated internal environment and its responses in health and disease. The illustrations provide a visual summary, the text a simple and direct description of the illustrated concepts. It does not pretend to be comprehensive and the student should *not* use it to replace more detailed and scholarly resources. The method of presentation is not conventional, but the underlying concepts are well established and lean heavily upon the influence of other authors; the references provided at the end of the book represent only a small selection of useful resources and do no more than point the way to wider reading. The lucid writings of the late L.G. Welt and R.F. Pitts have been especially valuable to me and I make no apology for the major influence they have had in the development of the concepts that are illustrated here. Many ideas for visual presentation have come from the enthusiasm of Dr. Elizabeth Brain and members of the Audio Visual Department, and thanks are due to those colleagues amongst students and faculty who have provided stimulation and advice over the past decade.

This text is based in part upon material that has been used in a slide/tape format for a number of years. It is now offered to a wider readership in the hope that it may prove useful in other medical, nursing and paramedical programmes and perhaps as a reminder for more advanced students.

Kinsey Smith

Chapter 1

The Cellular Environment

The Cell...

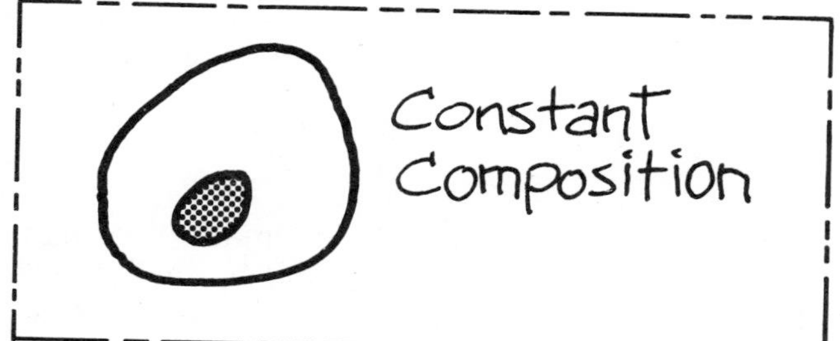

The unit of biological function is the single cell which requires a defined composition to function normally. It must be capable of acquiring those vital nutrients which may become depleted, and of rejecting those end products which are not required.

...and its Surroundings

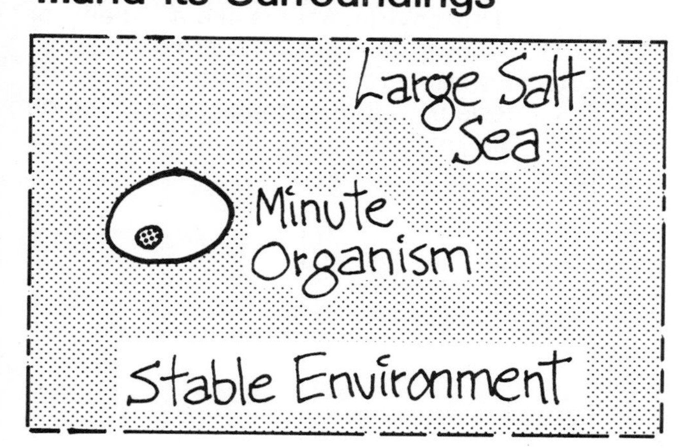

Simple, single-celled organisms evolving in an almost infinitely large salt sea may be considered as having an unvarying external environment. In this setting, regulation of their volume and composition is relatively simple. Surrounding osmotic pressures do not change appreciably and the concentration of solutes around them is constant. Waste products can leave the cell and become infinitely diluted in the surrounding fluid.

Small Freshwater Pond

Dilution

Rupture

Contractile Vacuole

A single-celled organism (such as an ameba) living in a small fresh water pond faces a more hostile environment. A vigorous rain storm may dilute the already solute-poor water in the pond and produce increasing osmotic pressure gradients which cause water to enter the cell which will swell up to the point of bursting. Adaptations which prevent this happening include the evolution of such devices as "contractile vacuoles" which can eject unwelcome volume.

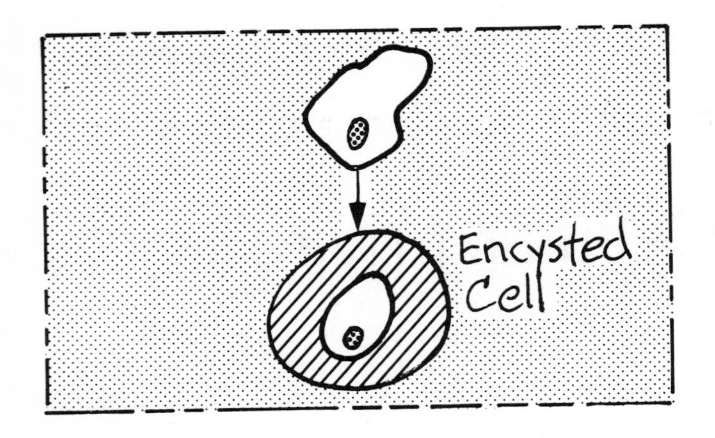

If the pond dries up, the reverse process occurs and the cell becomes shrivelled and dehydrated. The protection such an organism can provide for itself is to become encysted, surrounding itself with a ''waterproof coat'' which will allow it to withstand desiccation and become reactivated when conditions improve. The price paid for this survival is to undergo periods of immobility and inactivity.

"Milieu Interieur"

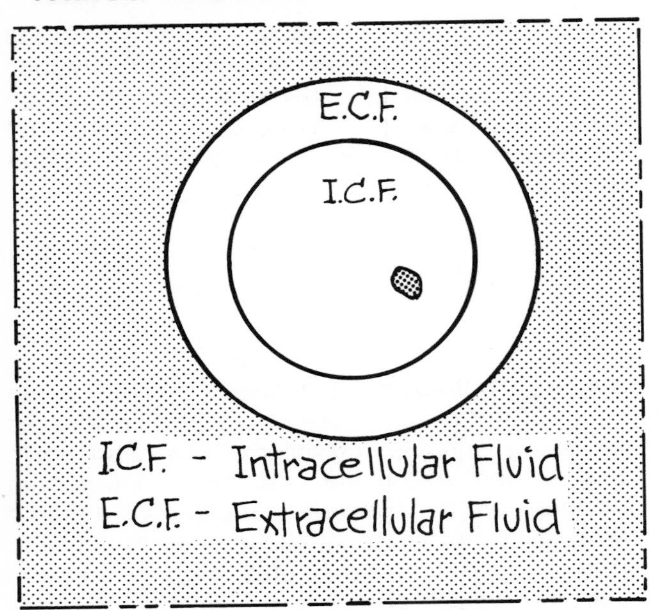

Warm blooded multicellular animals, like man, have adapted to life on dry land by taking with them a ''personal environment'', the composition and volume of which are precisely regulated even in the face of widely varying external conditions. This environment is like a ''shell'' of fluid which allows the cells to ignore changes in the outside world. This insulating shell is the ''milieu interieur'', first recognized by Claude Bernard. The internal environment is provided by the extracellular fluids.

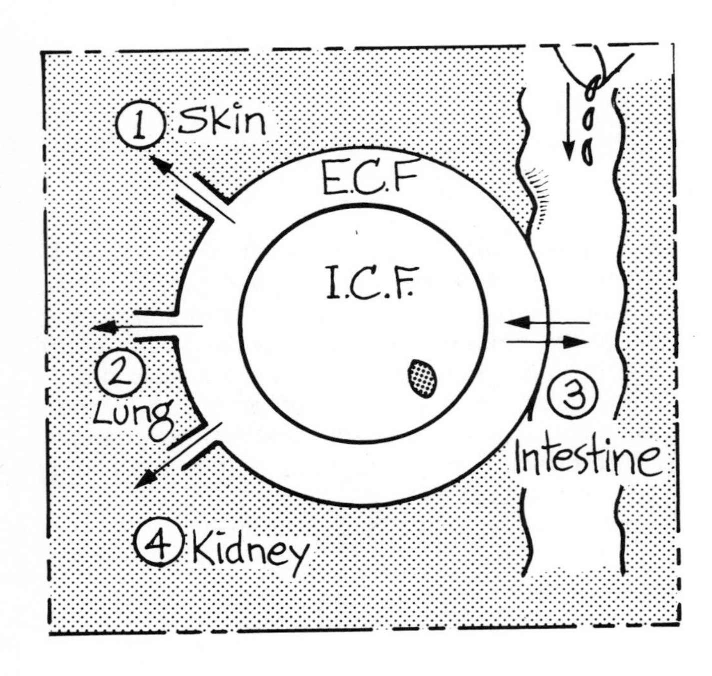

The internal environment is provided by the extracellular fluids. The only natural entry point to this fluid ''compartment'' is via the mouth and intestinal tract. There are four exit routes from the extracellular fluid space:

1) The skin, where loss occurs by perspiration and is related to temperature and humidity.
2) The lung, where loss is related to the need to moisten the air we breathe.
3) The intestine, where diarrhea or vomiting may cause abnormal losses.
4) The kidney.

Fluid Loss from the E.C.F.

Losses via the skin and the lung are dependent upon external factors which cannot be controlled. Thus men climbing high mountains in cold, dry air will have enormous water losses from the lungs as hypoxia demands increased ventilation; men suddenly transported to a hot climate may sweat unmanageably.

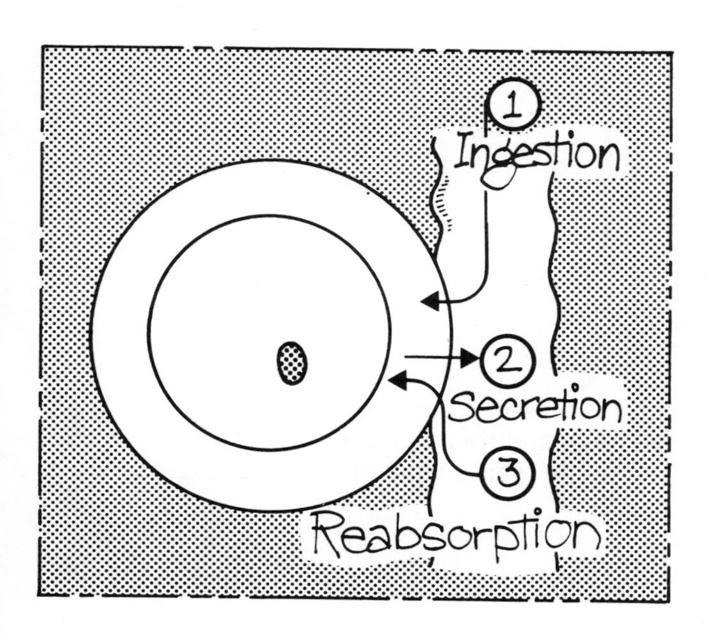

Intestinal losses are not ''physiological'' (as are losses from skin or lung), and are related to abnormal states where vomiting or diarrhea occur. The normal intestine absorbs fluids that are ingested (1); secretes large amounts of digestive fluids (2); and then reabsorbs virtually all that it secretes (3). The fluids in the intestine are separate from E.C.F., but because of the large surface area of the intestinal mucosa and its close contact with the E.C.F. compartment, potential losses from the E.C.F. into an abnormal intestine can be large. If mucosal function is disturbed the intestinal wall can cease to be a barrier to the E.C.F. with the result that diarrheal losses or losses into a dilated bowel, as in paralytic ileus, represent direct losses from the E.C.F.

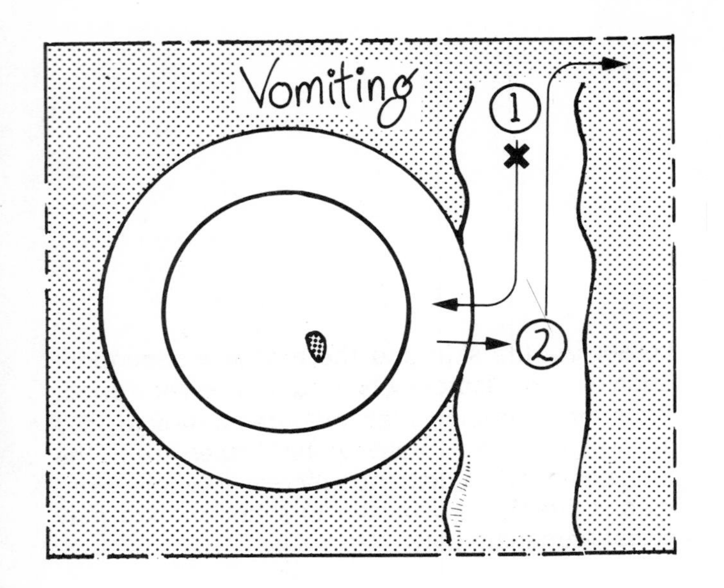

Vomiting causes fluid loss in two ways:

1) By failure to allow intake of normal fluid volumes,
2) By the ejection of fluids secreted into the upper intestine.

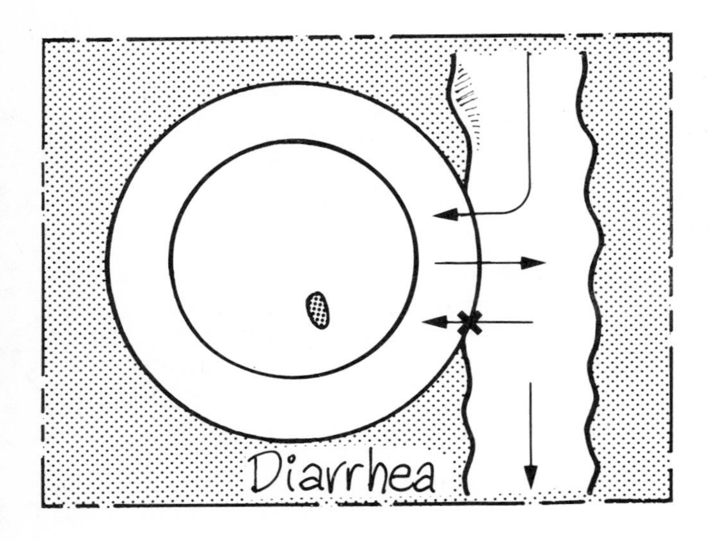

Diarrhea involves a failure to reabsorb the fluids secreted into the intestine and often a change in permeability of the mucosa. This allows the intestine to become the site of what, in effect, are direct losses of E.C.F. Thus most intestinal losses (either up or down) tend to be isotonic with respect to the E.C.F.

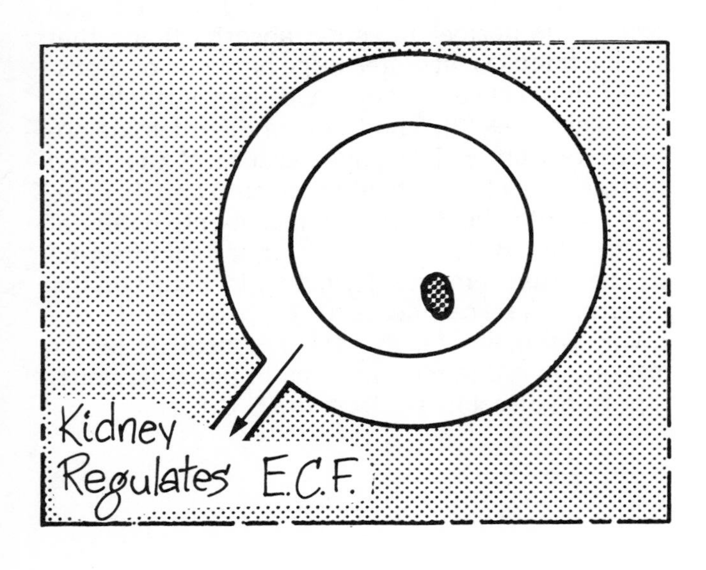

Only the losses from the kidney can be regulated precisely and this organ therefore plays a crucial part in the regulation of the E.C.F. volume.

Subdivision of E.C.F.

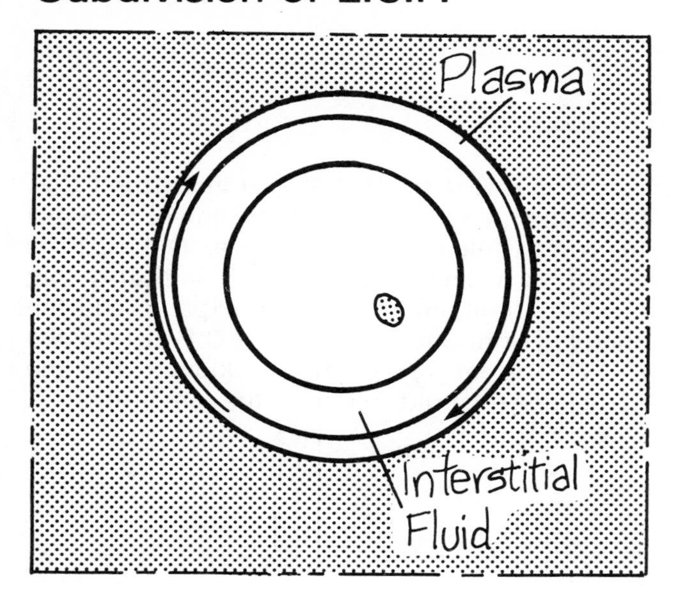

The E.C.F. is a little more complicated than we have indicated so far since it can be subdivided into two compartments. One, the larger part, contains the interstitial fluid which surrounds the cells. The second contains the plasma which, of course, circulates throughout the body and can therefore act as a "bulk transporter" of water and solutes in the E.C.F. It can also serve as a route for the even and rapid distribution of very small quantities of substances such as hormones and drugs.

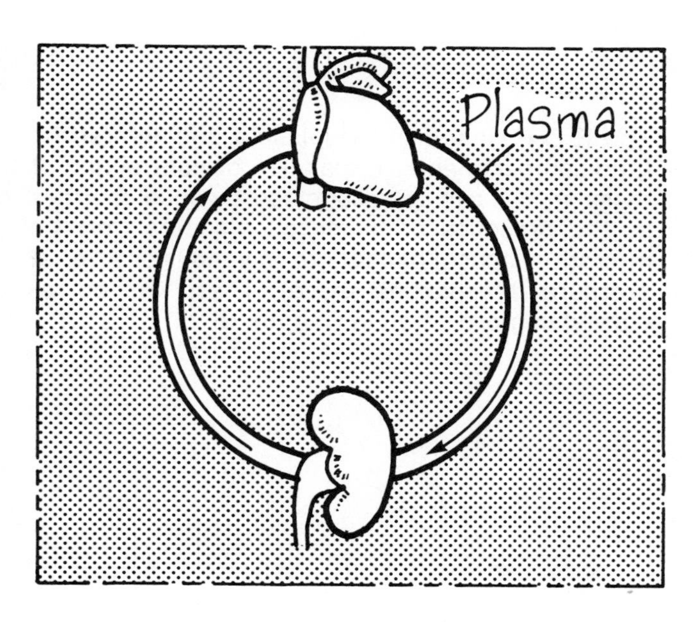

In some diagrams we will show the plasma compartment in this way, which indicates the importance of the heart as a pump for the plasma and the kidney as a sensor and regulator of its volume.

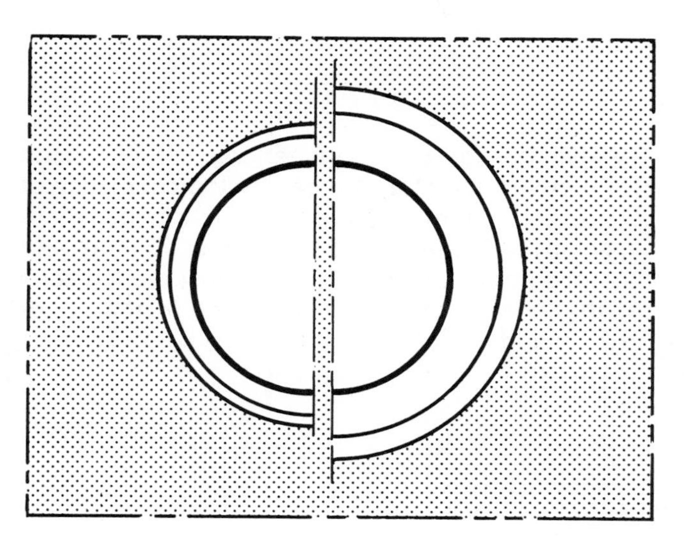

In normal health, changes in the volume and composition of the E.C.F. are within narrow limits. In most clinical situations the plasma and interstitial volumes vary in the same direction although sometimes they may change in opposite directions. Significant change in the E.C.F. can occur without any appreciable change in the I.C.F. because the E.C.F. provides a "protective shell".

The Tank Model

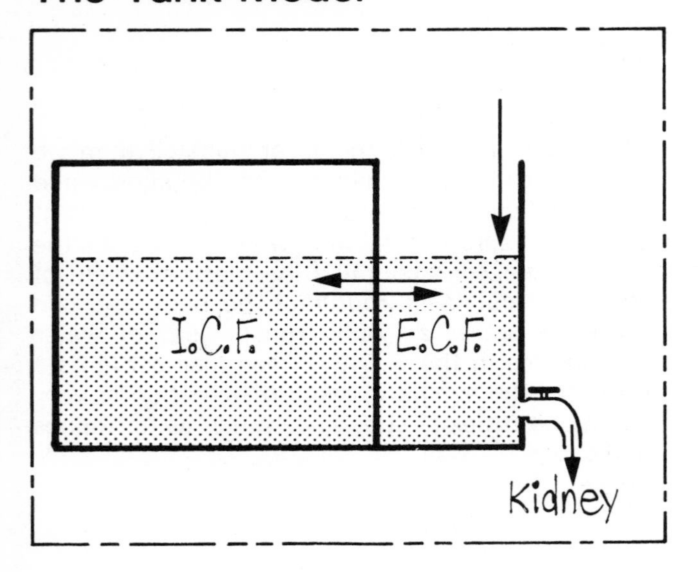

Whilst in reality the E.C.F. surrounds the I.C.F. like a shell, it is often more convenient to depict the compartments as shown here. The I.C.F. is drawn as a closed space which indicates that there are limits to its ability to expand, and the only way into and out of the I.C.F. is via the E.C.F. The tap represents the kidney, and the dotted line the fluid volume in the compartments.

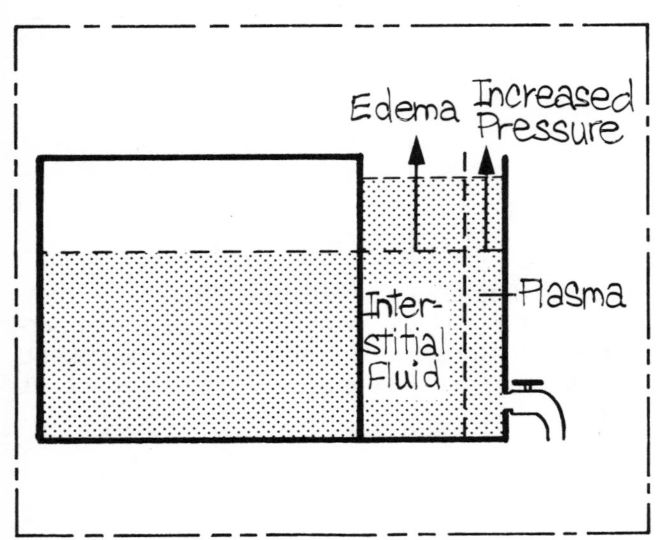

The E.C.F. includes the plasma and interstitial fluid compartments. Expansion or contraction of these spaces can be recognized by such changes as the presence or absence of edema (for the interstitial space) or changes in venous or arterial pressure (the plasma space). These will be reviewed in more detail in Chapter 3.

Cells as Regulators

Whilst the body cells in general are protected from environmental changes by the constancy of the E.C.F., some of the cells in the body are specially adapted to regulate the volume and composition of the E.C.F. These include the cells of the kidney, the lung, and some of the endocrine glands.

Purpose of this Book

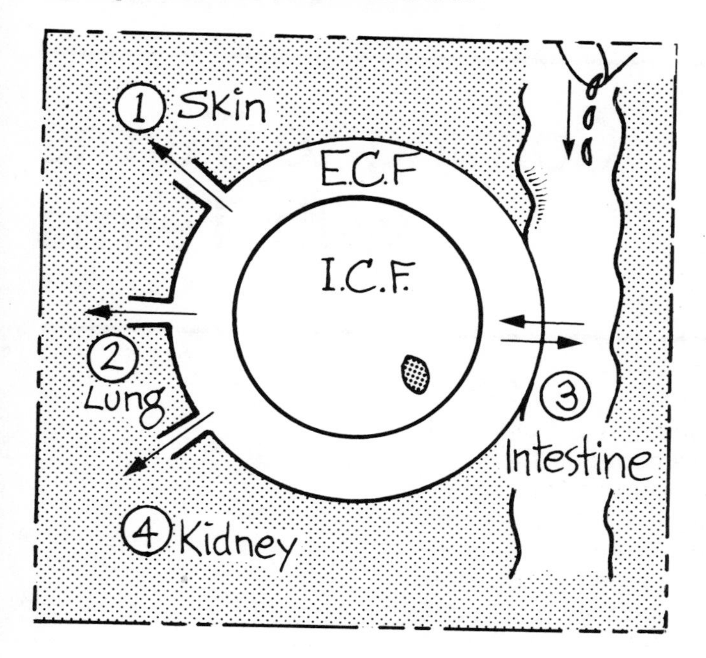

This book is about the body fluid compartments and the principles governing the regulation of both their volume and composition. It is concerned with *concepts,* and not with the details of distribution of every solute in every body fluid compartment.

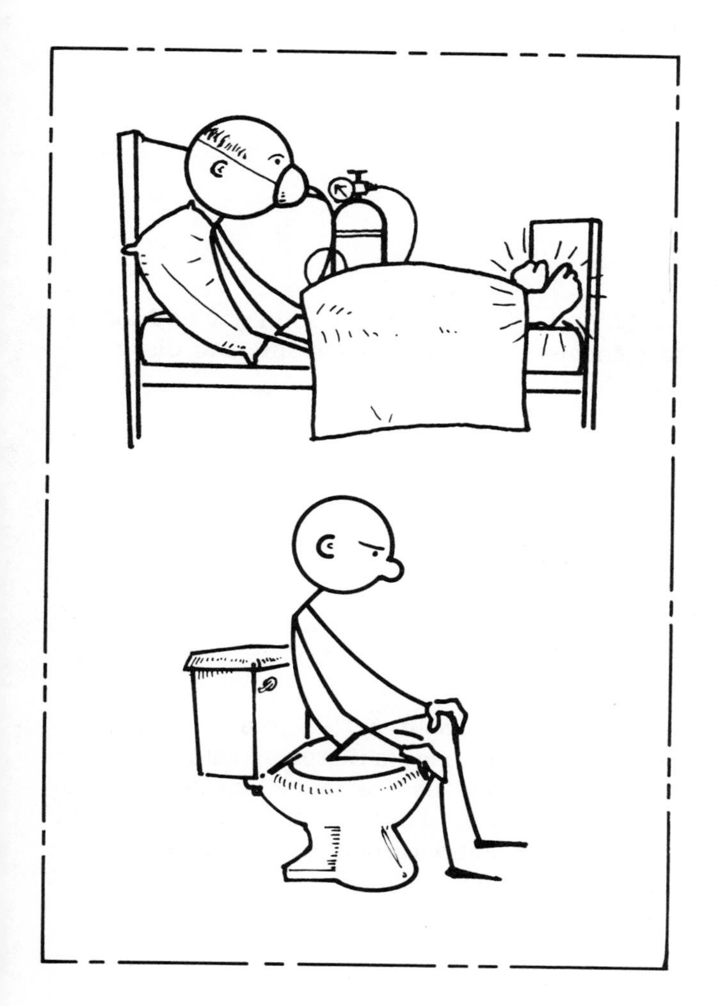

These concepts are fundamental to the understanding of the practical management of disorders of the body fluids. They explain such apparently different situations as heart failure and cholera.

They can be applied to the understanding of such things as diuretics, diabetes and drug distribution. They are, quite simply, basic building blocks for good medicine.

Chapter 2

Coming to Terms with Terms

Before going any further into the distribution and regulation of fluids and electrolytes it is important to understand their measurement and the units by which they are expressed. The problem here is that changes are occurring as the S.I. system of units is being introduced. "S.I." stands for Système International des Unités which attempts to relate all units of measurement to seven "base units." Where such units, together with the units derived from them, are applicable to clinical science they are being introduced as a rational advance. The rate at which these units are being introduced and accepted in clinical practice is more rapid in Europe than North America and the completeness with which they will be accepted is not yet clear. For these reasons the terms used in clinical medicine are not "pure" and for the time being this is likely to remain so. This chapter will attempt to clarify the concepts of current usage, with the understanding that the S.I. system is likely to become more widely used as time passes.

Body Fluids

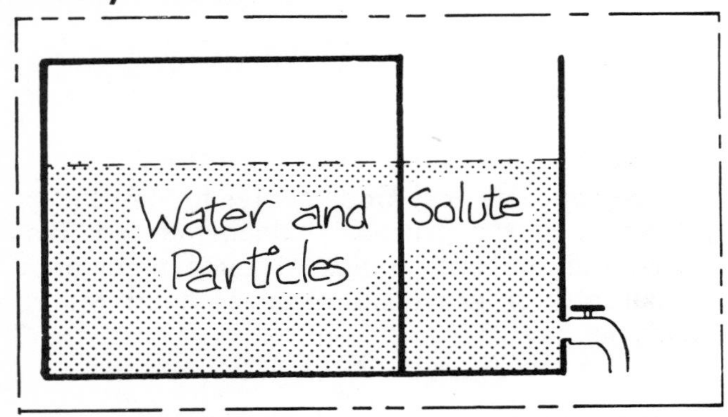

All body fluids are aqueous solutions and therefore consist of water (the solvent) and dissolved particles (the solutes).

Water - The Solvent

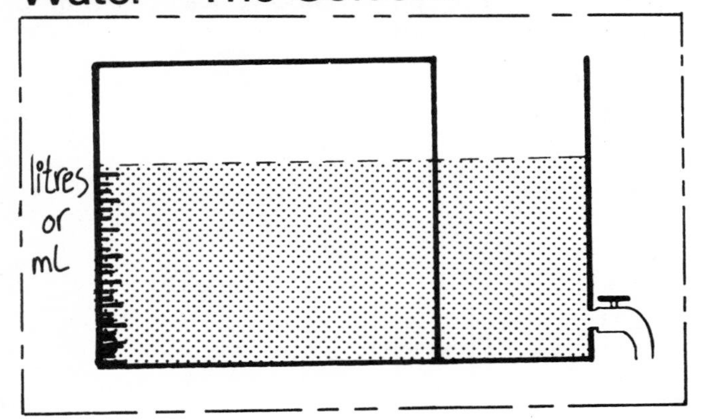

When we talk about a "volume" we mean a volume of water and its dissolved solutes. Water can be regarded as the major solvent for biological systems. The volume occupying any "space" or "compartment" is measured in litres (L) or millilitres (mL).

Solutes

The solutes are minute particles which may be molecules or fragments of molecules.

Molecules

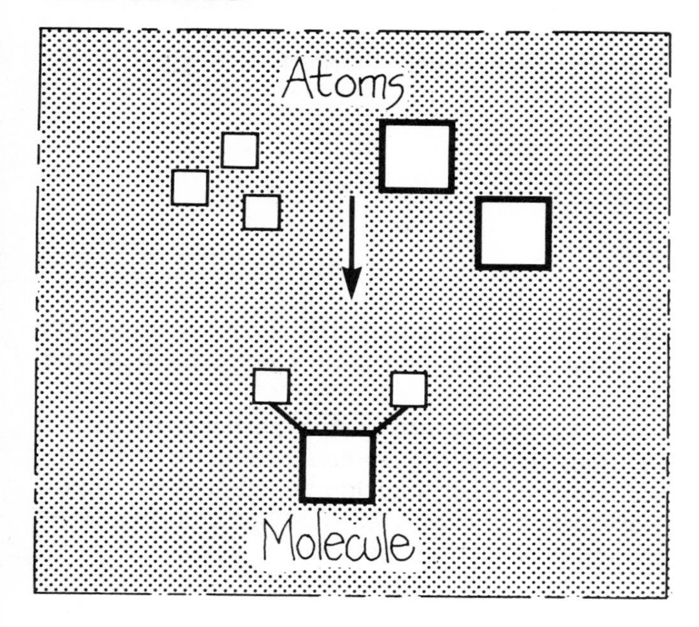

A molecule is the smallest particle into which a compound may be divided whilst still retaining its chemical identity. Molecules are made up of collections of *atoms* which are electrically neutral with a balance between positive charges in the nucleus and negatively charged electrons in the outer shell.

Ions

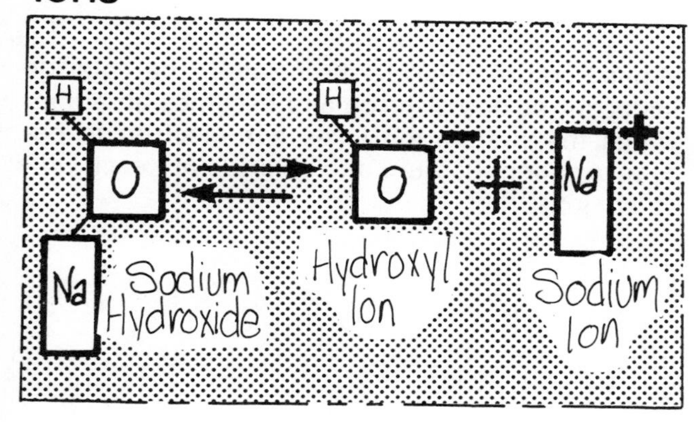

When molecules dissolve in water they may dissociate to a greater or lesser extent into their component parts which then carry electrical charges and are called *ions*. These ions may contain part of a single atom (such as sodium) or more than one atom (as in the hydroxyl ion).

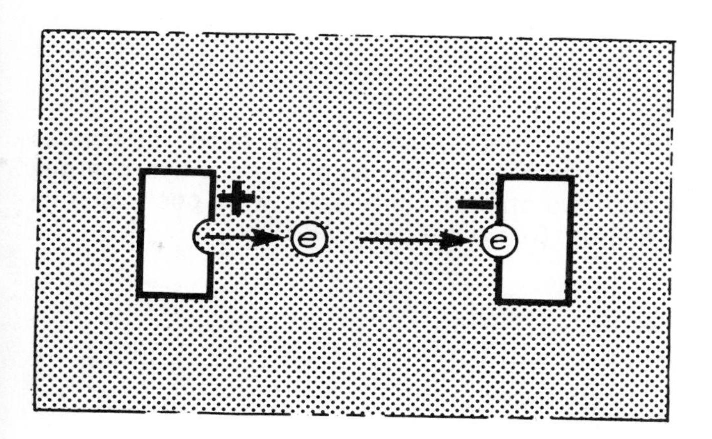

Positively charged ions have given up an electron, and negatively charged ions have gained an electron.

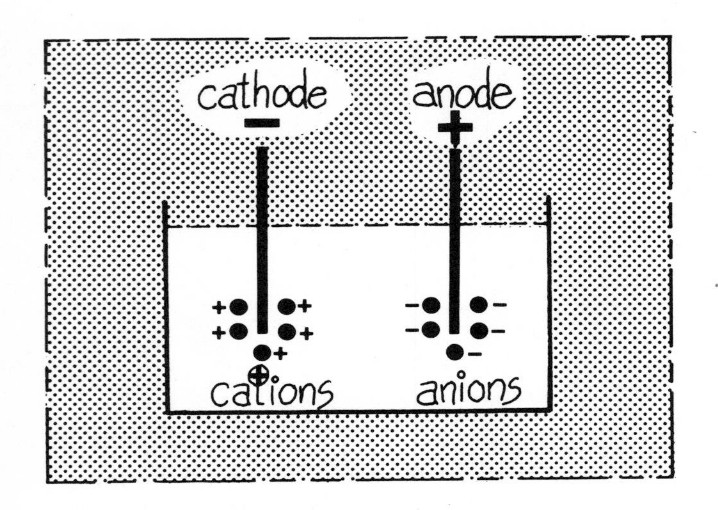

If electrodes are placed in a solution the positively charged ions migrate to the cathode and are called *cations*. Negatively charged ions migrate to the anode and are called *anions*.

Cations are positively charged.
Anions are negatively charged.

Electrolytes

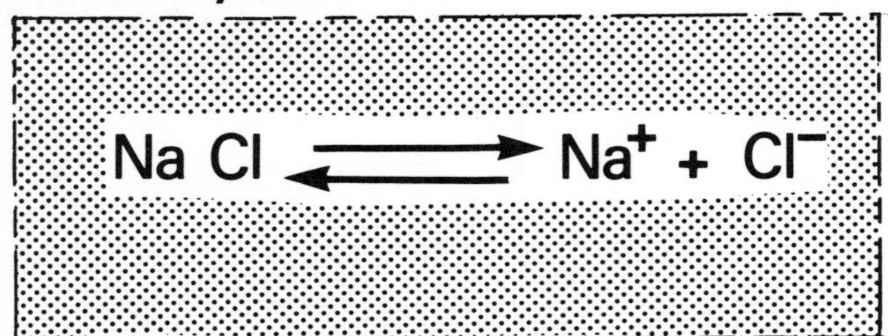

The interaction between the constituent ions of a substance in solution is an electrochemical one.

The charged nature of the particles in such a solution allows it to act as a conductor of electrical current. Such ionizing substances are therefore called *electrolytes*. Sodium chloride is an example of such a substance.

Valency

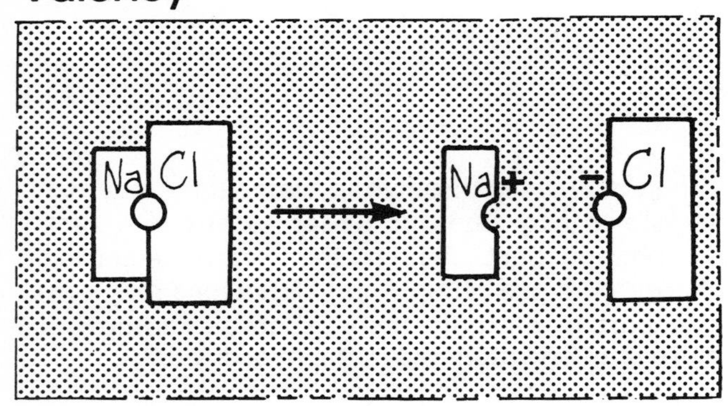

When sodium chloride dissociates, the chloride ion acquires one electron and becomes an anion, whilst the sodium ion loses *one* electron to become a cation. Sodium and chloride are each *univalent* ions.

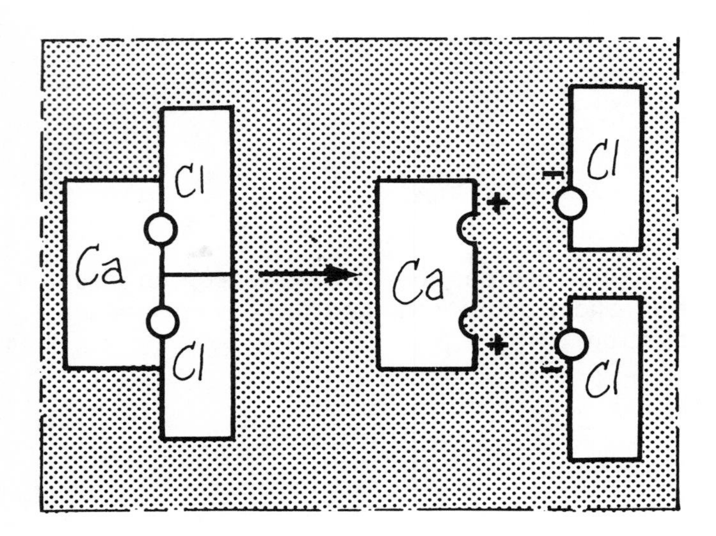

Calcium chloride, on the other hand, dissociates into one calcium ion with two positive charges and two chloride ions each with one negative charge. Calcium is a *divalent* ion.

Atomic and Molecular Weights

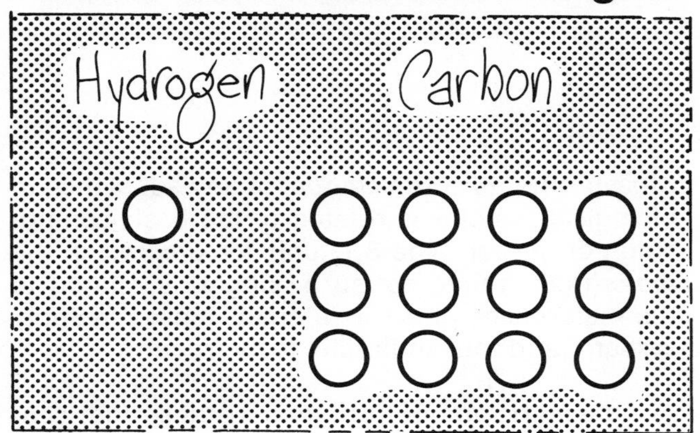

The atoms which make up a substance can be related to one another in terms of mass (*the atomic weight*). If the atomic weight of *hydrogen* is taken to be unity, then *carbon* has an atomic weight of 12, indicating that it is twelve times as heavy as hydrogen. Atomic weights are therefore *relative* masses to a reference atom and not actual weights.

Although the ''reference atom'' is, historically, hydrogen, *carbon,* with an atomic weight of 12, is now used as the absolute standard against which other atomic weights are compared.

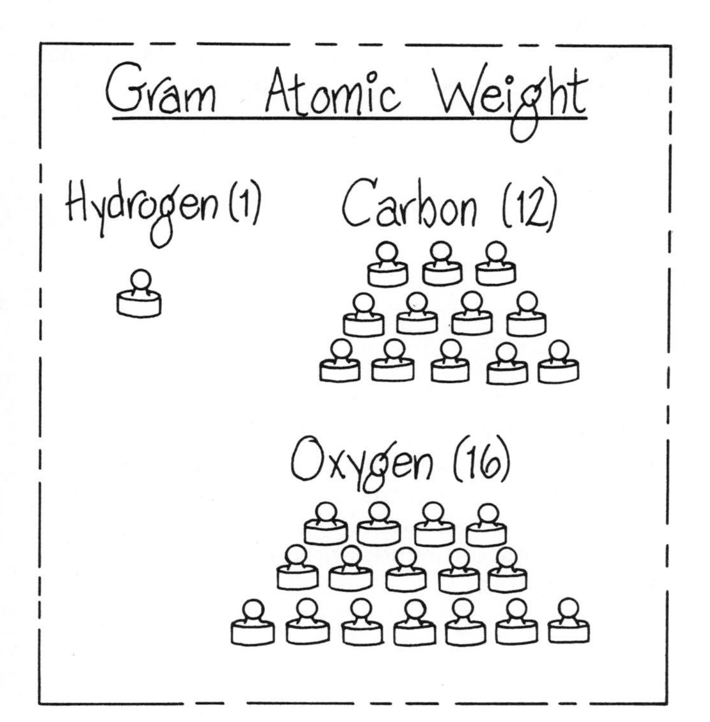

Atomic weights can be expressed in any unit of mass, but most commonly it is the gram. Thus the *gram atomic weight* of hydrogen is 1 gram, and of carbon 12 grams, but since atomic weights are relative to one another *any* unit of mass can be used for comparison so long as the two atoms being compared are expressed in the same unit.

Whatever unit of mass is used, an atom of oxygen will always be sixteen times as heavy as hydrogen, and one and a third times the mass of a carbon atom.

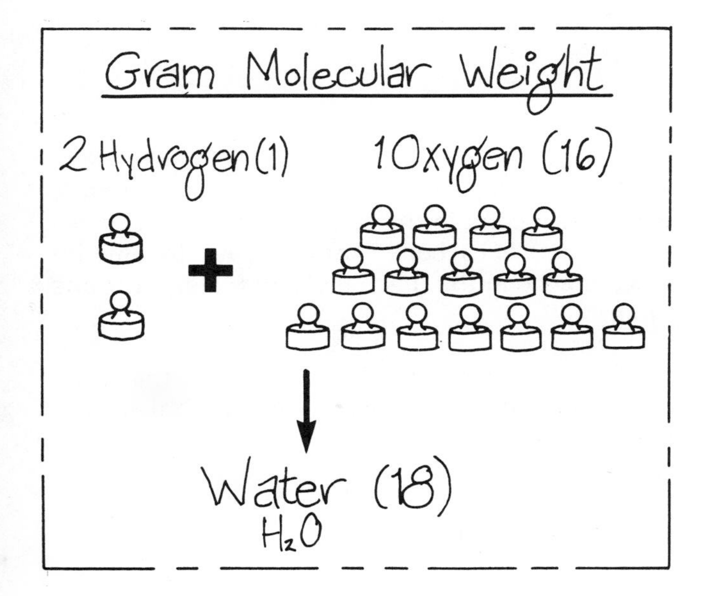

Since molecules are made up of collections of atoms their mass can be compared one to another as the *molecular weight,* which is the sum of the atomic weights of the constituent atoms. Just as atomic weights are relative to an absolute standard (hydrogen or carbon), so molecular weights are related to the *same* standard since they are made up of collections of atoms.

Thus a *molecule* of water has eighteen times the mass of an *atom* of hydrogen and one and one-half the mass of a carbon *atom*.

Mole - Amount of Substance

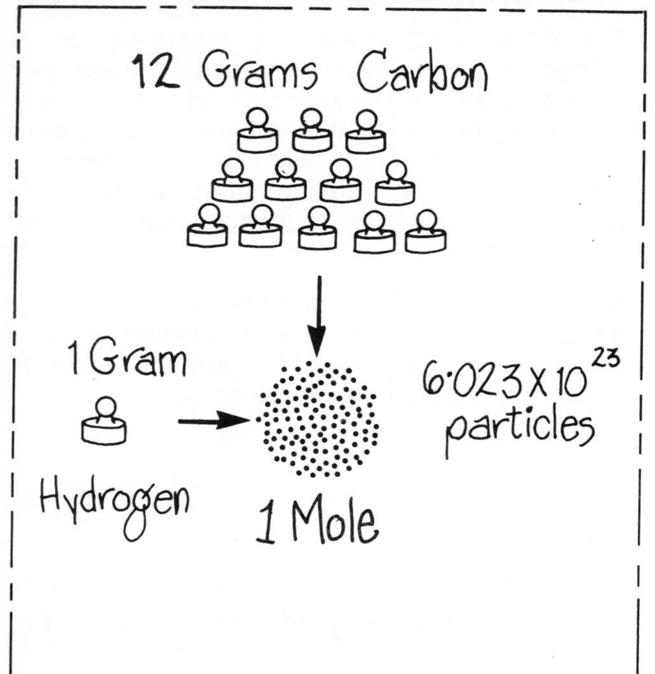

Measurements of the *amount* of any substance in biological systems is related to the basic composition of matter. The S.I. unit for *amount* is the *mole* (mol) which, not surprisingly, is related to the reference unit of atomic mass, namely carbon, (and thus indirectly to hydrogen).

A *mole* is the amount of any substance that contains the same *number* of elementary particles as there are atoms in 12 grams of carbon. (For the purist, this number is 6.023×10^{23} particles and is referred to as Avogadro's number.)

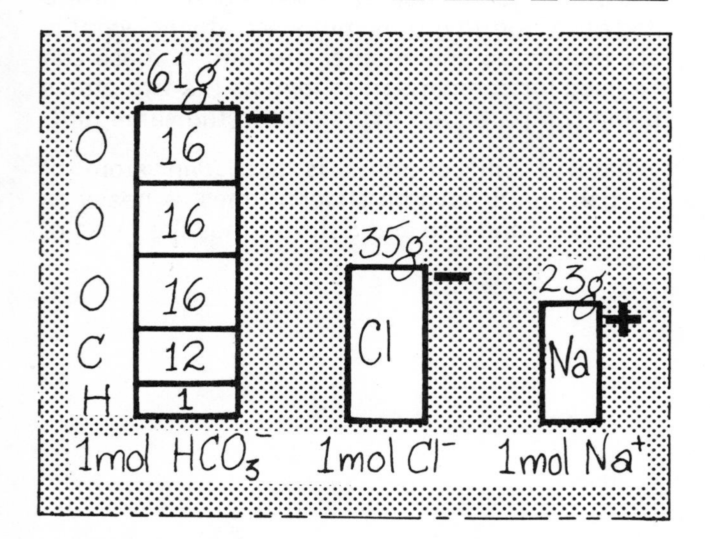

The particles may be atoms or molecules or ions, *all* of which can be measured in *moles*. Thus, 61 grams of bicarbonate ions will contain the same number of particles as 35 grams of chloride ions, 23 grams of sodium ions and (of course) 12 grams of carbon atoms.

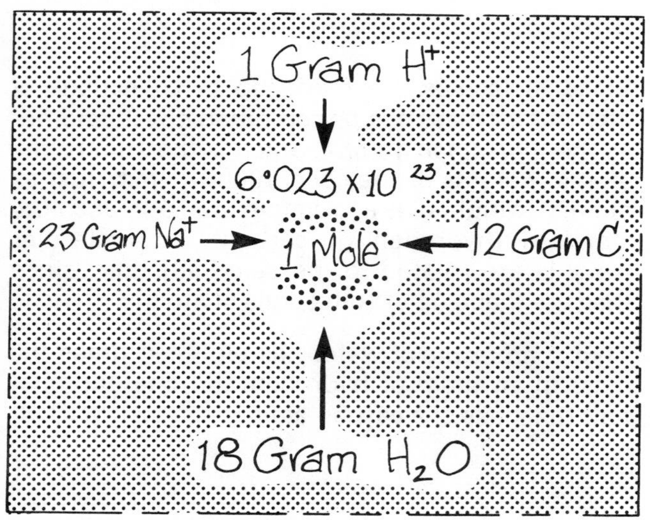

In practical terms, the gram molecular or atomic weight will contain the same number of particles as 12 grams of carbon. That is to say, the amount of substance in the gram molecular weight is one *mole*.

Similarly the amount of substance in the gram atomic weight or the gram ionic weight is also one mole.

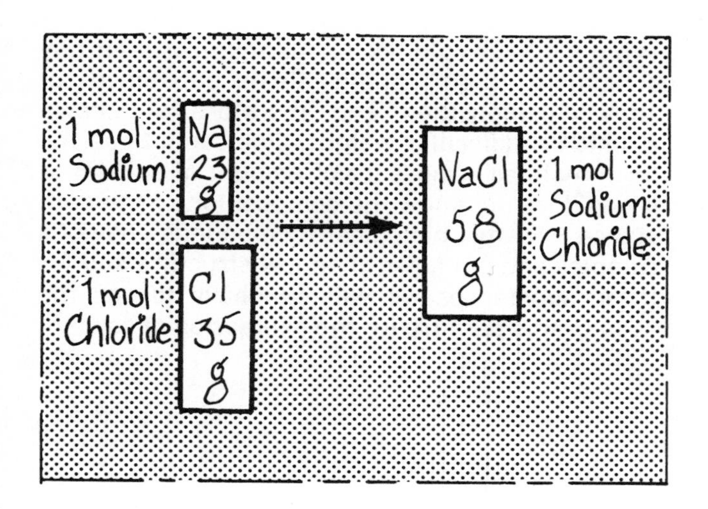

For example, one mole of sodium chloride is contained in 58 grams (the gram molecular weight) which is the sum of the atomic weights of sodium (23) and chloride (35).

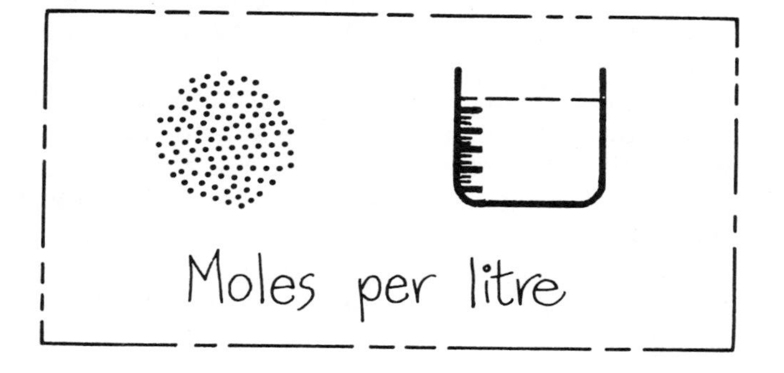

When substances go into solution their concentration is expressed as *amount per volume* and the units used are, of course, moles per litre (mol/L).

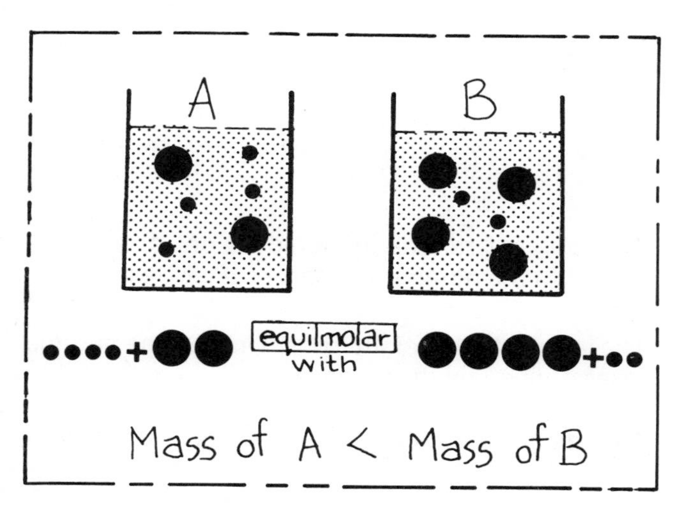

Because a mole is a number of particles rather than a mass of substance, solutions of *equal molar concentration* do not necessarily contain the same *weight* of solute. The two solutions shown in this diagram are equimolar because they contain the same number of particles, but it is apparent that the heavier particles predominate in solution B and so the mass of solute in solution B is greater than the mass of solute in solution A.

mole (mol)
millimole (mmol) 10^{-3} mol
micromole (μmol) 10^{-6} mol
nanomole (nmol) 10^{-9} mol
picomole (pmol) 10^{-12} mol

In biological systems amounts of substances are often fractions of moles, and they are commonly expressed as thousandths (10^{-3}) of a mole (millimoles or mmol), 10^{-6} mole (micromoles or μmol), 10^{-9} mole (nanomoles or nmol), or occasionally 10^{-12} mole (picomoles or pmol). In solution their concentrations will be expressed as fractions of a mole per litre (e.g. mmol/L).

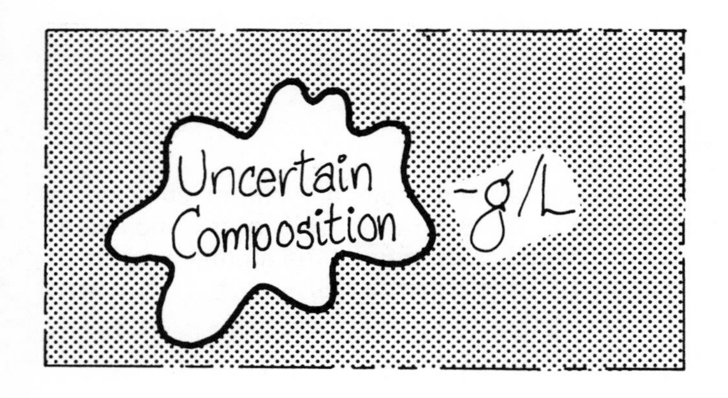

What we have said so far, although true of any substance, is only applicable to substances of known composition and therefore of known atomic or molecular weight. One cannot use the term mole as a measure of amount of substance if one does not know the composition of the substance. Many biological compounds have a complex and unknown chemical structure and for these substances we will still have to use the term grams per litre for the expression of concentration.

Equivalents

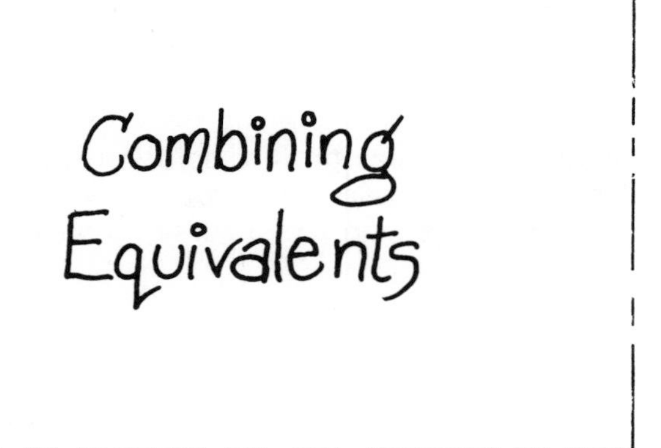

The concept of ''combining equivalents'' has been widely used in modern clinical practice. With the introduction of S.I. units the term ''equivalent'' will be replaced by mole (expressed as the gram ionic weight of a substance). You will still hear the term ''equivalent'' used because of its value in linking mass and charge in reactions between ions; although it is not an S.I. unit, it is so widely used and conceptually useful that it is likely to continue in use for the time being.

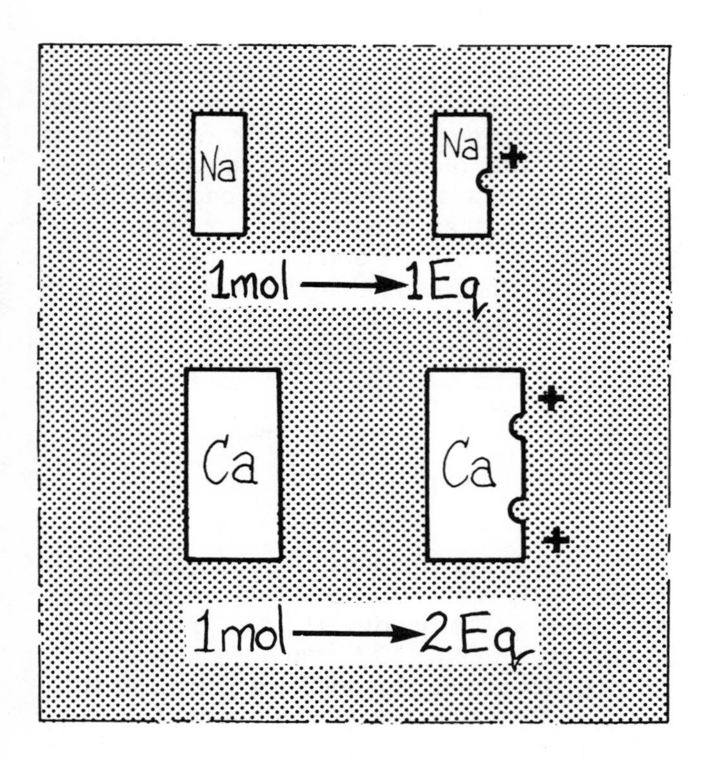

This concept relates the ionic weight of an electrolyte to the number of charges it carries.

The equivalent weight of an ion is the atomic mass divided by the valence.

This means that for a univalent ion like sodium

1 mole contains 1 equivalent.

For a divalent ion such as calcium

1 mole contains 2 equivalents.

S.I. Units in Summary

In the S.I. system of units, the concentrations of substances of known and exact composition will be expressed as moles per litre (mol/L).

Substances of ill defined composition, such as proteins, will continue to be expressed by weight and their concentration as weight/L. So the terms gram/litre (g/L) and milligrams/litre (mg/L) will be used. Terms such as ''mg%'' and ''mg/100 ml'' are more accurately stated as ''mg/dL'', where ''dL'' is decilitre.

This means that we will have to become accustomed to changes in the units for a number of substances, although some that we already express in molar terms will not alter.

Plasma Concentration	OLD	NEW
Calcium	10mg/dl	2·5mmol/L
Magnesium	1·0mEq/L	0·5mmol/L
Sodium	140mEq/L	140mmol/L
Chloride	100mEq/L	100mmol/L

Amongst the electrolytes, sodium and chloride will not change.

Calcium and magnesium, whch are now often expressed as mg/dl, or mEq/L, will be expressed in molar terms (mmol/L).

Plasma Concentration	OLD	NEW
Bilirubins	1mg/dl	18µmol/L
Urea Nitrogen	10mg/dl	3·6mmol/L
Creatinine	1mg/dl	88µmol/L
Glucose	100mg/dl	5·5mmol/L

Pure substances other than electrolytes will also change to molar terms and the ranges we are used to dealing with clinically will change very strikingly.

A number of non-S.I. terms are likely to continue in use. We have already mentioned the continued use of equivalents as one of these. Another is the concept of pH as a way of expressing hydrogen ion activity rather than using molar concentrations of hydrogen ions. This will be discussed further at the end of this chapter.

Osmotic Activity

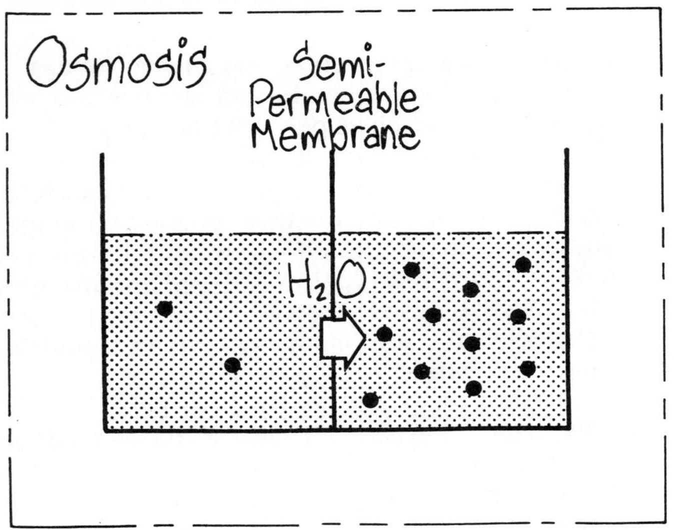

Osmosis refers to the movement of water (the solvent) across a membrane from a solution of lower concentration to a solution of higher concentration. The membrane concerned must be permeable to water but effectively impermeable to the solute.

Where osmotic effects (as opposed to chemical or electrical effects) are being considered, the units osmole (osmol) and milliosmole (mosmol) are used.

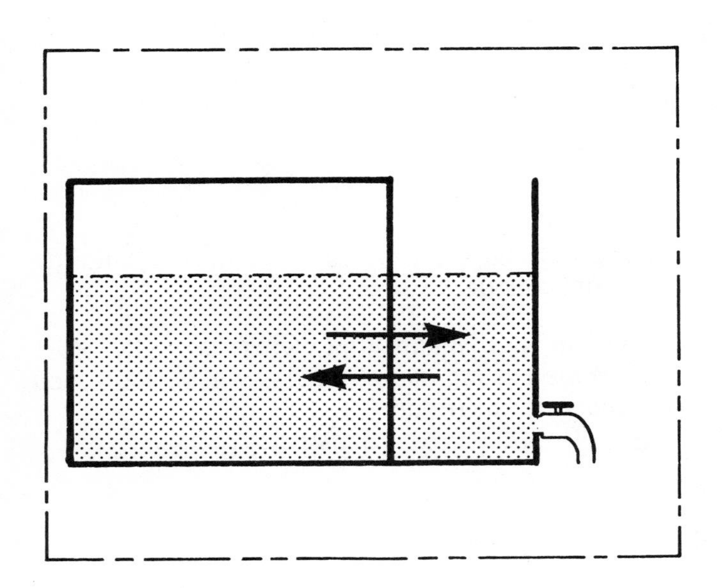

Water can move freely throughout the body compartments and its distribution is dependent upon osmotic forces which ensure that the concentration of particles on each side of a membrane is equal. The distribution of particles (and thus of water) depends on the permeability of the membrane and the size of the particles concerned.

The tendency for water to move along concentration gradients can be measured in terms of the hydrostatic pressure required to prevent it. Thus, osmotic pressure can be measured in millimeters of mercury.

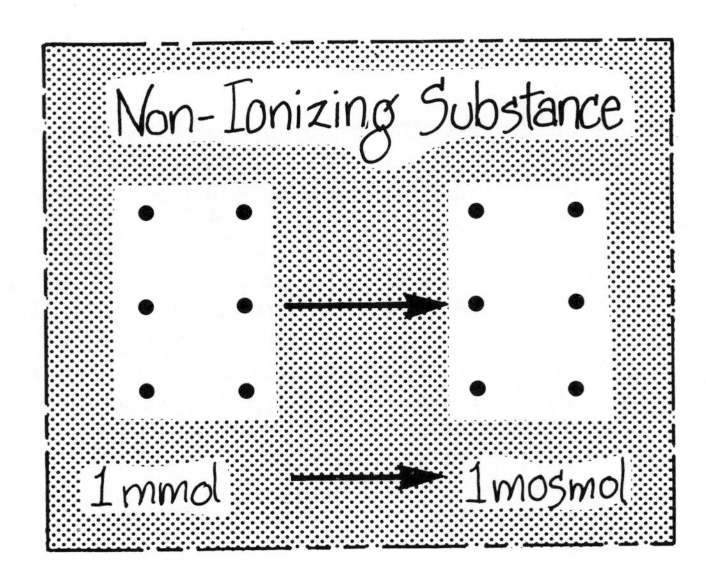

The number of particles in a solution determine the osmotic activity of the solution and can be expressed in terms of osmoles or milliosmoles and measured by determining the degree to which the freezing point of the solution is depressed. For a compound that does not dissociate, such as glucose or urea, the number of osmotically active particles will equal the number of molecules and therefore 1 millimole is equivalent to 1 milliosmole.

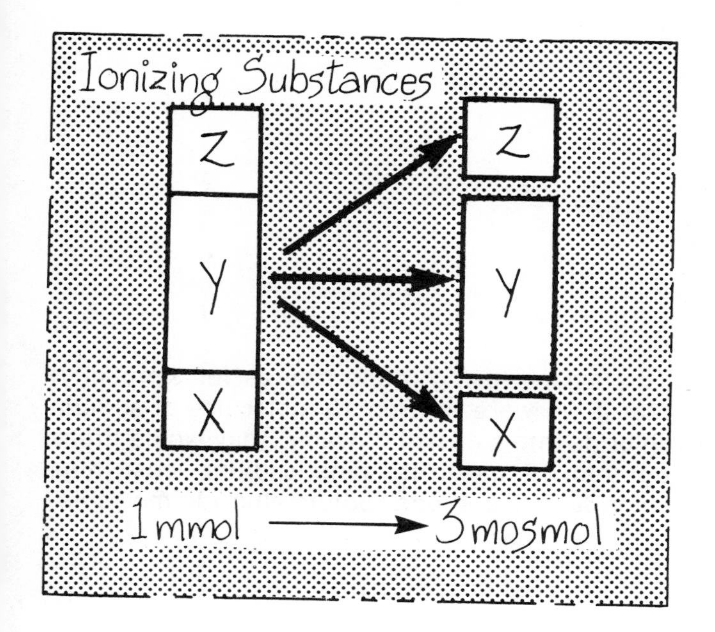

For a compound that does dissociate, one molecule will break up into more than one particle. Thus, in solution, 1 millimole will produce *more* than 1 milliosmole, in this case 3 milliosmoles.

It is important to recognize that measurement of the number of particles in a solution (its *osmolal concentration*) tells nothing about its osmotic *activity* which only takes effect when the solution is exposed to an adjacent solution across a semipermeable barrier.

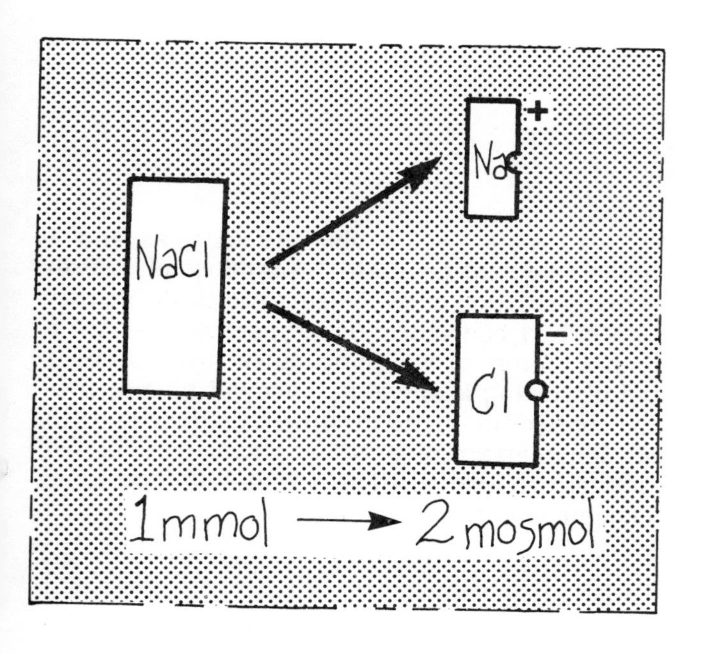

Thus for a substance like sodium chloride which breaks up into a positively charged sodium ion and a negatively charged chloride ion, 1 millimole would generate 2 milliosmoles in solution (provided complete dissociation occurred).

In practice the dissociation of sodium choride is often accepted as being complete. In fact it dissociates about 90% completely and has an *osmotic coefficient* of about 0.9. This value will vary to a small extent depending upon the concentration of the solution.

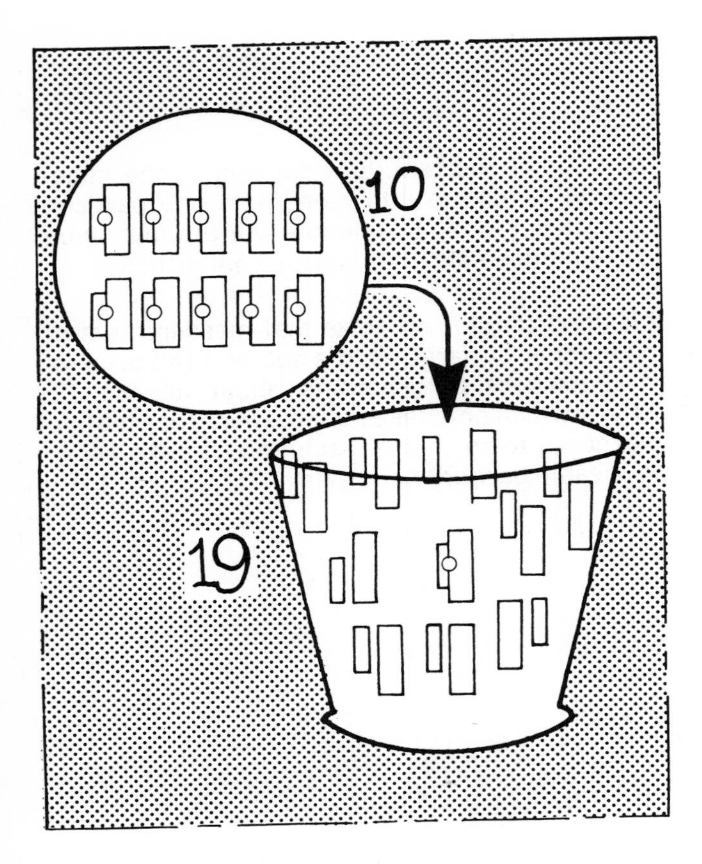

This means (using round figures for simplicity) that if we were to drop 10 molecules of sodium chloride into a bucket of water, they would break up into 19 particles: nine sodium ions, nine chloride ions and one sodium chloride molecule.

Thus 10 moles would yield 19 osmoles and 1 millimole would generate 1.9 milliosmoles. Preparing a solution of an ionizing solute with a predetermined concentration of particles is not a simple matter of weighing out a given amount and dissolving it in water. This will give a solution containing a predetermined *amount* (mol/L) but the number of particles depends upon an osmotic coefficient which varies with the ionic concentration of the solution.

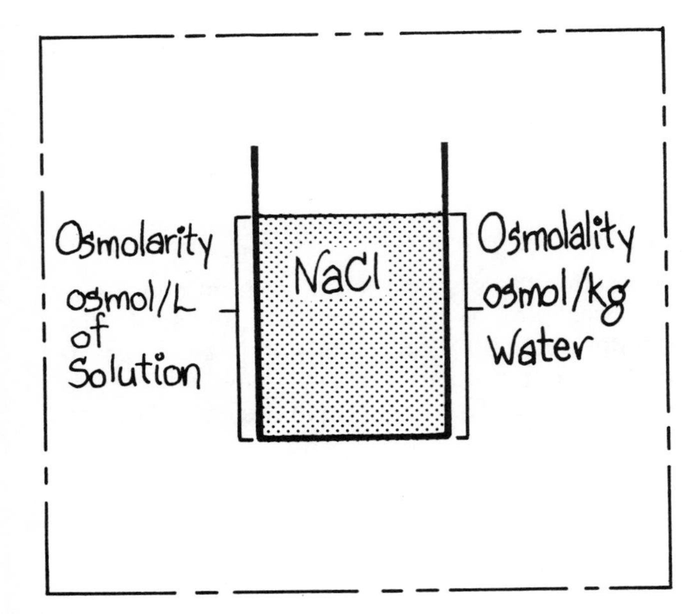

The concentration of particles in a solution can be expressed either as its *osmolarity* or as its *osmolality*.

An *osmolar* solution contains one osmole dissolved in water and made up to one litre of solution. Osmolarity thus refers to a concentration of active particles in a litre of solution.

An *osmolal* solution contains one osmole dissolved in a kilogram of water. Osmolality thus refers to a concentration of active particles in a kilogram of water.

For a pure solution of sodium chloride, for instance, the difference is so small that the two terms are effectively interchangeable.

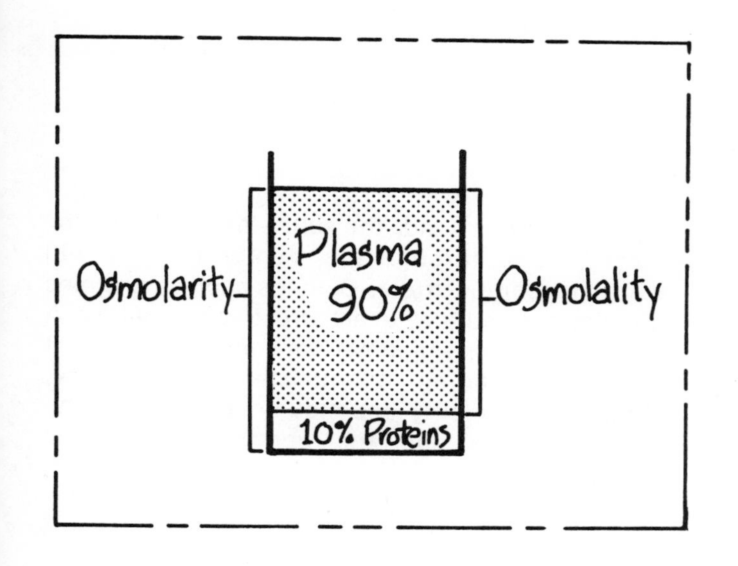

However, for a more complicated solution, such as plasma, the situation is less simple. Of a given volume of plasma, 10% is made up of solids (including proteins, lipids, urea and glucose, etc.) rather than water. As a result, osmolality and osmolarity are appreciably different; the osmolality is greater than the osmolarity because of the smaller amount of water.

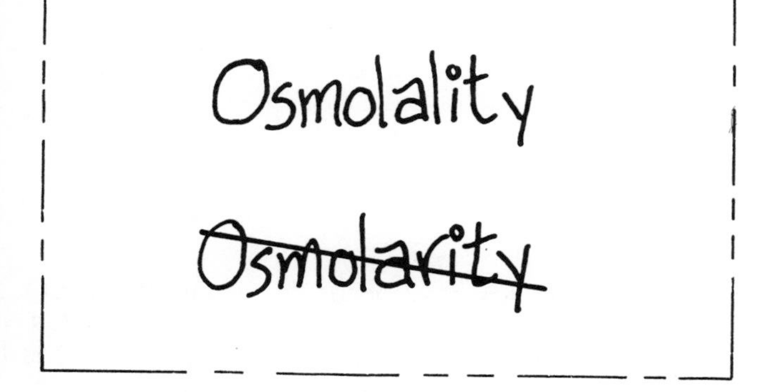

Since osmotic activity depends upon the concentration of active particles per *kilogram of water* the term *osmolality* is the correct one to use and is the quantity measured by osmometers. Osmolarity is therefore a technically incorrect and inaccurate term.

Effective Osmolality

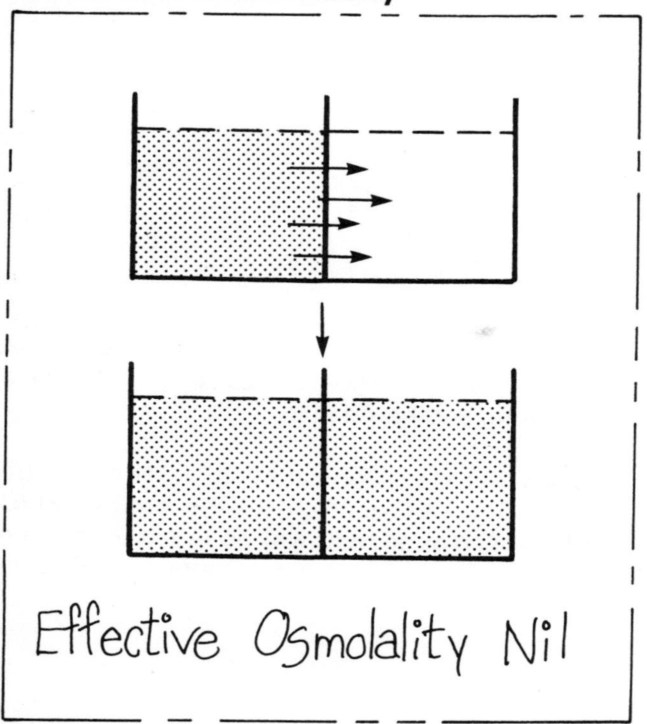

If two compartments are separated by a membrane with water on either side and a solute is put into one compartment it will only exert an osmotic effect if the particles stay on one side. If they are small and very permeant (like urea) they will rapidly diffuse across the membrane and abolish any osmotic gradient. Thus the *effective osmolality* of such a solution rapidly becomes nil; the calculated osmolality of such a solution produces no long-term effect because biological membranes are permeable to urea.

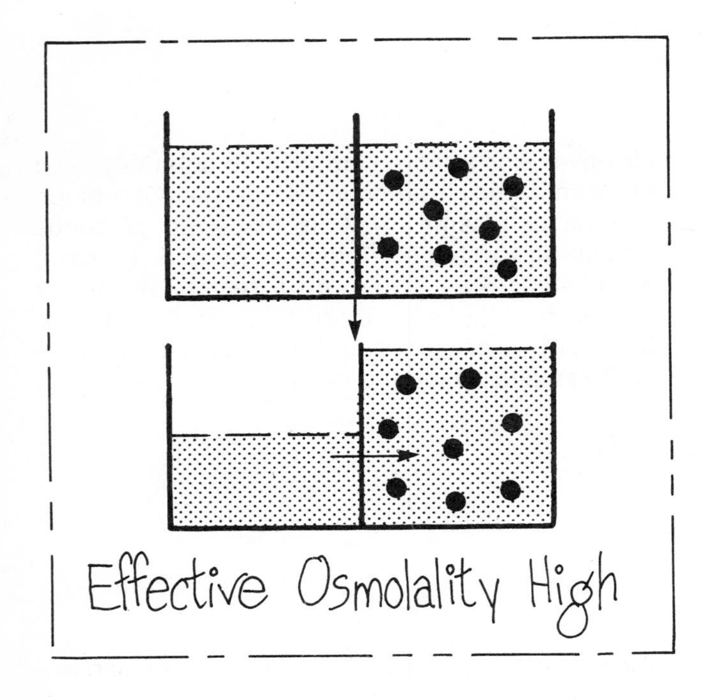

If, on the other hand, the solute particles are too large to pass through the membrane they will exert an osmotic effect. Colloids such as proteins are like this and they exert *effective osmolality.* A volume of water will move from one compartment to another tending to equalize the osmolality. Effective osmolality depends on the size of the solute particle and the permeability of the membrane.

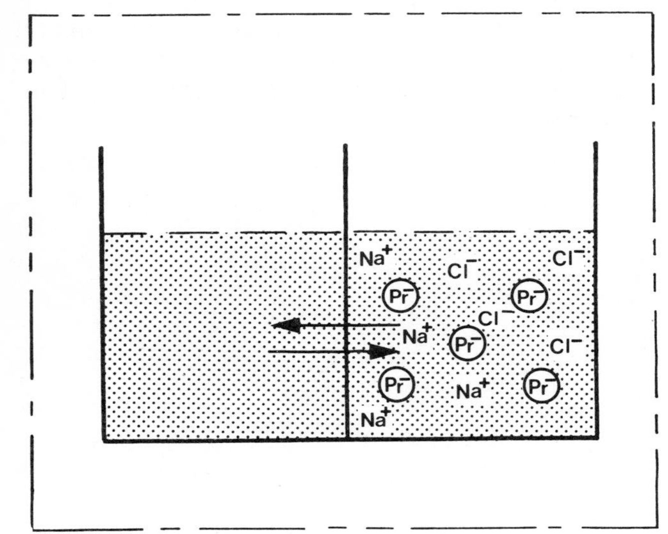

A special effect may occur in situations where fluid in one compartment contains small *diffusible* ions together with large *non-diffusible* ions like protein, which is anionic and can be designated Pr⁻. Without the presence of protein, electrochemical and osmotic forces would result in equal distribution of the small diffusible ions across the membrane.

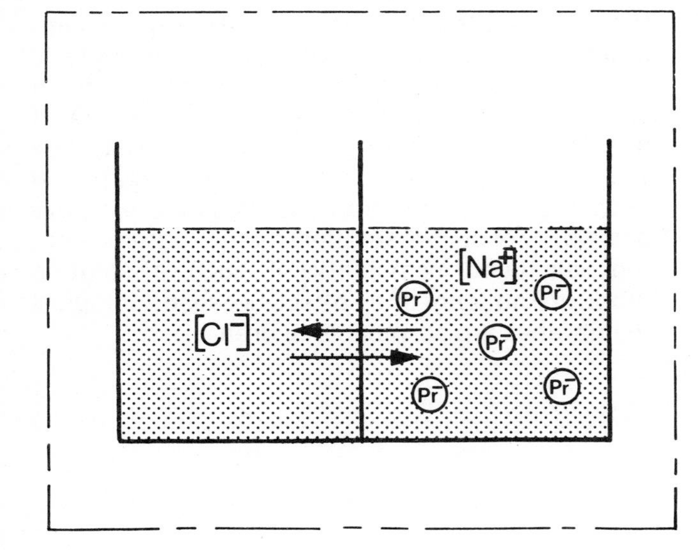

But because the large protein ions cannot diffuse across, their presence causes an asymmetry of distribution of the small ions.

Anions, like chloride, will be in larger concentration on the side opposite to the protein whilst cations such as sodium will be in higher concentration in the compartment containing protein.

Gibbs-Donnan Equilibrium

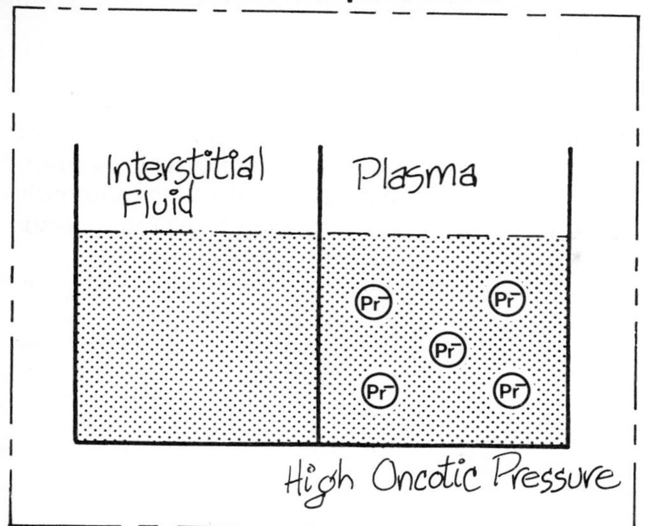

This kind of asymmetry is an example of the *Gibbs-Donnan Equilibrium*.

It explains in part the difference in effective osmotic pressure between the plasma and the interstitial fluid; since plasma albumin is a protein which cannot diffuse out of the capillaries it thus produces an effective osmotic pressure partly due to its own small osmolar contribution and partly due to asymmetry of distribution of small diffusible ions that it produces. This total osmotic effect of a non-diffusible colloid is called the *oncotic* pressure.

Tonicity

Tonicity = Effective Osmolality

The term *tonicity* refers to the *effective osmolality* of a solution. The two terms are interchangeable.

Isotonic Solutions

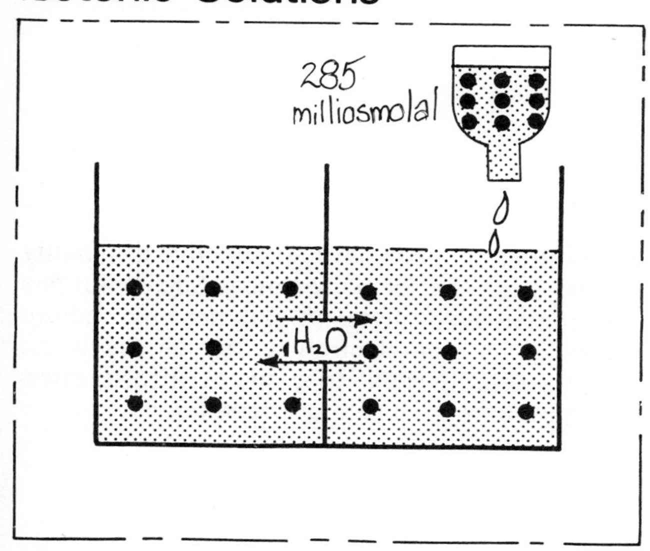

Isotonic solutions have the same effective osmolality as body fluids, that is they are close to 285 milliosmolal. Thus *isotonic* sodium chloride must also be close to 285 milliosmolal and this osmotic activity will be provided by sodium chloride solution that is 154 millimolal.

This has been calculated as:

$$(154 \times 2)\ 0.93$$

where 0.93 is the approximate osmotic coefficient for sodium chloride at this concentration.

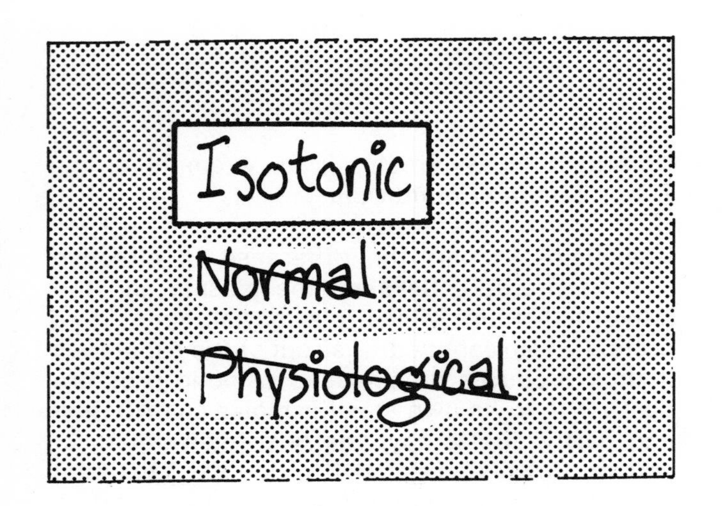

Isotonic sodium chloride is also known as ''normal saline'' and ''physiological saline.'' These latter two terms are obsolete and only serve to confuse, but you may hear them used.

Hypotonic Solutions

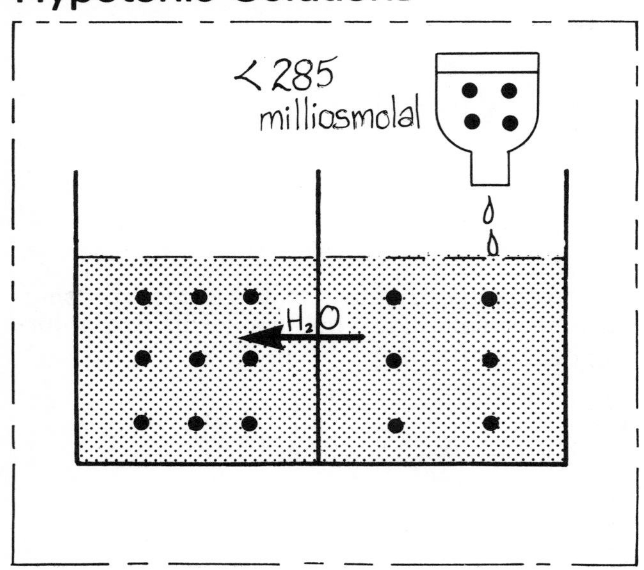

Hypotonic fluids have a lower osmolality than body fluids. One such solution, sometimes called ''half normal saline,'' contains 77 millimoles of sodium chloride per litre.

Hypertonic Solutions

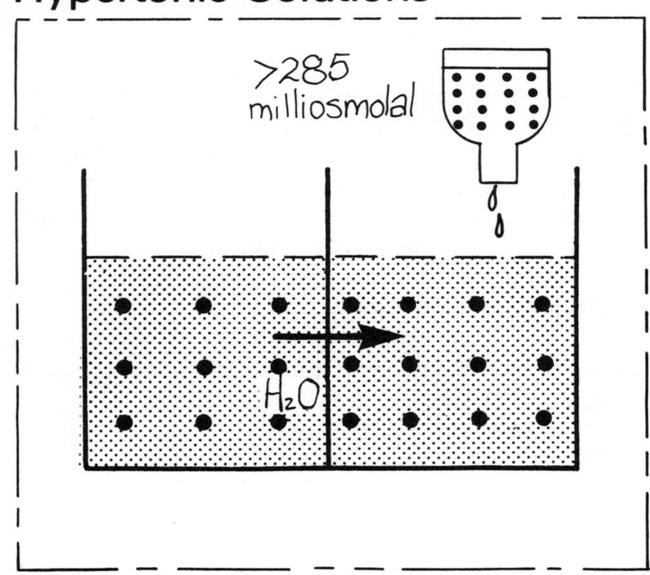

Hypertonic solutions have an effective osmolality greater than that of body fluids. Such solutions may contain electrolytes (e.g. hypertonic sodium chloride) or non-electrolytes which have an osmotic effect because they do not diffuse across biological membranes (e.g. Mannitol).

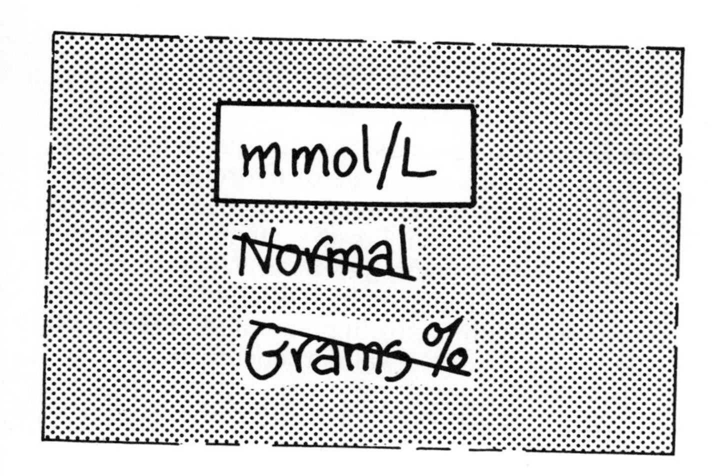

You will hear clinicians use such terms as "half normal saline" (77 mmol/L NaCl) or "five per-cent saline" (855 mmol/L NaCl), but it is obvious that expressing all concentrations in terms of *millimoles per litre* is both consistent and eminently rational; it also avoids confusion which, in clinical care, is an important consideration.

Hydrogen Ions and pH

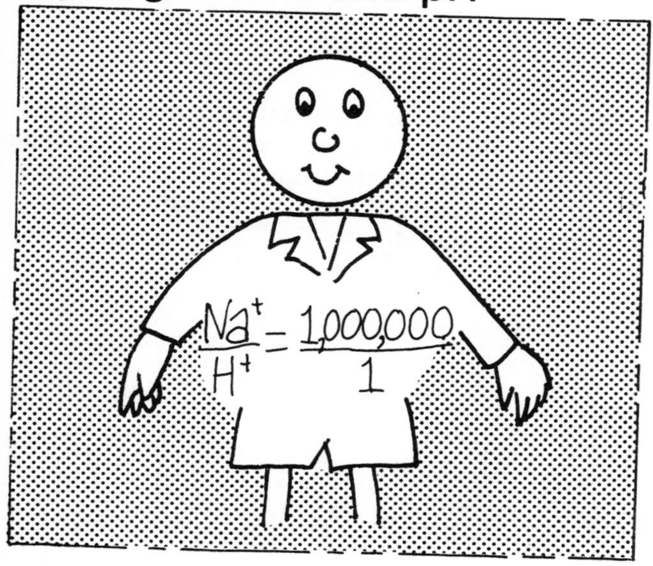

Free hydrogen ions are present in most body fluids in very small concentrations. Blood plasma contains about 40 nanomoles of hydrogen ions per litre (i.e. 40×10^{-9} mol/L). This is roughly one millionth of the concentrations of ions such as sodium and chloride.

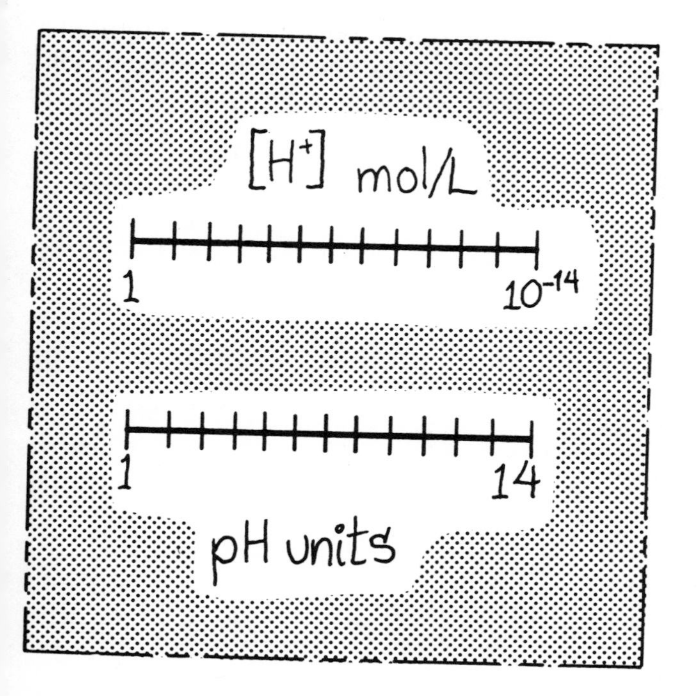

The concept of "pH" was developed by Sorenson to express hydrogen ion concentration more compactly. Hydrogen ion concentration might range from 1 molar to 10^{-14} molar, an enormously wide range. pH is defined by:

$$pH = \log \frac{1}{[H^+]}$$

and is a measure of hydrogen ion activity in a solution; usually it is measured by a hydrogen ion sensitive glass electrode and a "pH meter." The same range of hydrogen ion concentration is thus described by a range of pH from 1 to 14.

Fluid	pH*	H^+
0·1molar HCl	1	10^{-1} mol
Water	7·0	10^{-7} mol
Plasma	7·4	4×10^{-8} mol
0·1molar NaOH	13	10^{-13} mol

* At 20°C

The concept of pH has been useful since it expresses such a wide range of hydrogen ion concentrations so conveniently; it also expresses hydrogen ion activity rather than merely the concentration of hydrogen ions. Whether or not pH is officially replaced by hydrogen ion concentration as SI units are implemented, pH is still likely to continue as a valued concept.

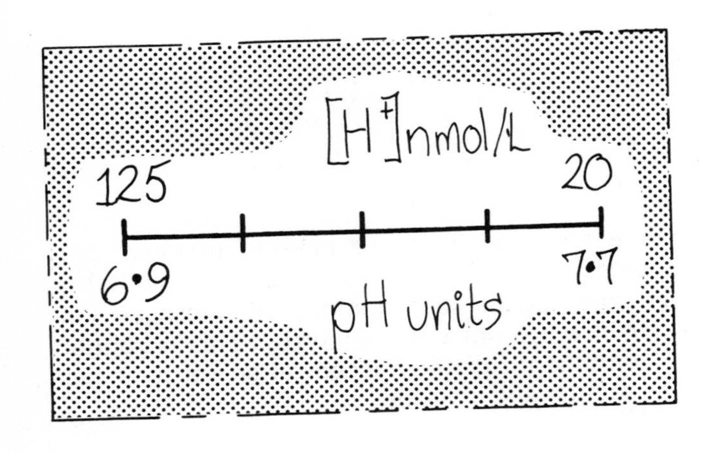

In clinical situations hydrogen ion concentrations can vary from 125 to 20 nanomoles per litre. This represents a pH range from 6.9 to 7.7 and is a much wider range than the normally tolerated physiological variations.

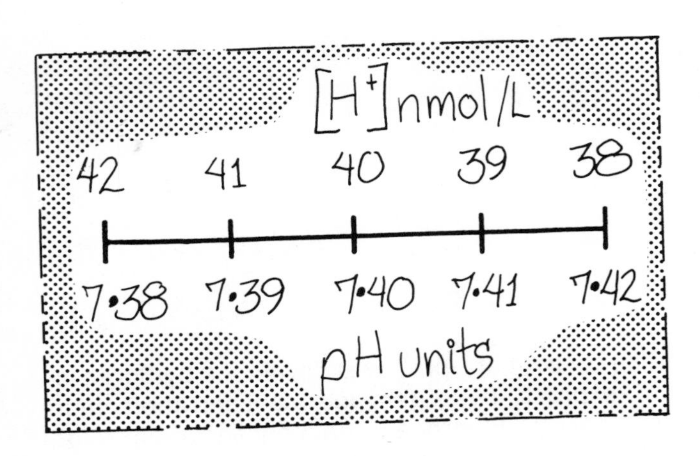

The physiological range lies close to pH 7.4 (a hydrogen ion concentration of 40 nanomoles per litre). Near this range each change of 0.01 pH unit is equivalent to a change of hydrogen ion concentration of 1 nanomole per litre.

Chapter 3

The Distribution of Body Fluids

1.— Body water content
2.— Distribution between E.C.F. and I.C.F.
3.— Distribution of Solutes—electrical and osmotic activity
4.— Distribution between plasma and interstitial fluid

This chapter outlines the way in which water is distributed throughout the body, and explores some of the basic rules governing its transfer from one compartment to another.

Body Water Content

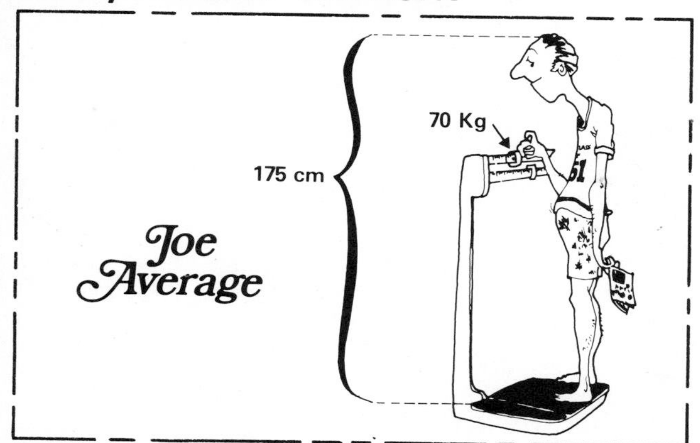

When we talk about the average man, we are talking by convention about a male who is about 175 centimetres tall and weighs about 70 kilograms. This average man contains about 60% water by weight. In many clinical situations, you will not be dealing with "average" people; they may be fatter or thinner, and this will influence the size of their fluid compartments.

The degree to which they are fat or thin has a significant effect on how much water they contain. A thin man contains less fat and more water. A fat man contains proportionally less water because adipose tissue contains only about 20% water. The fatter a person is, the less water he contains in proportion to his body weight.

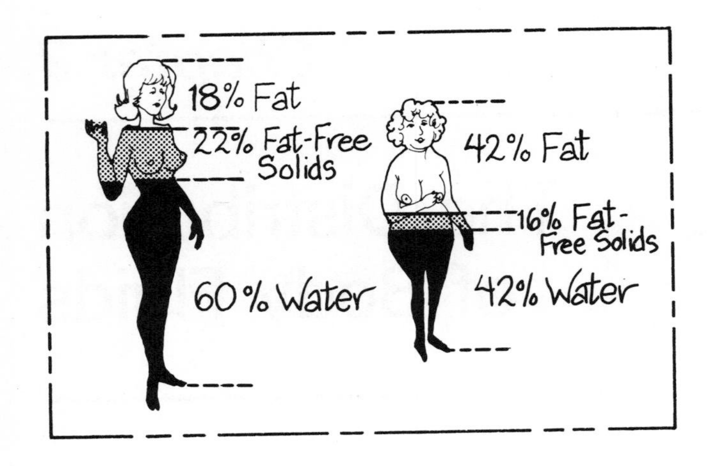

Similarly there are differences between fat and thin women.

There is a sex difference, too, because women have more body fat than men. This means that, by and large, women contain less water in proportion to their body weight than men do. This is not usually a very major consideration in clinical medicine, but there are circumstances where these variations can be very significant.

	Total Body Water *		
	Infant	Male	Female
Thin	**80**	**65**	**55**
Average	**70**	**60**	**50**
Fat	**65**	**55**	**45**

*As % of Body Weight

This table shows how big the variation can be between infants, and adult males and females. An infant may have as much as 80% of its body weight as water, and a malnourished infant who is depleted of fat may have even more than that. This becomes very important because a child is much more vulnerable to the effects of volume depletion than an adult, and what may be trivial fluid loss for a normal adult may be a catastrophic loss, in percentage terms, for a child.

Distribution of Water Between E.C.F. and I.C.F.

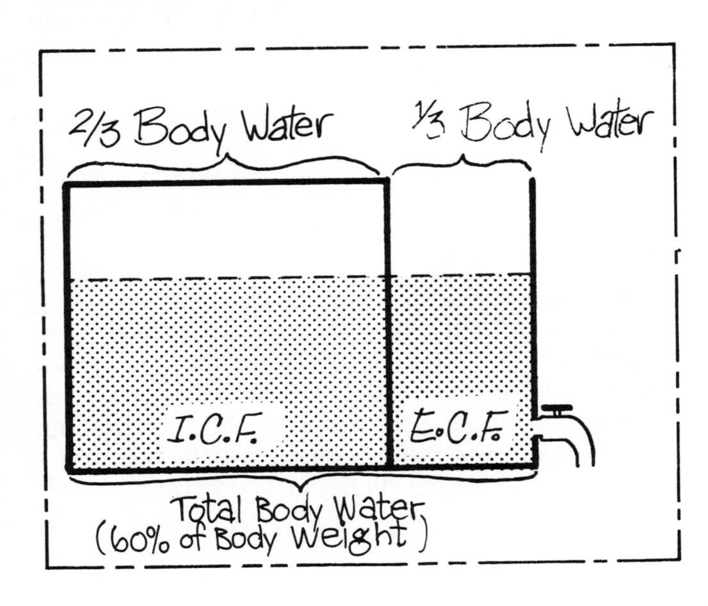

As described in chapter 1, the body fluids can be divided between an intracellular compartment (I.C.F.) and an extracellular compartment (E.C.F.). Approximately two-thirds of body water will be inside the cells and one-third outside the cells.

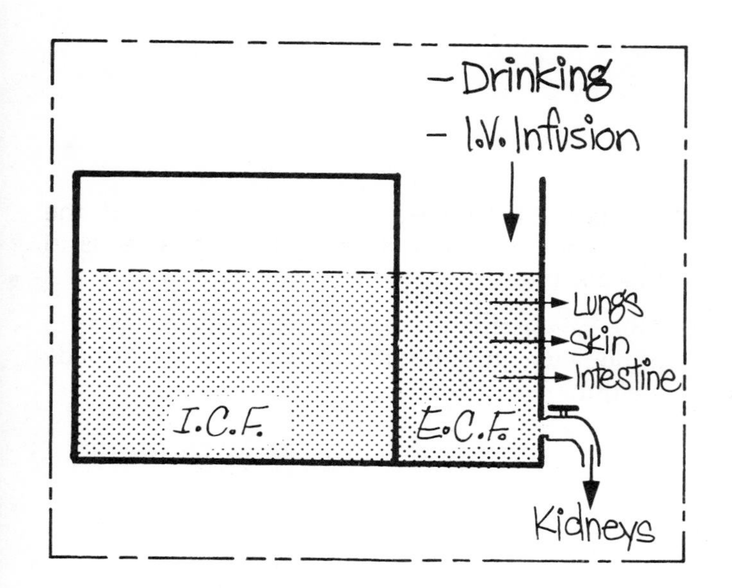

Normally, *additions* to the extracellular compartment are made by drinking, but physicians often add to it by unphysiologic routes, such as intravenous infusion. Fluid can be *lost* from the extracellular compartment via a number of routes; these include losses via the lungs, the sweat, and the intestine, but the route of loss that is physiologically most important is from the kidney.

All such losses occur from the E.C.F., and the intracellular compartment can only lose or gain volume via the extracellular compartment.

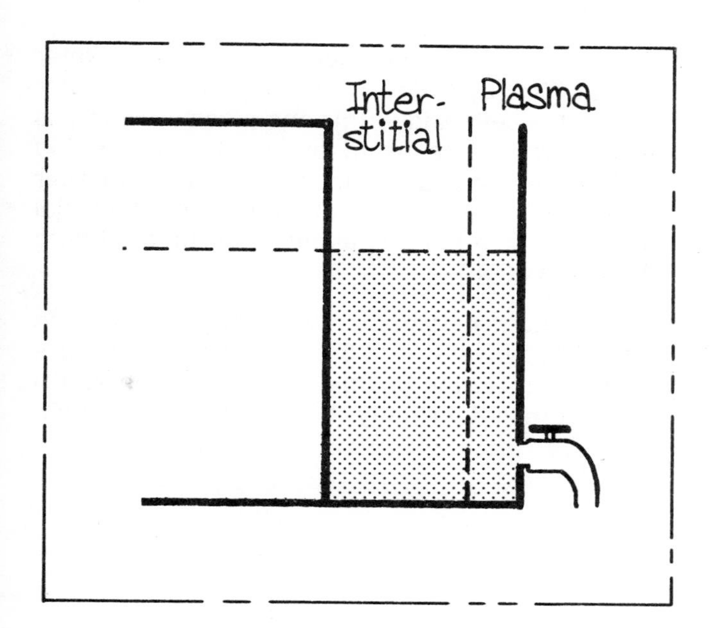

About a quarter of the total E.C.F. is confined to the vascular space and is referred to as the plasma volume. The remaining three-quarters is in the interstitial space outside the blood vessels and bathing the tissues. Once again these proportions are what you expect in an ''average'' adult and will vary with age, sex and fatness.

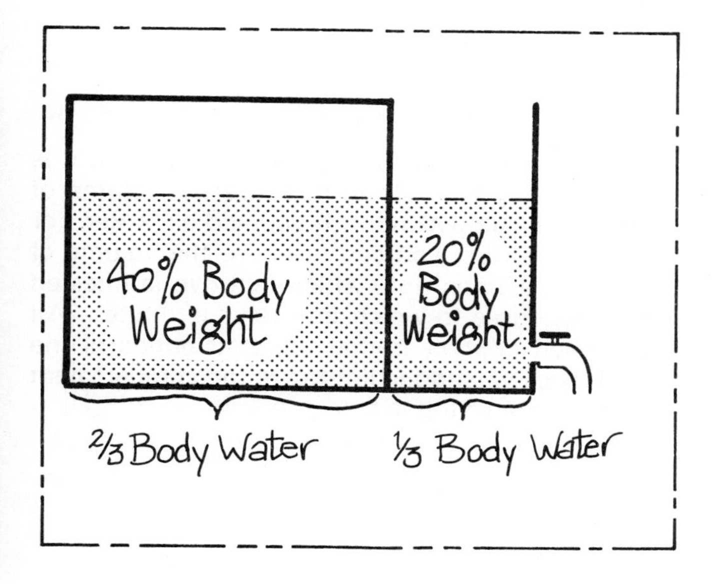

In our diagrams we have shown that two-thirds of total body water is in the I.C.F. and one-third is in the E.C.F. In terms of total body weight this means that for a normal adult I.C.F. water constitutes 40% of body weight and E.C.F. water makes up 20% of body weight.

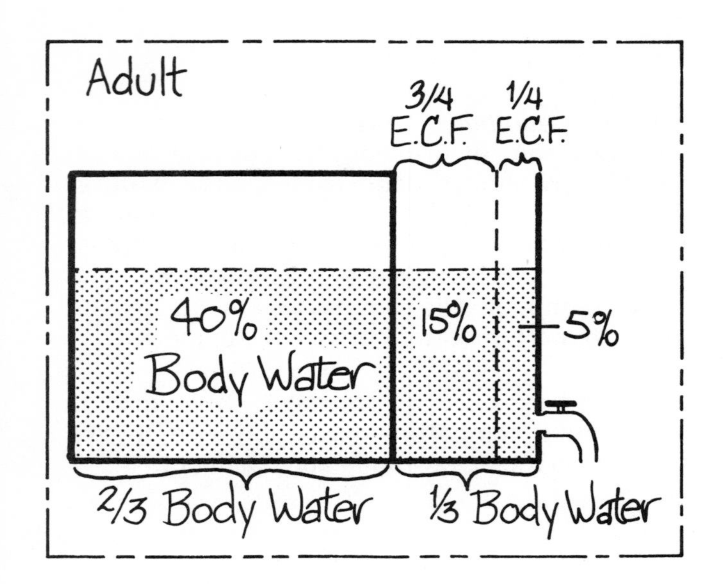

Similarly, in the E.C.F. three-quarters of the fluid volume is interstitial and one-quarter is in the plasma.

This means that interstitial water makes up 15% of body weight, and plasma water is 5% of body weight.

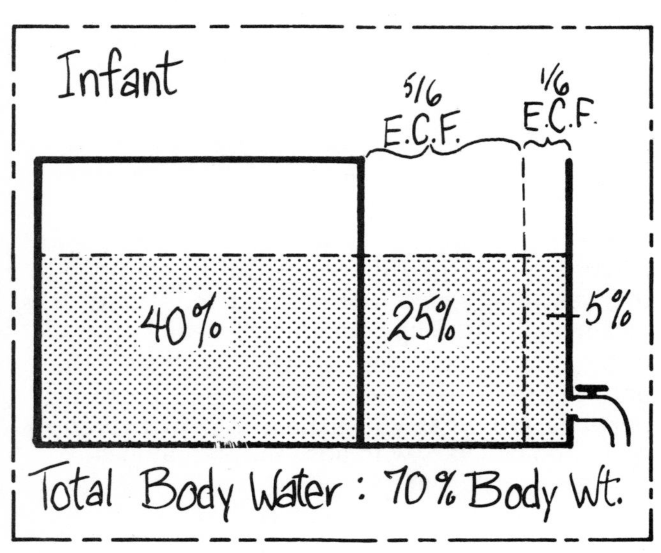

In an infant, however, there is much more water (about 70% by weight). The proportion of water within the cells and within the plasma volume is exactly the same as in an adult, but the big difference is that an infant has a relatively larger interstitial volume. The ratio between interstitial and plasma volume is about 5:1 compared with 3:1 in the adult.

For an infant weighing only 5 kilograms and with a total body water of 3.5 litres a loss of one litre from the intestine (a small amount for an adult) may represent virtually the total E.C.F. volume.

Measuring Volume of Compartments

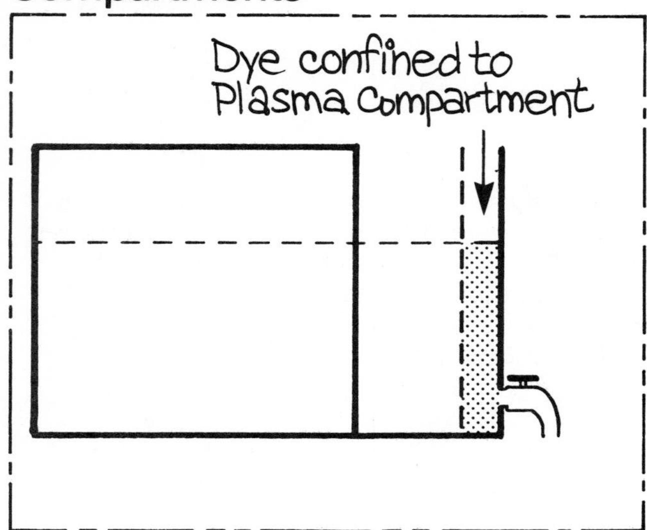

Given a compartmental system of this kind, it is theoretically possible to measure the volume of any given compartment. If you add a marker, for example a dye, you can measure the volume of the compartment into which that dye is injected by the degree to which its colour is diluted. All you need to know is that the distribution of the substance you add is limited to the compartment that you are trying to measure.

Markers:
Dyes
Radioactive Isotopes
Chemicals

Markers may be radioactive isotopes such as ^{125}I-albumin which is distributed in the plasma volume. Others may be chemical markers, such as thiocyanate, which is distributed throughout the E.C.F. None of these markers is *absolutely* confined to a single compartment, but they do provide reasonably close approximations which are of practical use.

Distribution of Solutes

1. Electrochemical Activity
2. Osmotic Activity

In looking at the composition of the fluids within the body's compartments we are interested predominantly in substances present in large amounts which have a major impact on either the electrochemical or the osmotic activity within each compartment.

Positive Charges = Negative Charges

In all body fluids, whatever their composition, anions and cations will *always* be present in equal amounts since positive and negative charges must be equal.

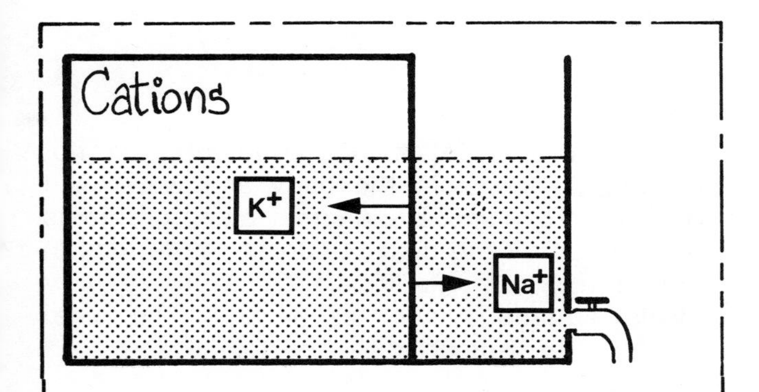

Sodium is the predominant cation in the extracellular space whilst in the intracellular space potassium is the predominant cation.

In complex animals like man, the cell wall has a pump system which pushes *out* the sodium that tends to leak in and pumps *in* the potassium that tends to leak out. This means that as long as the extracellular fluid compartment is maintained at a constant composition, the cells can maintain a constant composition by mechanisms within their own cell walls which utilize the energy provided by cellular metabolism.

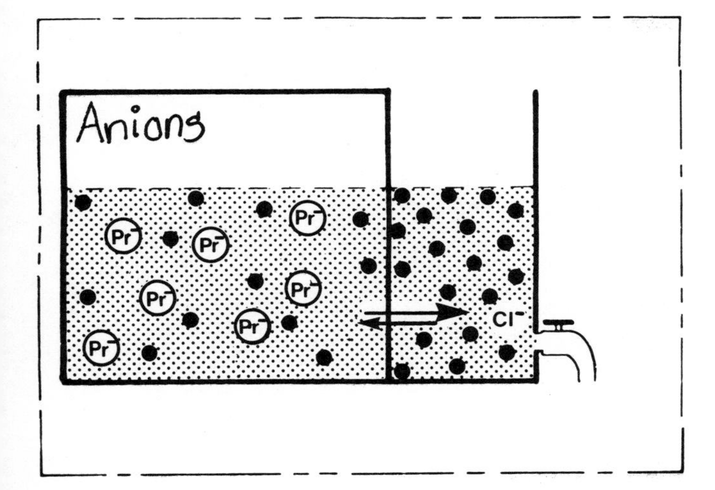

Within the cells the important anions are large organic molecules such as proteins together with phosphates. Proteins remain within the cell and are too large to diffuse out.

The main extracellular anions are chloride and bicarbonate. Chloride is not exclusively outside the cells and distributes according to electrical forces. Since many of the cellular anions cannot diffuse out chloride tends to end up predominantly outside the cells.

The Donnan Effect:

The asymmetry of distribution of a diffusible ion (chloride) due to the presence within the cell of a non-diffusible ion (proteinate) is another example of the Donnan effect.

Distribution of Ions determines

1. Electrical activity on cell surface
2. Osmotic pressure in compartments

The presence of small, charged, ionic particles in body fluids is vital to the function of the body, determining factors such as the surface charges on the cell walls (largely due to a leak of potassium from the cell), and the osmotic activity of each compartment.

As a general rule osmotic gradients are not allowed to persist in the body and water will move from one space to another along gradients determined by the distribution of solute ions to correct any osmotic imbalance. Only in very special situations (such as parts of the renal tubule) is this rule broken.

Osmolality of Body Fluids

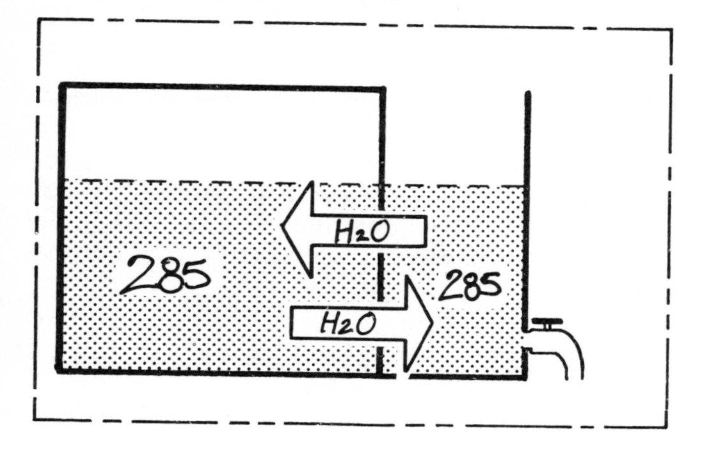

In man, therefore, the intracellular and extracellular fluids are of comparable osmotic composition and normally this is about 285 milliosmoles per kilogram. Water, therefore, will move equally in either direction across the cell wall since the random movement of water molecules will be the same in each direction.

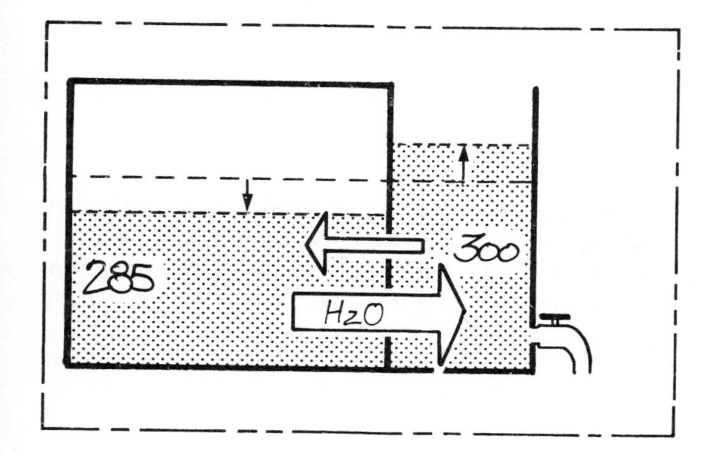

But suppose that the osmolality of the E.C.F. should become 300. There is now an osmotic gradient, and water moves predominantly from the cells into the E.C.F. until there is equalization of the osmotic pressure. Thus osmotic factors exert control over the distribution of volume between compartments.

Distribution of Volume in E.C.F.

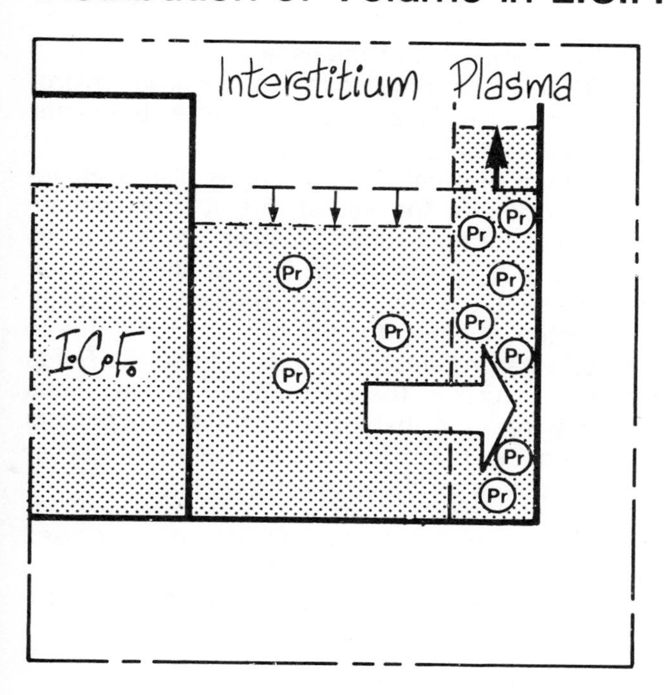

As we have seen, the E.C.F. is composed of two compartments, the plasma and the interstitial volumes. These are separated simply by the capillary walls which are freely permeable to both water and solutes, but only slightly permeable to proteins. The distribution of sodium, chloride, and other ions through the whole of the E.C.F., whether it be plasma or interstitial space, would be expected to be uniform, but because within the capillaries there is a higher concentration of protein there is a relative osmotic gradient between the interstitial and the plasma space; water and solutes tend to move constantly from the interstitial space into the plasma space in an attempt to balance the osmotic effect of the plasma proteins.

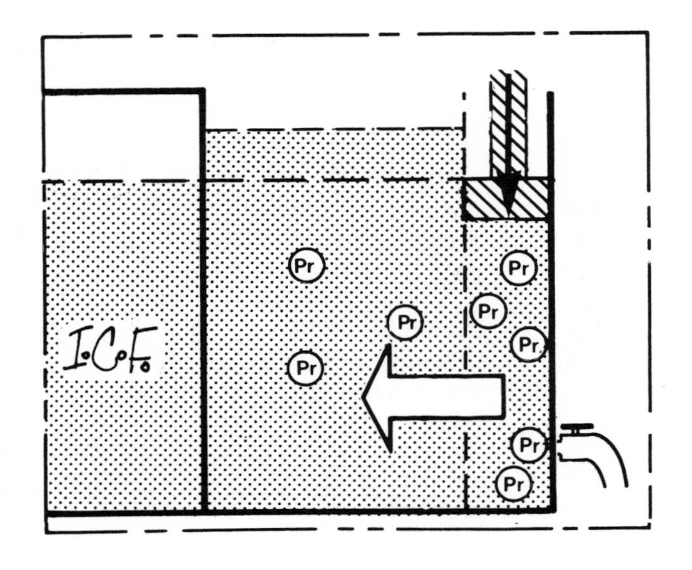

But another thing happens in the capillaries, especially at the arterial end. Here there is a positive hydrostatic pressure inside the vessel which tends to push water and solutes, but not protein, into the interstitial space, tending to increase its volume at the expense of the plasma volume.

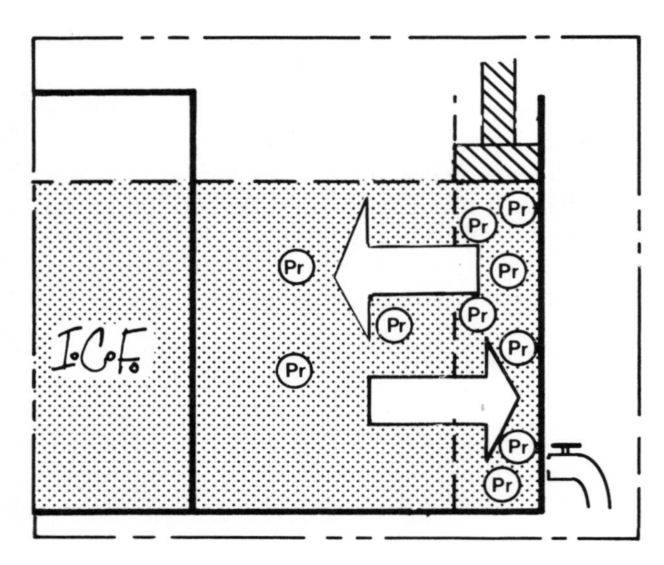

In normal circumstances these two processes balance out and a steady state develops. Osmotic factors tend to move fluid from the interstitial space into the plasma volume, and hydrostatic pressure tends to move them in the opposite direction. The balance of these forces varies as the plasma flows along the smallest capillary blood vessels.

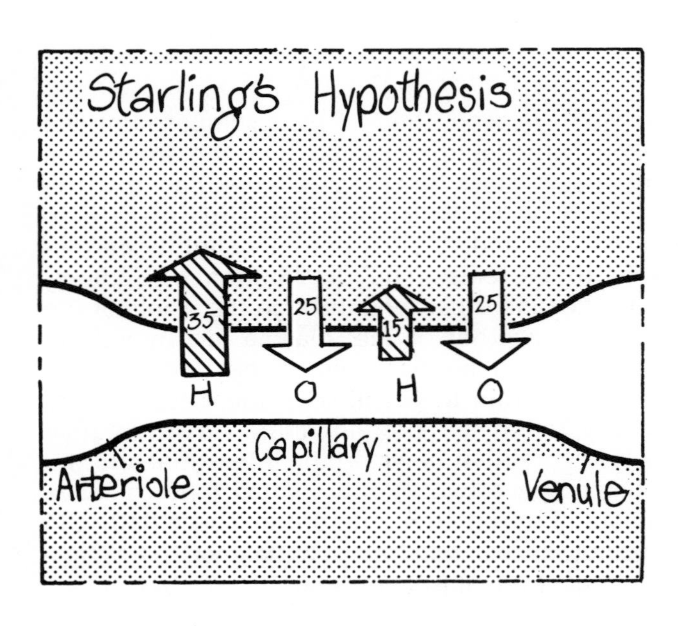

At the arterial end of the capillary fluids tend to leave the plasma under the influence of hydrostatic pressure. This is not just water, but water and all the small solute particles it contains, such as sodium and chloride. There is also a movement back into the vessel due to the osmotic pressure, but the hydrostatic pressure is greater, leading to a net loss from the capillary. At the venous end the hydrostatic pressure has been dissipated, and now the osmotic pressure, due to the plasma proteins, is greater and so the fluids tend to return to the circulation. This explains how there is a balance of fluid movements between the plasma and the interstitial space. This is known as *Starling's Hypothesis.*

The pressures in this diagram are all expressed as millimetres of mercury and are representative of an ''average'' capillary in the systemic circulation.

Edema

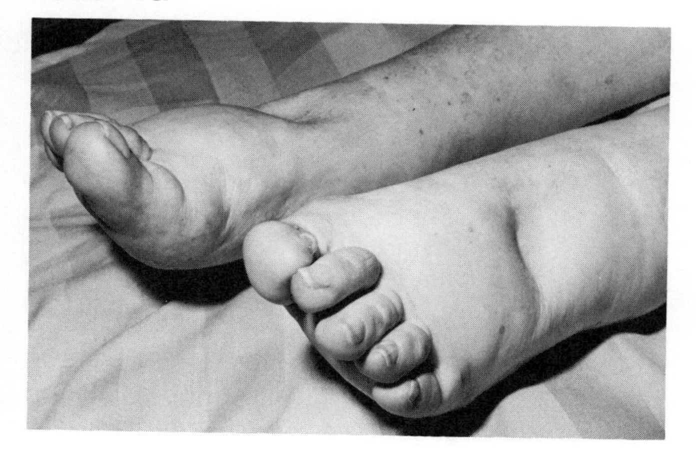

If this equilibrium becomes disturbed, water and solutes may accumulate in the interstitial compartment, producing visible swelling (edema).

Firm pressure on such a swollen area will leave an impression—hence the clinical sign of ''pitting'' edema.

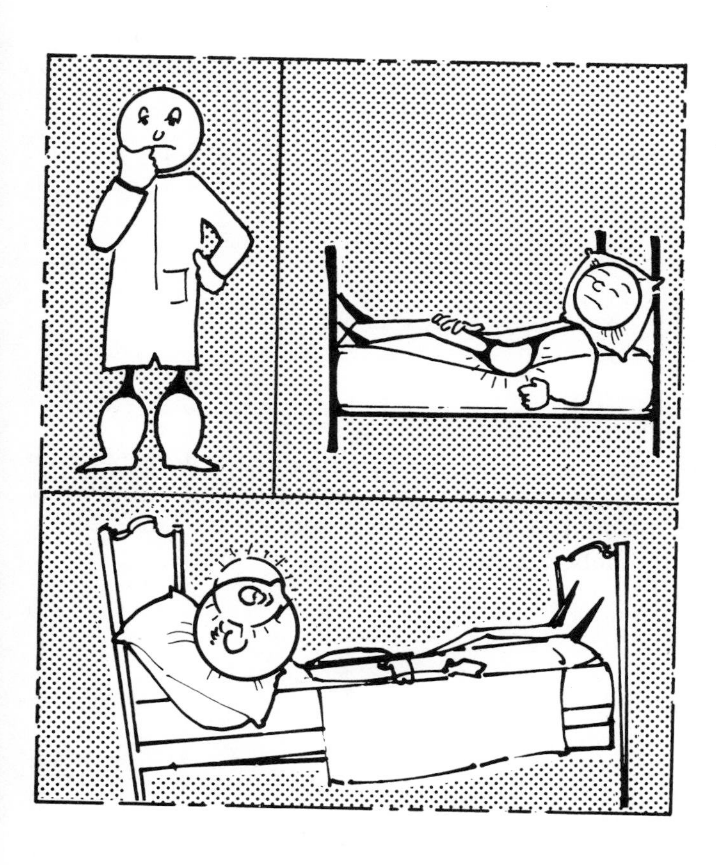

In states of generalized edema the swelling is in the most dependent part; it is in the feet if you stand up, but of course it will end up in the face if you lie down. This explains why many people with edema tend to have puffy eyes in the morning and puffy feet in the evening. Patients lying in bed will accumulate edema over their sacrum. This is simply due to the effects of gravity added to the hydrostatic pressure in the capillaries.

Generalized Edema means:

1. Expanded Interstitial volume
2. Increased total body sodium

One of the most important rules about edema is that anybody who has generalized edema must have an expanded interstitial volume. Since this volume is made up of water and the solutes present in the interstitial space, anyone who has edema of a generalized kind also has an increased total body sodium because sodium (and accompanying anions) are the major E.C.F. solutes.

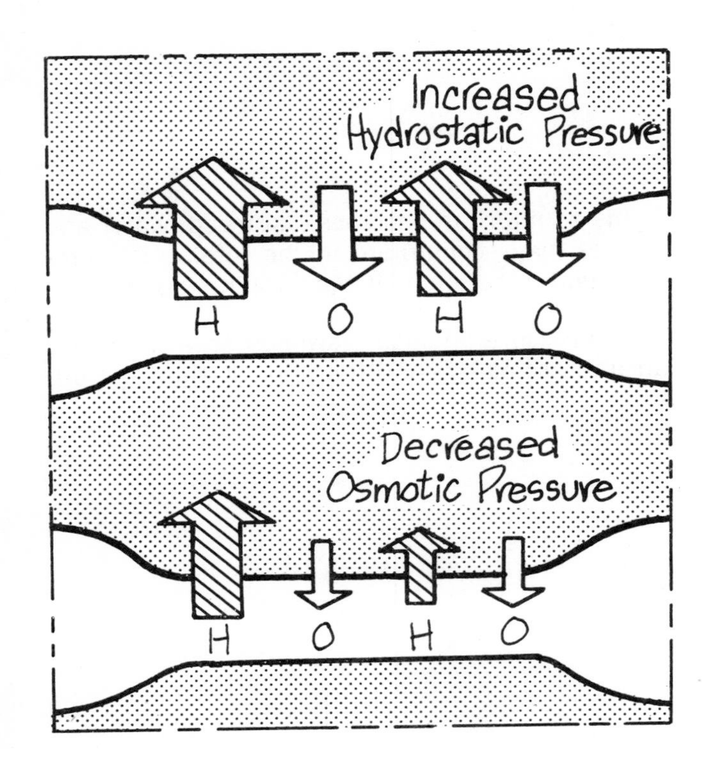

If you think of the balance involved in the Starling Hypothesis it becomes obvious that edema can be caused by an increased hydrostatic pressure in the capillaries, (such as in heart failure), or by a decreased osmotic pressure due to low plasma proteins, (such as in the nephrotic syndrome). In the former, the plasma volume is also expanded, whilst in the latter the plasma volume will tend to be contracted.

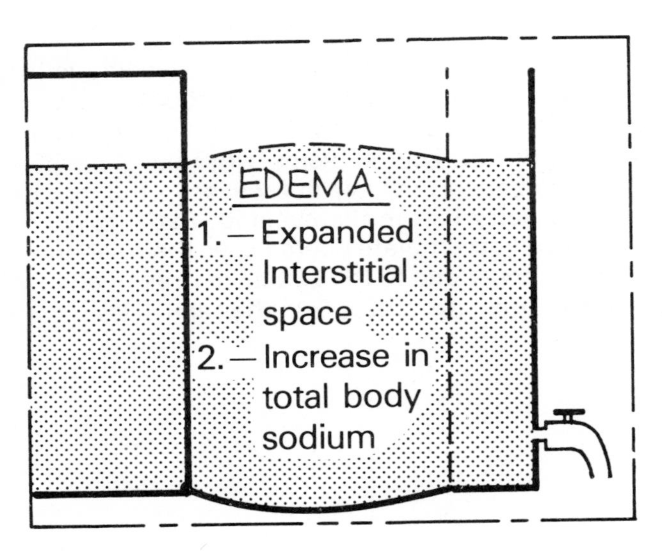

But in either case, edema indicates an expanded interstitial space and, whatever the size of the plasma volume, it also implies an increase in total body sodium.

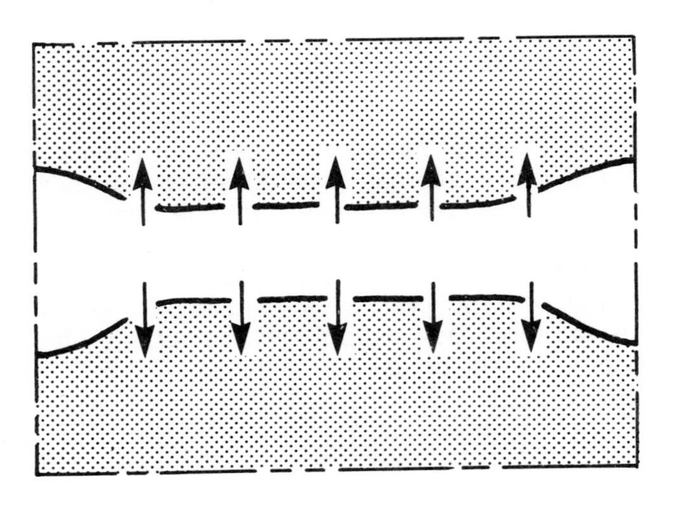

Increased permeability of capillary walls will also favour the formation of edema, but this seldom occurs on a generalized basis. It is, however, a cause of local edema, in, for example, inflammation.

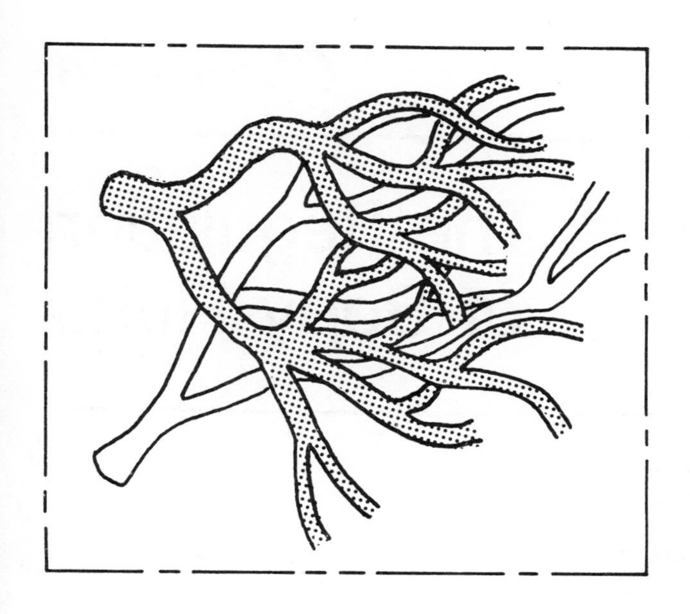

It must be remembered that running parallel to the capillary blood vessels are the lymphatic vessels which are capable of transporting interstitial fluid back into the plasma compartment. Obstruction of the lymphatics can give rise to localized edema which is usually non-pitting. In generalized states of edema lymphatic flow may increase markedly.

The lymphatics are also capable of taking up stray protein molecules that leak out into the interstitium and returning them to the plasma compartment via the central lymphatics and the thoracic duct.

Chapter 4

The Regulation of Body Fluids

The regulation of fluid distribution within the body involves a series of co-ordinated responses to change, which all tend to maintain the constancy of volume and composition of the E.C.F. The more constant the environment provided by the E.C.F., the more will the cells be protected from changes in the external world.

Changes in Plasma Volume

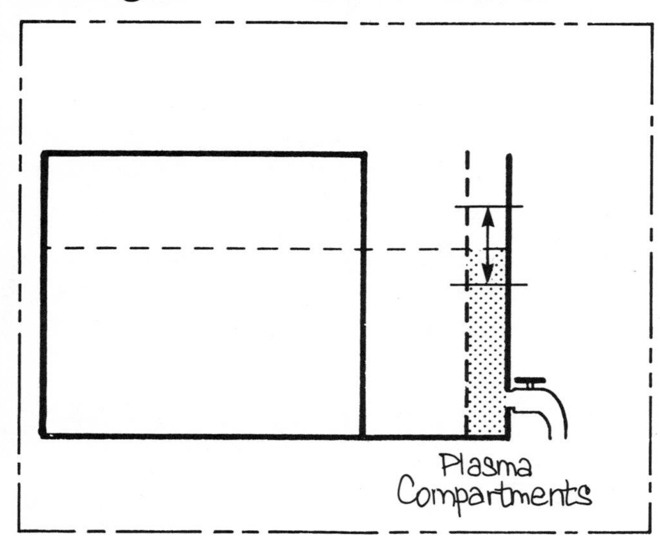

The plasma compartment has a relatively small volume, but the pressures and flows within it may be high. Shifts of volume and pressure within it may be rapid, for example in response to the gravitational effects caused by the simple expedient of rising from a lying to a standing position. Direct loss of volume, such as occurs with hemorrhage will also be rapid especially from the arterial side where pressure and flow are highest.

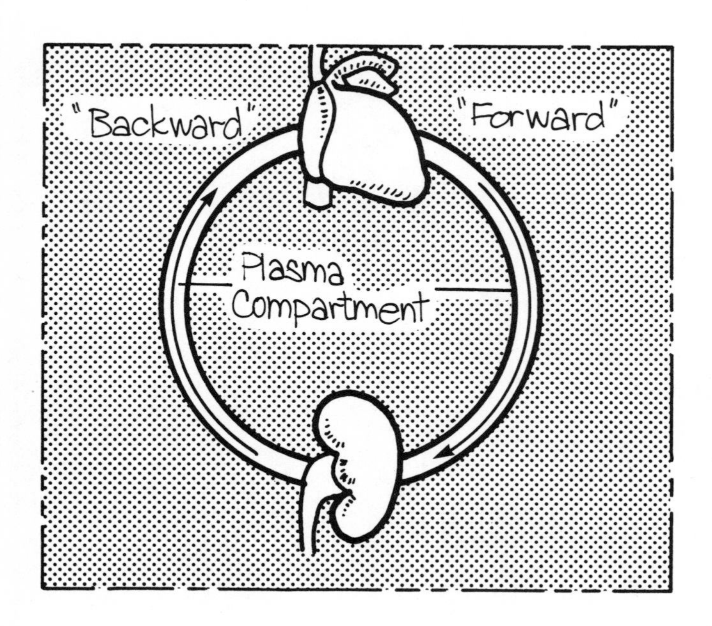

At this point it is important to modify the circuit diagram that was developed in previous chapters. We have drawn the ''forward'' and ''backward'' parts of the plasma compartment as being equal. This equality applies to *rate of flow* since cardiac output and input must be the same over a fixed time period; however, it does not apply to *volume* and *pressure* which are markedly different.

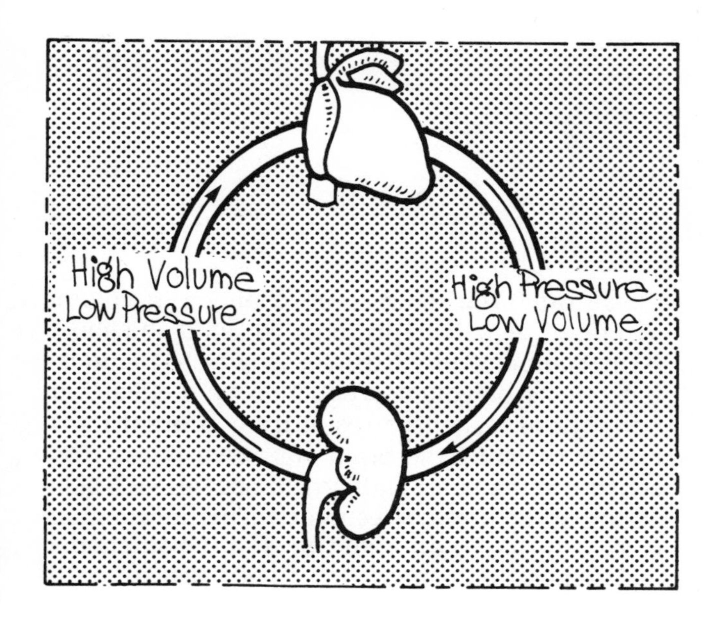

In fact, the "forward" circulation is a high pressure, low volume system, whilst the "backward" circulation is a high volume, low pressure system.

The smaller vessels "in front of" the heart (small arteries and arterioles) provide a resistance to blood flow and are called *resistance vessels.* Changes in their caliber in response to nervous and humoral stimuli cause changes in peripheral resistance which influence pressure and flow.

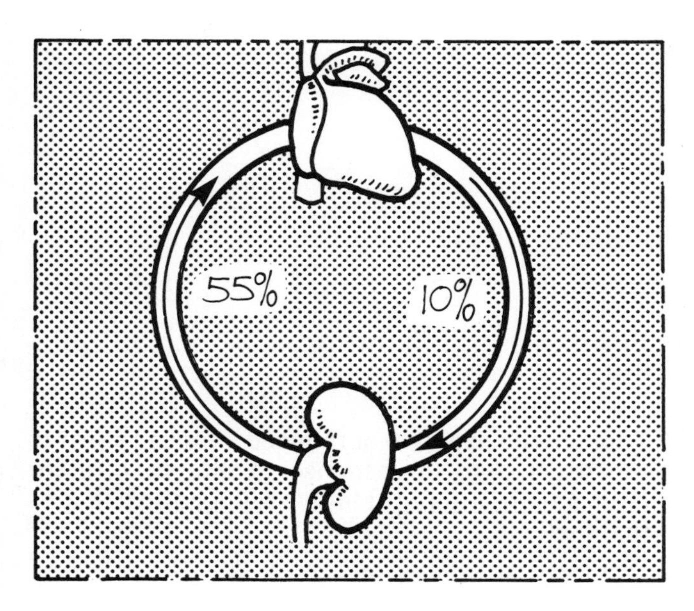

In the normal situation about 55% of the plasma volume is in the venous system whilst 10% is in the arterial system at any particular time. The remaining 35% is distributed in the heart, the lungs, and the capillary bed.

The large capacity of the venous system "behind" the heart allows it to accommodate an expanded volume with relatively much less change of pressure and flow than in the arterial system. For this reason veins are sometimes called *capacitance vessels.*

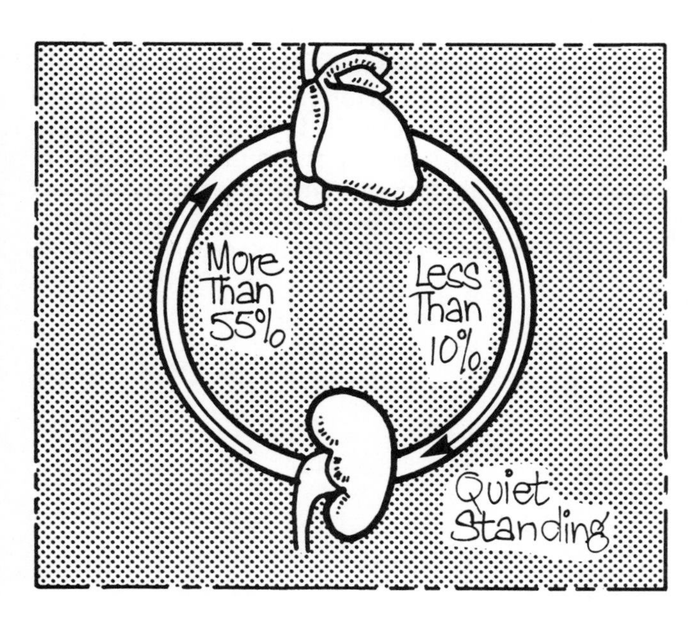

Prolonged, quiet standing can produce an increased venous pooling (due to gravitational forces) with a resultant reduction of the volume available to the arterial side of the circulation and thus a diminished blood flow to key organs. This is what is happening in the soldier who faints on the parade ground. It is an example of the relatively rapid short-term change that can alter volume and flow in the plasma compartment and alter the distribution of blood between resistance and capacitance vessels.

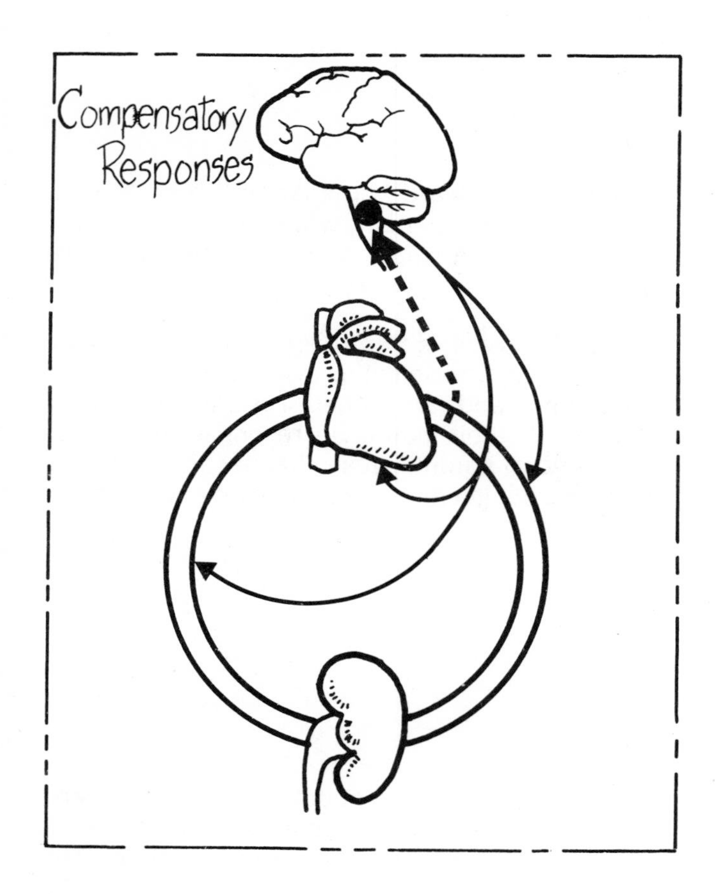

These rapid and short-term changes in volume and flow are matched by equally rapid and short-term compensatory responses which minimize the effects of volume reduction or maldistribution. These responses result from afferent information reaching the brain via nerve fibres from sensors in the heart and large blood vessels, such as the aortic stretch receptors. This information is transmitted by efferent fibers in the autonomic nerves back to the heart and blood vessels.

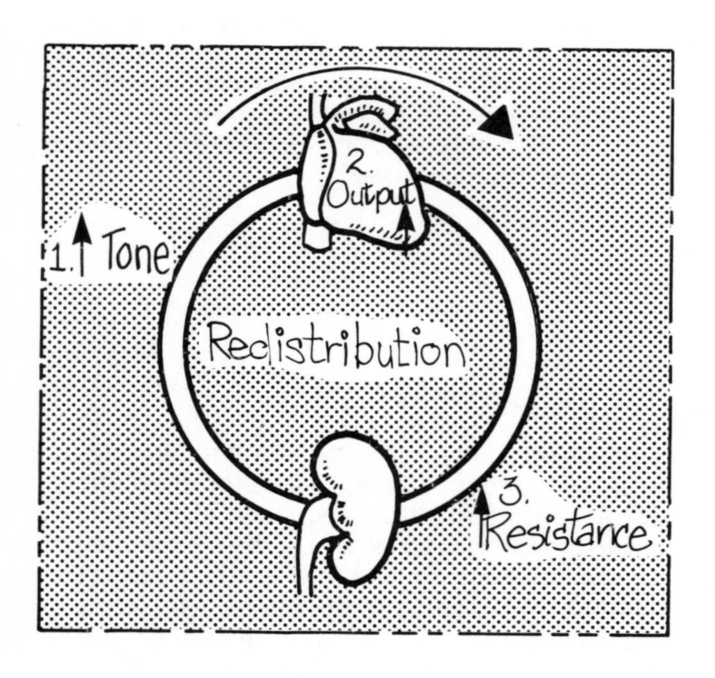

Thus, in the example of reduced venous return to the heart, and in response to altered afferent information, efferent stimuli will lead to changes that include:

1) Increased venous tone,
2) Increased cardiac output, and
3) Increased arteriolar resistance.

All of these tend to allow blood to be redistributed from veins to arteries. These changes all result from the direct effects of autonomic nerve stimulation and possibly the humoral release of vaso-active substances. They effectively correct the short-term changes in volume and flow that set them in motion in the first place.

Changes in Interstitial Volume

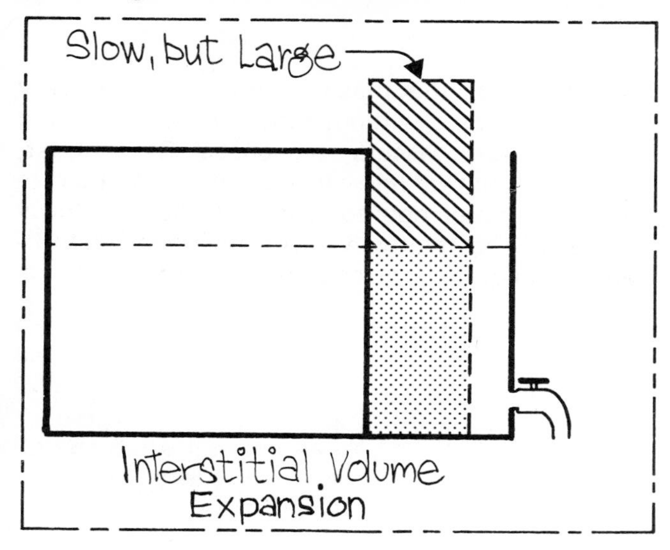

Changes in interstitial volume may be, by contrast, much slower than changes of plasma volume; however, they may eventually be much larger. For example, the interstitial space may expand by many litres over a long period of time without producing any major functional disturbance within the plasma compartment or the I.C.F.

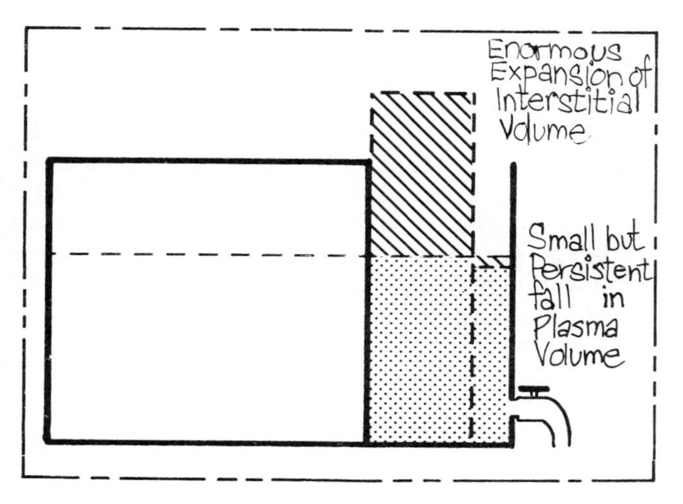

Such a change of interstitial volume may occur when a reduction of plasma albumin leads to a fall of oncotic pressure in the plasma with a consequent "shift" of volume into the interstitial compartment. This results in a small but persistent fall in plasma volume which activates a series of regulating systems that result in retention of salt and water by the kidney. So long as the plasma albumin remains low, however, volume will continue to shift out of the plasma compartment and produce ever-increasing interstitial expansion. The persisting small change in plasma volume may produce no apparent effects of itself but the interstitial edema which results may be overwhelming to the patient.

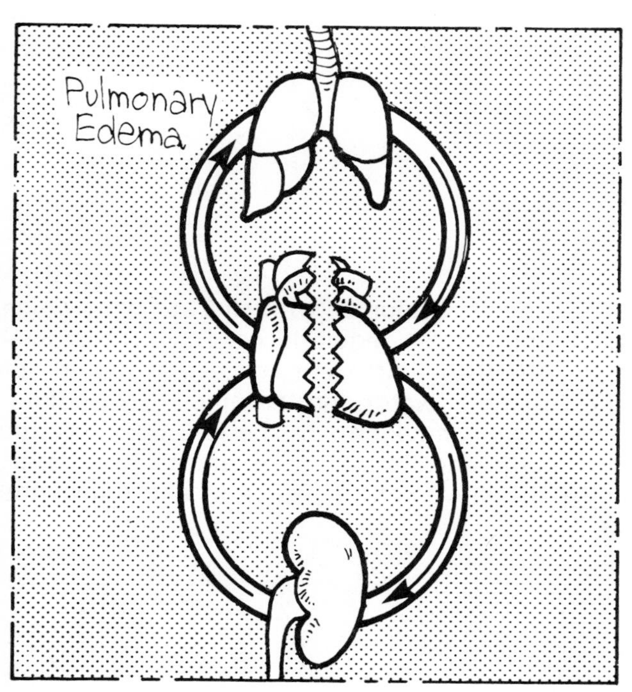

The balance between the plasma and interstitial compartments is more delicate in the pulmonary circulation. Our single circuit model is an oversimplification, and it can be redrawn to include a systemic and a pulmonary component of plasma volume.

The pulmonary circulation accomodates the same cardiac output as the systemic circulation but is more distensible and has a lower capillary hydrostatic pressure (about 10 mm Hg.). In addition there is periodic subatmospheric pressure in the chest cavity, associated with breathing, which will be transmitted to the distensible blood vessels in the chest. Increases in pulmonary interstitial volume result in breathlessness and extravasation of fluid into the alveoli (pulmonary edema) which may be an early sign of E.C.F. volume expansion in some clinical situations.

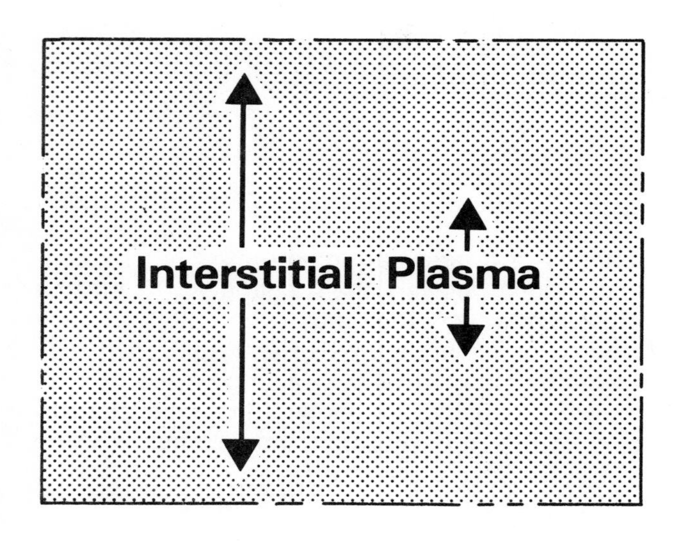

In summary, the plasma cannot change very much in volume without inducing fairly striking compensatory vascular responses. By contrast, the interstitial volume can expand (or contract) to a much greater degree, and if it does so slowly there will be relatively little functional disturbance until quite large changes have occurred.

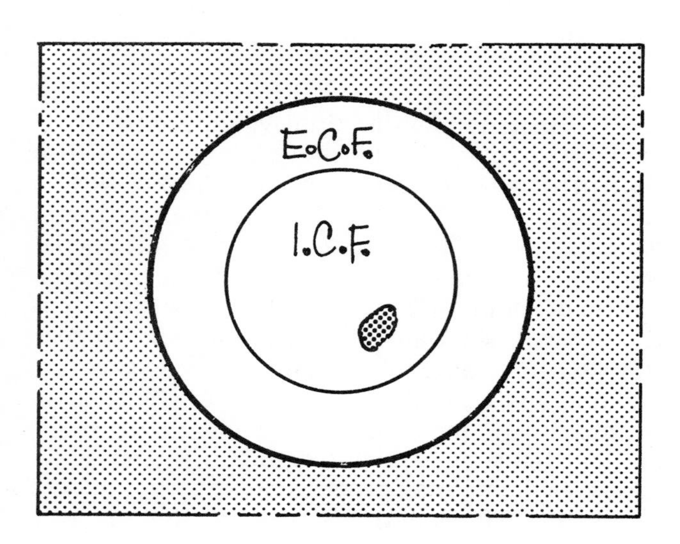

The important concept is that the E.C.F., as a whole, may vary within fairly wide limits as the body reacts to its environment, but the I.C.F. remains extremely stable. The E.C.F. interfaces with the outside world and becomes modified by it; the degree to which its composition is regulated by such organs as kidneys and lungs determines the degree to which the E.C.F. provides an optimum "bath" for the cells.

The Regulation of E.C.F.

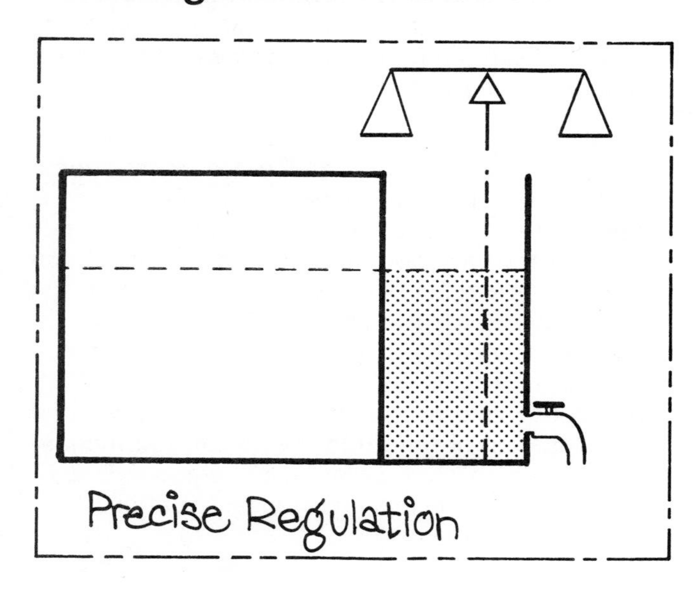

The body is equipped with mechanisms which allow the precise regulation of E.C.F. composition. The most important concern regulation of the volume of water and the amount of sodium. These two components of the E.C.F. determine its volume and its osmolality and in turn determine the movement of water in and out of the cells.

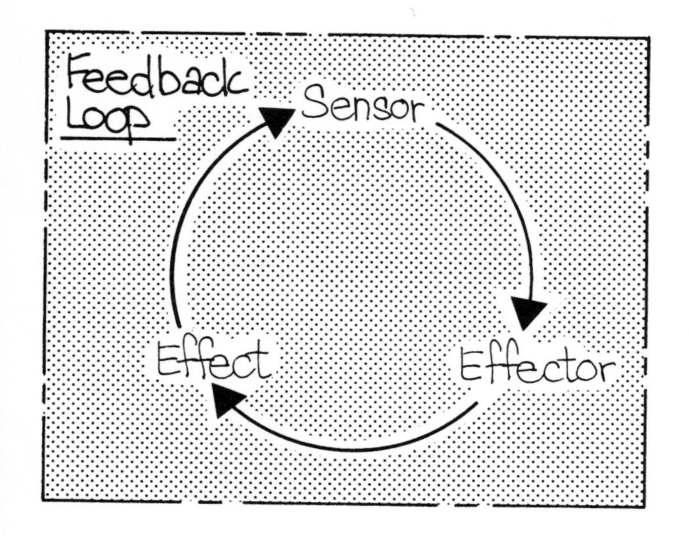

These regulatory mechanisms will be introduced here and described in more detail in later chapters. Both involve a "sensor" which recognizes changes, and an "effector" which corrects the changes. This implies a "feedback loop" which allows a continuous linkage between the output of a sensor and the effect which it produces.

Water Regulation -Antidiuretic Hormone

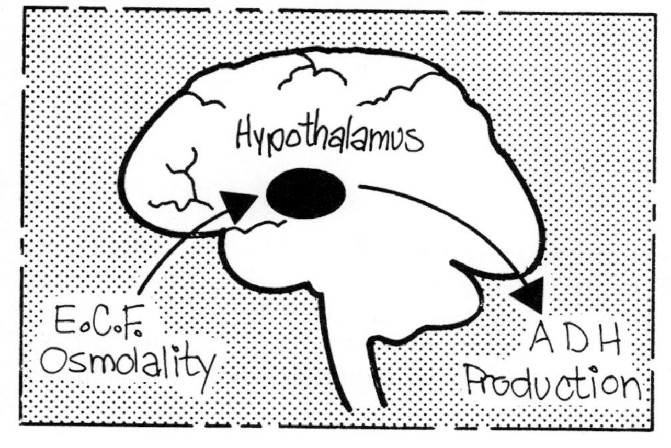

One sensor is a group of specialized cells in the hypothalamus which recognize changes of osmolality in the surrounding E.C.F. As a result, they regulate the amount of the peptide antidiuretic hormone (ADH) released from the hypothalamus. Minor changes in osmolality (which are constantly occurring) lead to minor changes in ADH production.

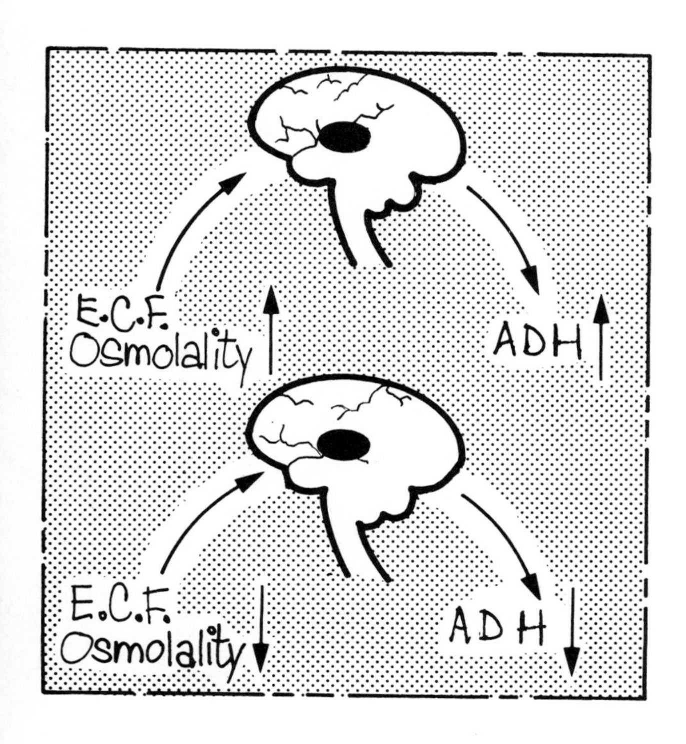

A rising osmolality increases ADH production, a falling osmolality reduces it. In normal living, with intermittent salt and water intake, these two components (osmolality and ADH) are in constant interaction, both moving within limits which we call "physiological."

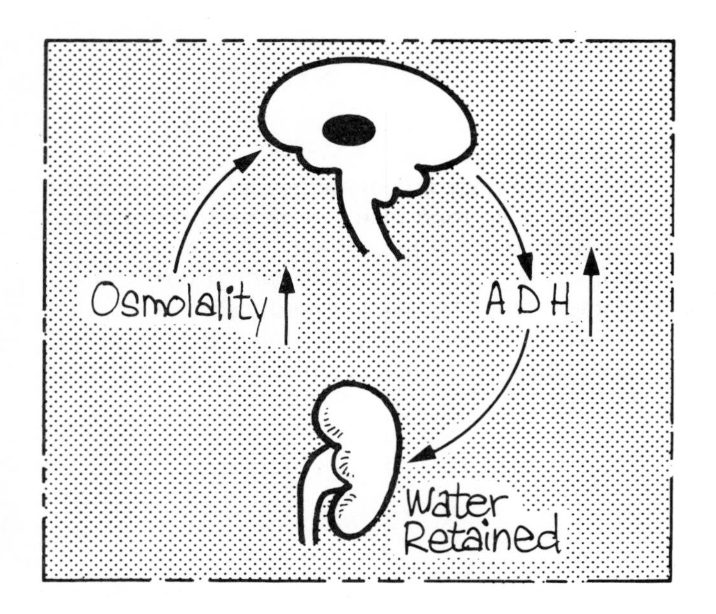

ADH regulates the amount of water retained by the kidney; the more ADH that is secreted, the more water that is retained.

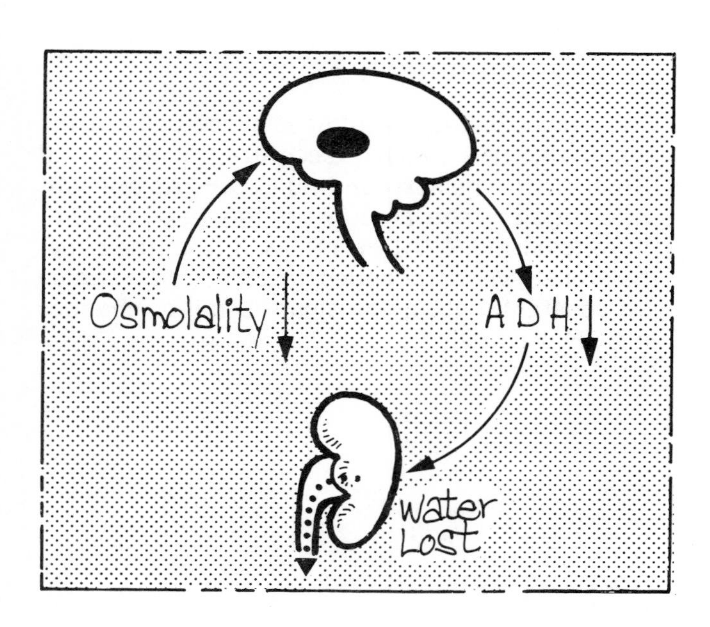

Conversely, the less ADH that is secreted, the more water is lost. Thus, varying ADH release results in a variation of water loss that precisely matches the variations of osmolality in the E.C.F. The details of this mechanism will be discussed in Chapter 7.

Volume Regulation-Aldosterone

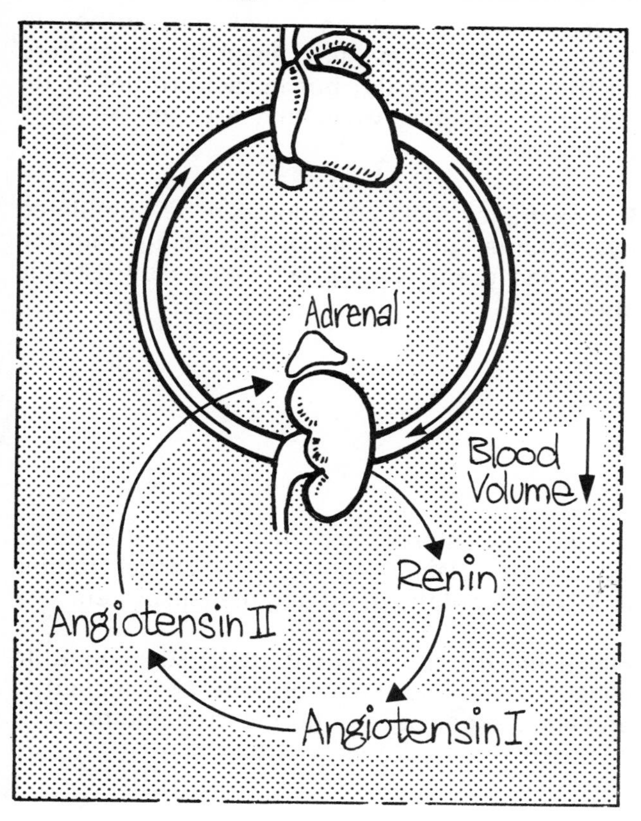

Other sensors are concerned with changes in volume rather than changes of osmolality. One of these activates the "Renin-Angiotensin-Aldosterone system." This involves a series of steps in a cascade that begins with diminution of blood volume or flow being sensed by specialized cells within the kidney and results in the secretion of aldosterone from the adrenal gland.

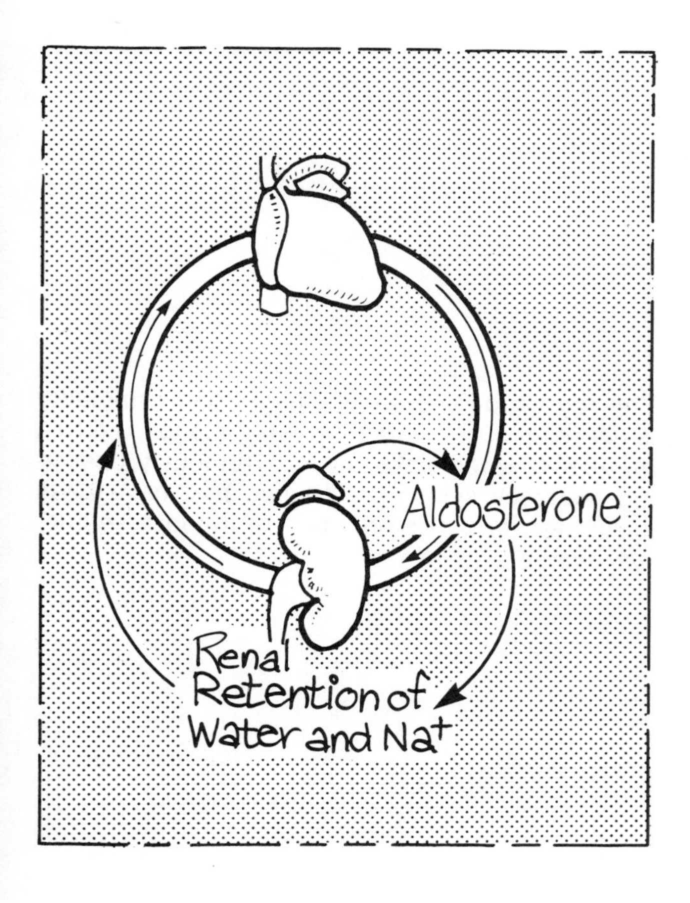

The loop is completed as aldosterone causes the kidney to retain sodium (and with it, water) to correct the volume deficit. The reverse process results in the loss of salt and water by the kidney in states of volume expansion.

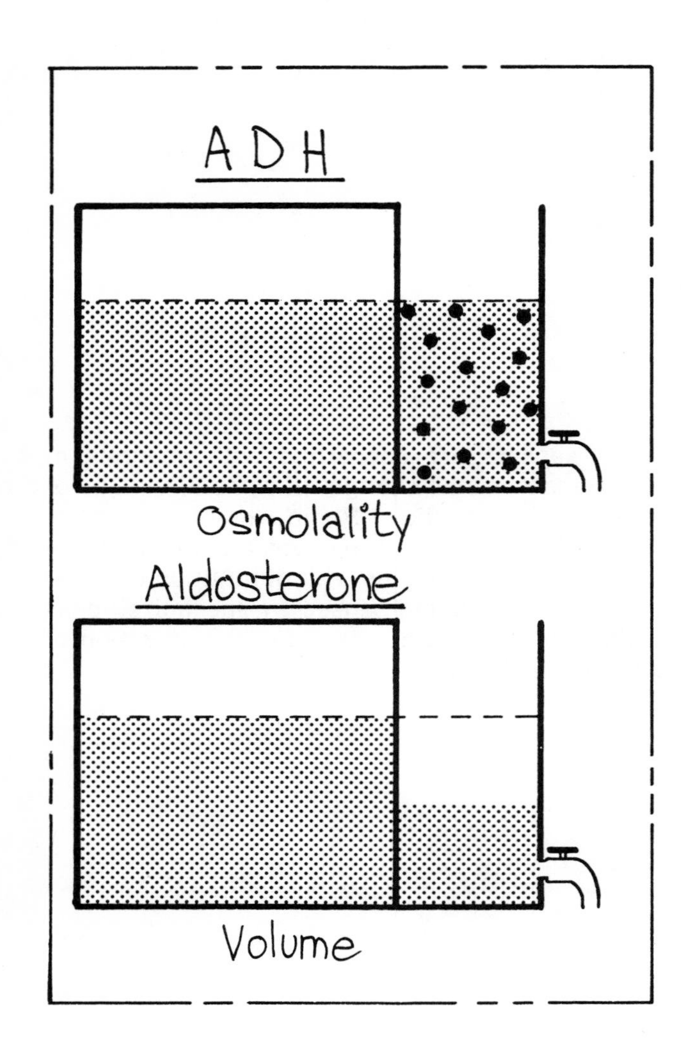

These concepts will be developed further in chapters 6 and 7. They have been simplified here for the sake of clarity, but in fact are linked together in complex ways that are not yet fully understood.

For example, ADH secretion can be stimulated by ''volume'' stimuli, probably arising from the heart and great vessels, as well as by osmotic stimuli.

Similarly aldosterone secretion can be stimulated by rising plasma potassium concentrations independent of the renin-angiotensin system.

However, for the moment, it is convenient to consider ADH as the ''hormone of osmolality'' and aldosterone as the ''hormone of volume.''

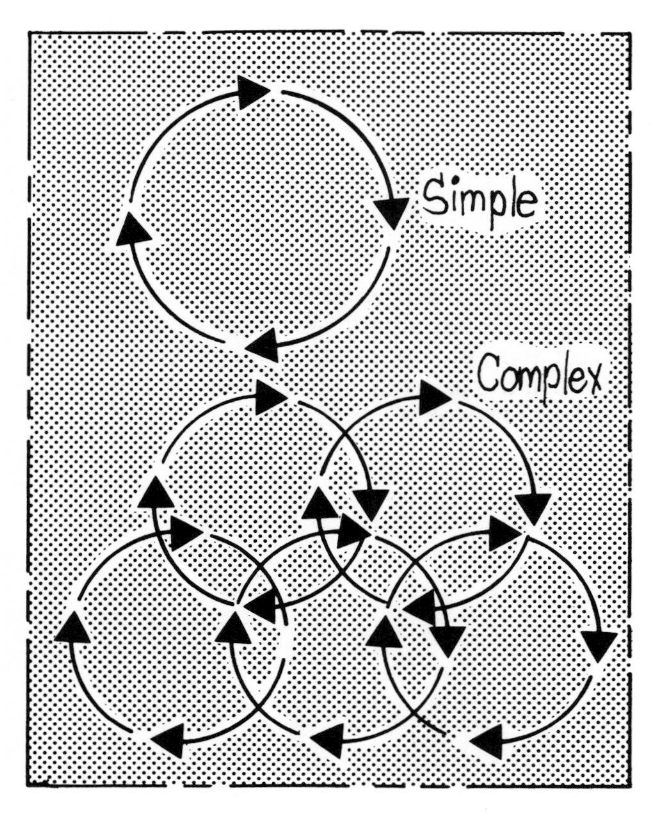

The regulation of E.C.F. volume and osmolality is not related to a *single* ''feedback loop'' but to a *number* of interrelated and continuously operating feedback systems. The two that have been described here form only part of the total regulatory system.

Clinical Assessment of Changes in Fluid Compartments

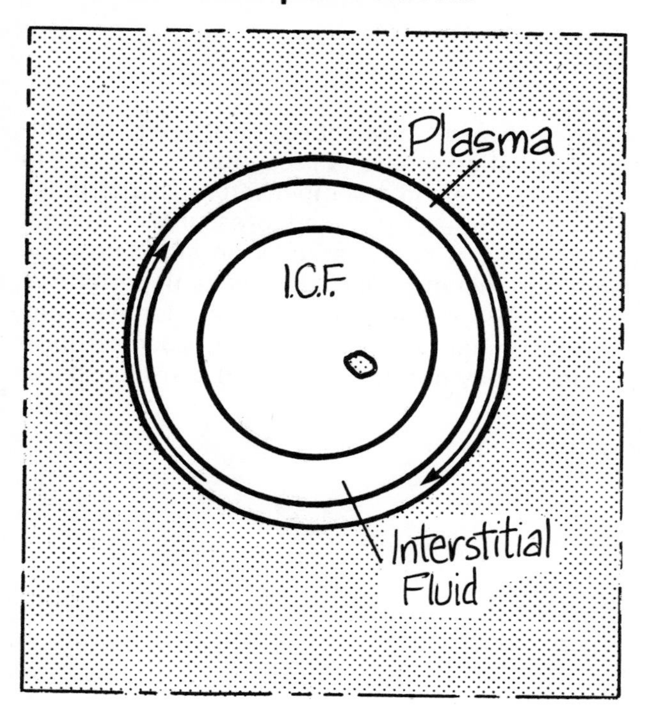

Changes in the distribution of fluid volume between the major body compartments can be assessed by an observer at the bedside.

Because of its relative accessibility, the E.C.F. is open to clinical examination more readily than the I.C.F. which is "hidden" within the protective shell of the E.C.F.

Clinically, changes in the plasma compartment are the most obviously visible whilst minor changes in the interstitial space are often more subtle.

Plasma Compartment

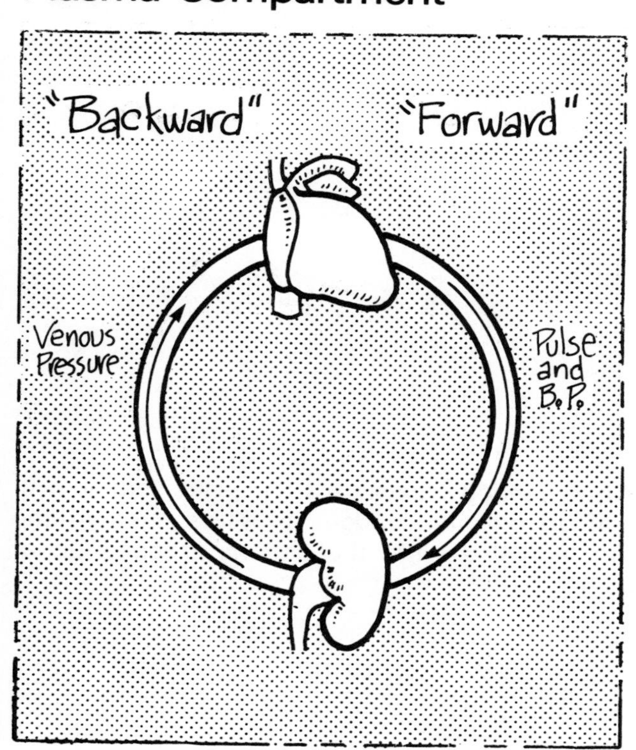

We have divided the plasma compartment into "backward" and "forward" components where changes in volume and flow go hand in hand. "Forward" volume and flow are mirrored by the pulse and arterial pressure, whilst "backward" volume will be reflected by changes in venous pressure and interstitial volume.

Diminished "Forward" Volume and Flow

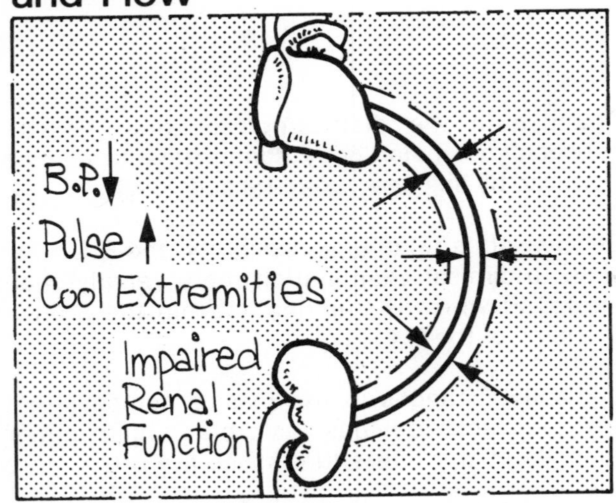

A low "forward" flow causes diminished peripheral circulation which is recognized by a fall in blood pressure and a compensatory rise in pulse rate. Sometimes the blood pressure is normal while lying but falls markedly on standing. The hands and feet may be cold, often with peripheral cyanosis. Poor perfusion of the kidneys may result in impaired renal function; poor cerebral perfusion may produce disturbed consciousness.

Expanded "Forward" Volume and Flow

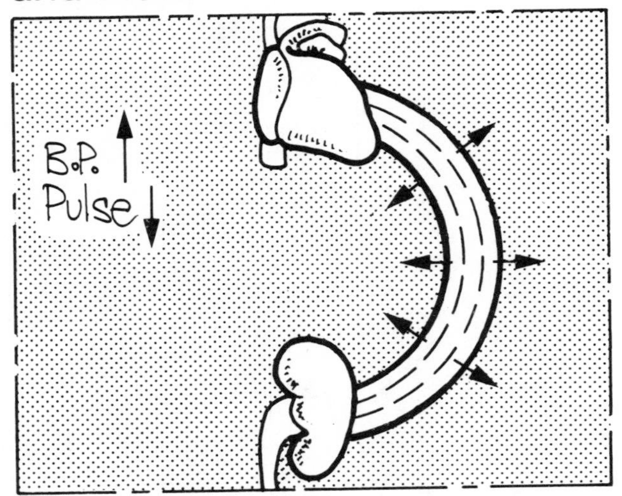

An expanded "forward" flow will result in a rise in blood pressure and compensatory cardiac responses which include a reduced pulse rate.

Expanded "Backward" Volume and Flow

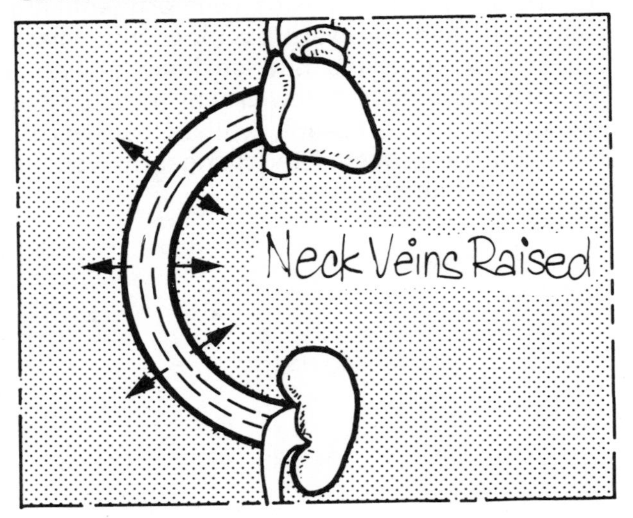

An expanded "backward" plasma volume will lead to elevated venous pressure seen in the jugular veins (with the patient propped up at 45°).

Diminished "Backward" Volume and Flow

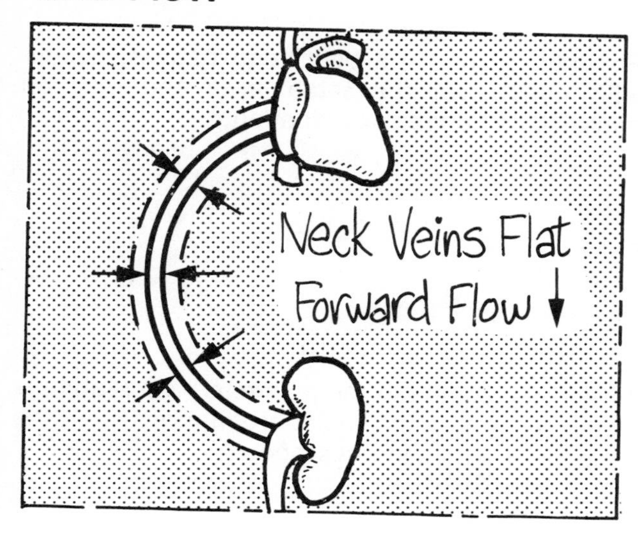

A reduction in "backward" volume rapidly leads to a fall in cardiac filling and the signs of a low forward flow. In practice, therefore, there are few clinical features specific to a contracted venous volume, although the neck veins will not be visibly engorged.

Interstitial Compartment

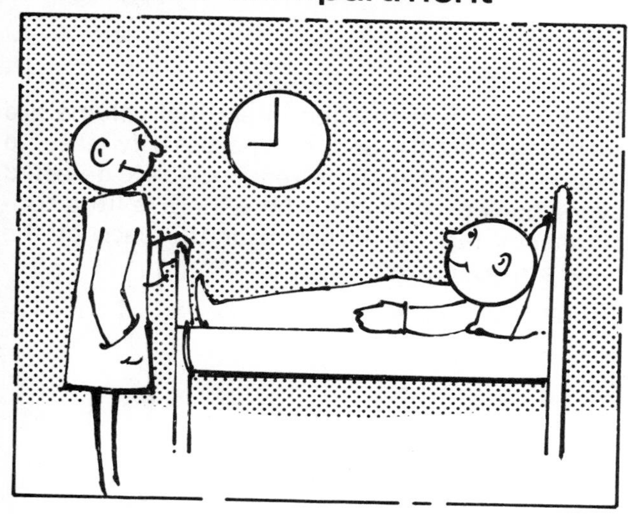

The interstitial volume can also be assessed at the bedside, and careful examination will be well worth the time.

Expanded InterstitialVolume

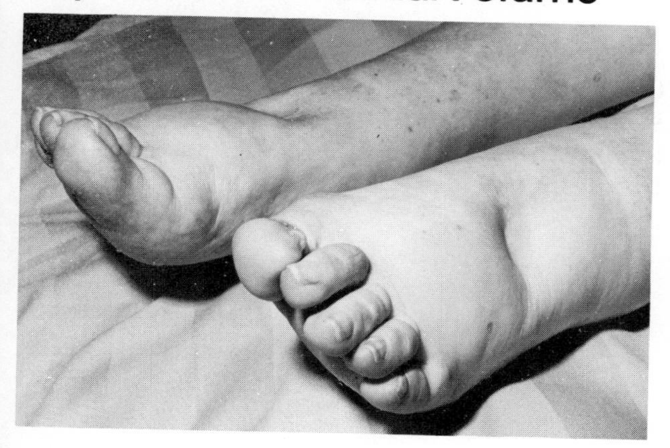

An expanded interstitial volume will result in edema, revealed by the pitting produced by firm and sustained pressure with the thumb over the ankle. Gravitational effects and the degree of laxity of subcutaneous tissues determine the distribution of generalized edema.

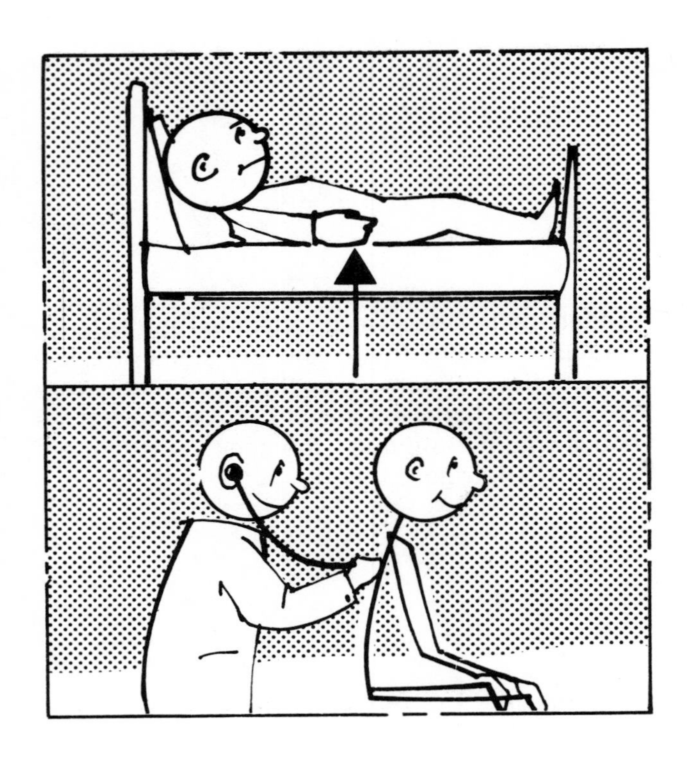

For this reason, in a bedridden patient, edema will collect over the sacral area and may be missed by a hurried examination.

Interstitial edema in the pulmonary circulation is detected by hearing moist crepitations over the lung bases with a stethoscope.

Contracted Interstitial Volume

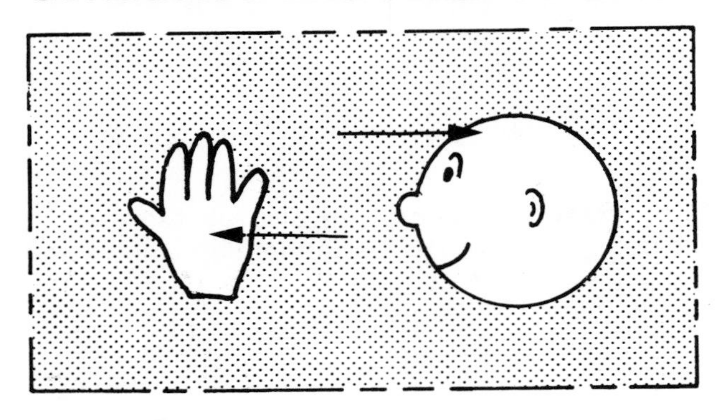

The signs of a contracted interstitial volume are sometimes subtle and not necessarily easy to detect. Skin elasticity is reduced, best detected over a bony prominence where there is little subcutaneous fat; the forehead or the back of the hand are often used.

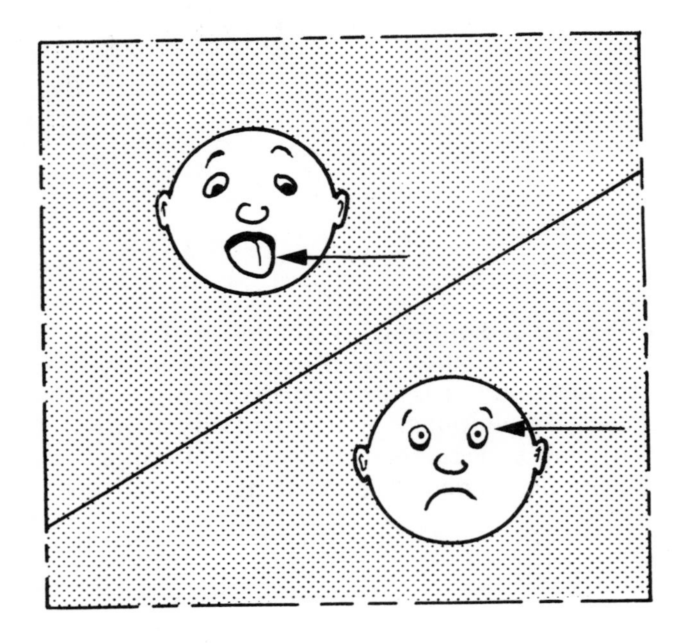

Other signs are dryness of mucous membranes, such as the tongue (unreliable in a mouth-breathing patient) and reduced eyeball tension (difficult to determine).

In the elderly these signs may be difficult to assess, particularly because of the loss of elasticity of the skin with age. The complete *absence* of edema may be as important as the *presence* of reduced skin turgor in an elderly patient whose interstitial volume is reduced.

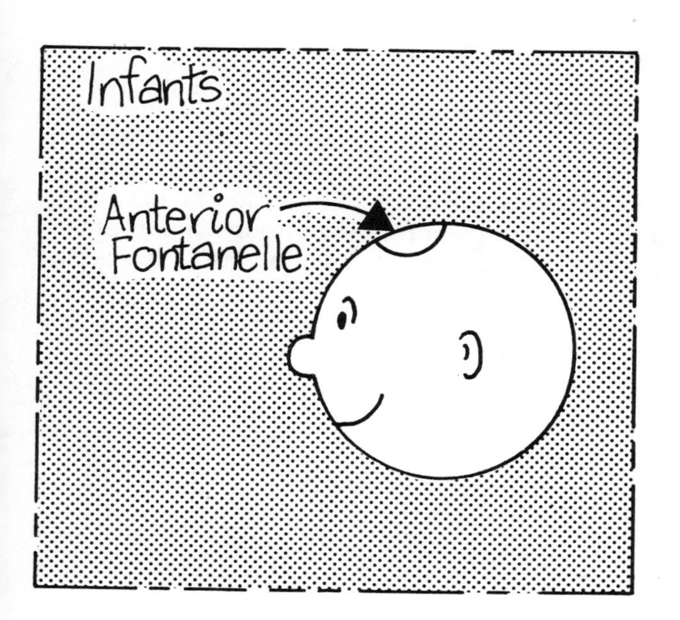

In the infant the tension of the anterior fontanelle is a good indicator of interstitial volume. In contracted states it will be more deeply depressed than usual, whilst in expanded states it may be slightly bulging.

There are many highly sophisticated ways of assessing pressures and flows (using special central venous catheter recorders, accurate measurements of cardiac output and so on.) These reflect changes in volume in both systemic and pulmonary circuits.

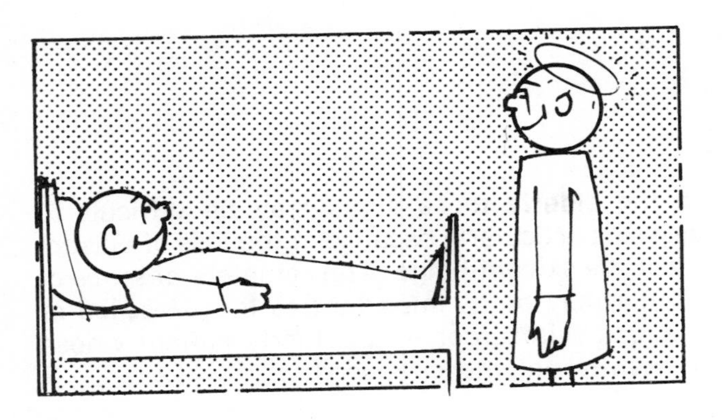

However, measurements of pressure and flow may be influenced by factors other than changes in volume. Whilst they can be very helpful in special situations, remember that nothing can replace good bedside clinical skills in the initial assessment and follow-up of patients with volume problems.

The Intra-cellular Compartment

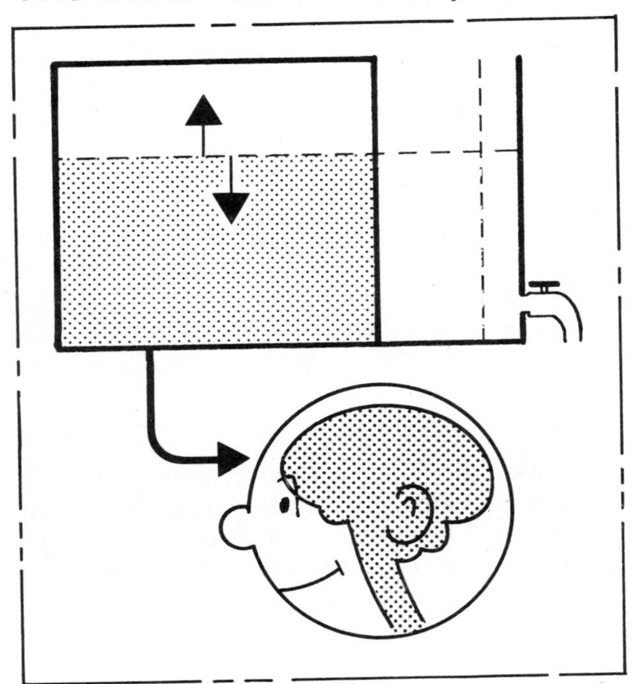

Clinical assessment of the intracellular space is very difficult. The brain, because of its rigid bony constraints, is the organ whose function may reflect changes in intracellular volume in a way that can be clinically apparent. Headache and confusion may occur with either expansion or contraction of the I.C.F. Expansion of the brain can only occur by herniation through the foramen magnum or in the presence of a hole in the skull.

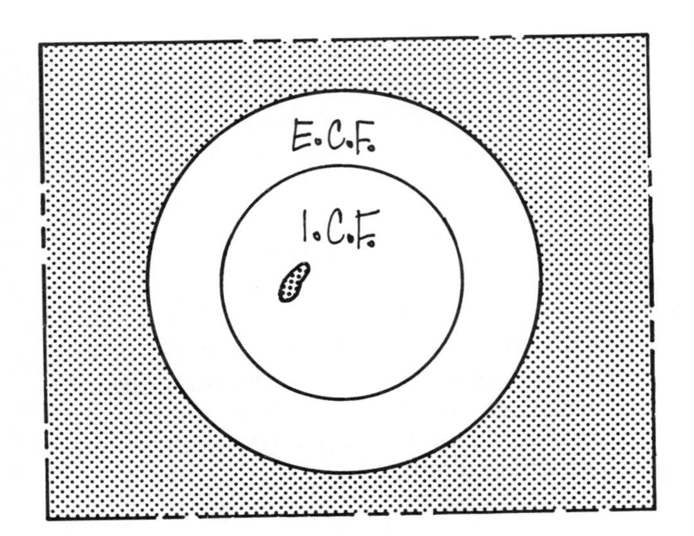

It is, of course, because the I.C.F. is surrounded by an E.C.F. ''shell'' that the I.C.F. is clinically inaccessible. Clinical assessment of body fluids therefore means, for practical purposes, assessment of the E.C.F.

Chapter 5

Body Fluids in Abnormal States

Using the models built up in the previous chapters it is now appropriate to discuss what happens when the system is stressed. The examples dealt with here are, of necessity, simplified, and provide only a skeleton upon which to hang the concepts of disordered regulation. Many details are left out in the interests of clarity, and not because they are unimportant.

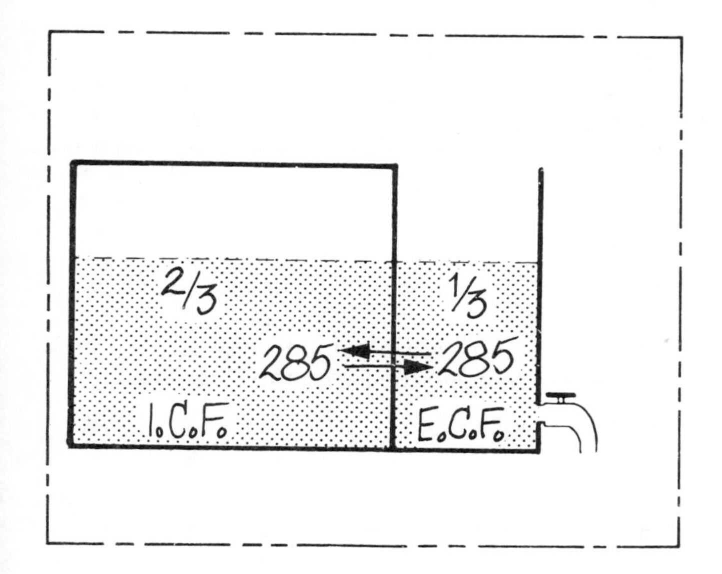

Remember that normally the relative volume of the intracellular and extracellular compartments is 2/3 :1/3. There is an osmotic balance between the two compartments, both having an osmolality of 285 milliosmoles per kilogram. Sodium ions and chloride ions predominate in the E.C.F. and are the major determinants of osmotic pressure.

A Man in the Sahara

Now we can consider what would happen to this system if water was lost. Suppose we take the extreme example of a man stranded in the Sahara desert at high noon: Water *intake* will be reduced to zero, but water *loss* will continue via the skin (as sweat) and the lungs. He will also lose water as the obligatory solvent for those solutes that *must* be excreted in the urine. There will, of course, be a loss of sodium and chloride (as well as water) in the sweat, but to simplify the present example we may consider him as losing primarily water.

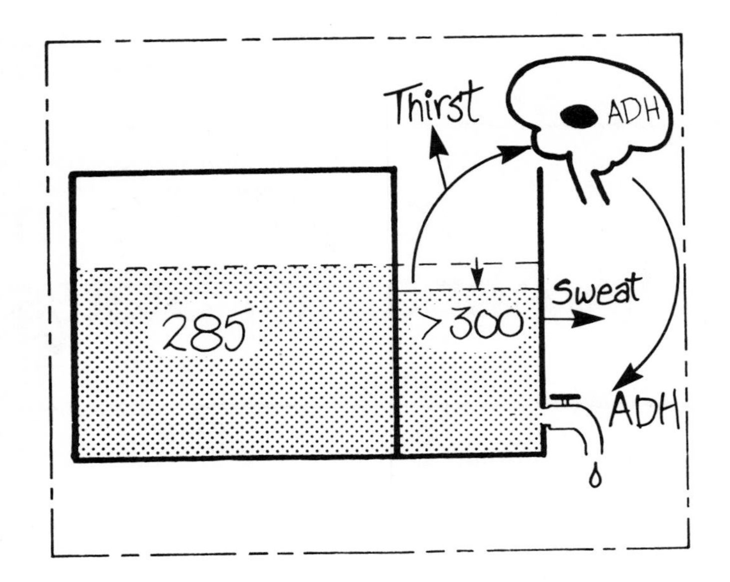

At first this will result in a loss of water from the E.C.F. and a rise in its sodium concentration and thus its osmolality. This will lead to thirst and the release of ADH which will minimize water losses by concentrating the urine maximally. Man is not well adapted to desert conditions and even under the severest stress an adult will still excrete about 400 millilitres of water in the urine daily.

This continued loss of water from the E.C.F. leads to a rising sodium concentration and osmolality, as well as a fall in the E.C.F. volume.

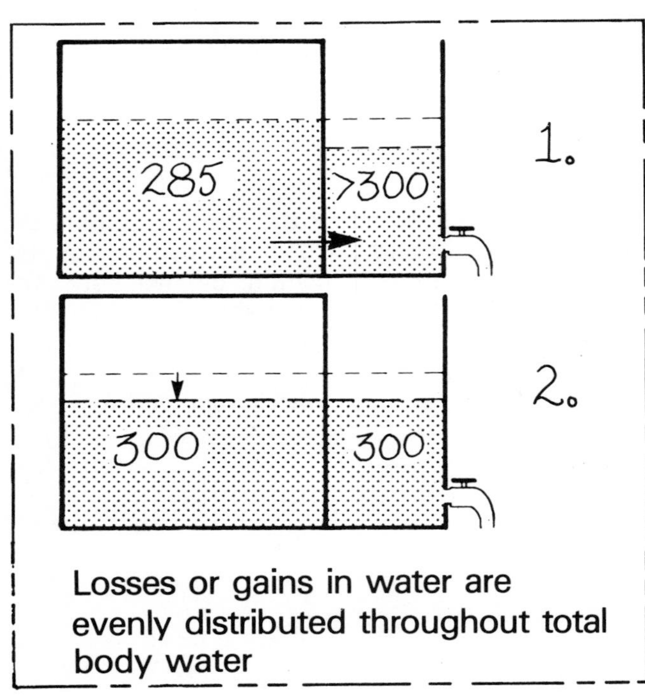

Normally he would drink water to correct matters but since he has no water, the osmotic gradient that has now been established between E.C.F. and I.C.F. leads to movement of water from the I.C.F. into the E.C.F.

This movement will continue until osmolality is equalized between the two compartments.

With each upward step of E.C.F. osmolality there is a compensatory upward step in the I.C.F. osmolality. So long as no external water is available this process will go on, resulting in a continuing fall of volume and rise of osmolality in both body compartments.

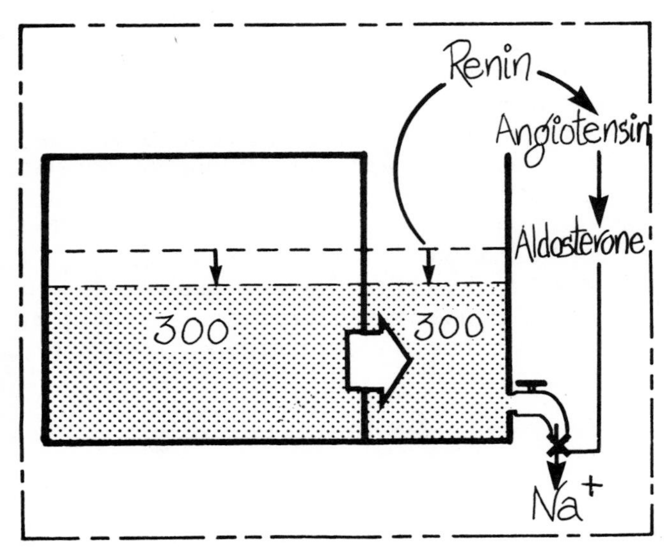

Because E.C.F. volume is reduced, the Renin-Angiotensin-Aldosterone system will be activated and in addition to retaining water the kidney will also retain sodium which, in normal circumstances would expand the plasma volume by retaining more water. In this case, however, no additional water is available.

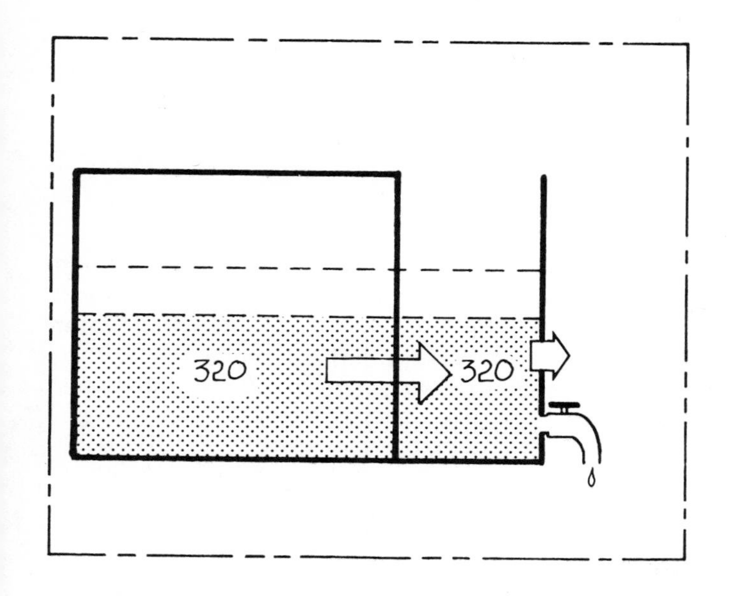

Each body compartment in the man in the desert becomes equally water-depleted. As the day gets hotter and the water gets scarcer, he continues to lose water. The osmolality of his E.C.F. goes on rising and so the osmolality of the I.C.F. also goes on rising, and of course the fluid volumes gradually diminish, with equal distribution of fluid contraction between all body compartments.

Signs of ICF Volume Depletion

As time goes on there is continued intense stimulation of hypothalamic osmorecepters, but progressive shrinkage of the I.C.F. volume. The effect of this is most obviously seen as a disturbance of function in the cells of the brain with the result that our deserted friend ''goes mad with thirst.''

Signs of E.C.F. Volume Depletion

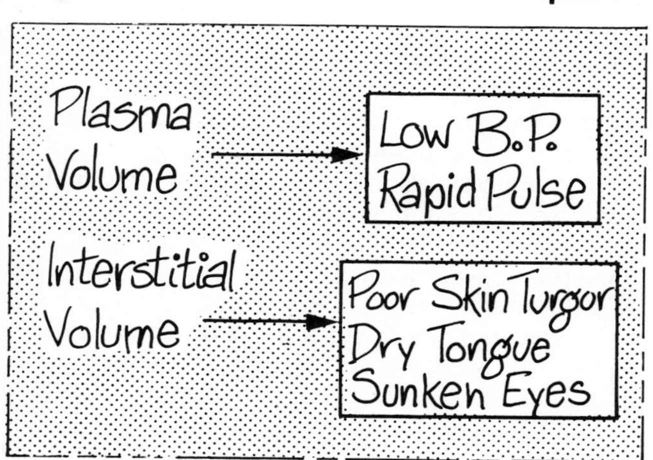

If you were to examine him at this time, it would be apparent that both his E.C.F. and I.C.F. had shrunk. He would have the signs of a reduced plasma volume, that is low blood pressure, rapid pulse, and a reduced interstitial space as you would see by his poor skin turgor, dry tongue, and sunken eyes. Because he has lost water, but only a relatively small amount of sodium and chloride, his serum sodium and chloride concentration will be elevated.

A Man in the Pacific Ocean

But now let's alter the situation and put the man on a raft in the Pacific ocean. He is stili exposed to heat and sunshine and still has no fresh water but he is now surrounded by sea water with a sodium concentration of about 450 millimoles per litre.

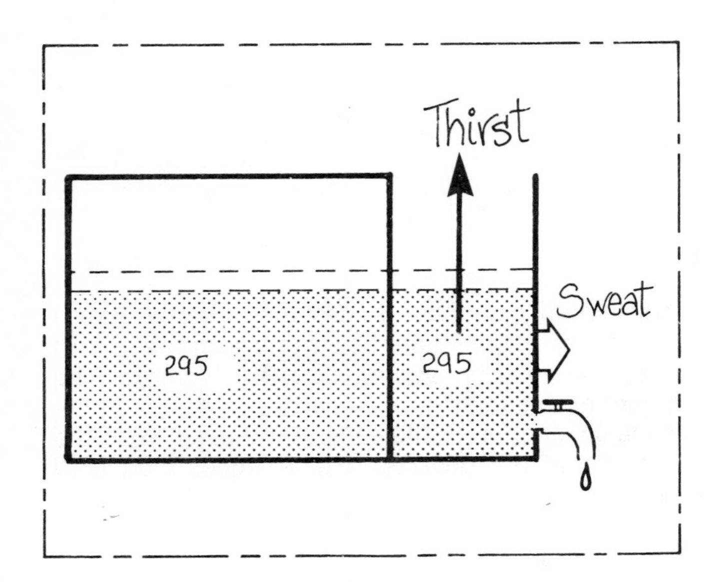

What happens at first is exactly what happened to him when he was in the desert. He develops an increased osmolality in both compartments and a water deficit spread throughout the total body water.

This leads to thirst which may become severe enough to drive him to drink sea water in an attempt to relieve it.

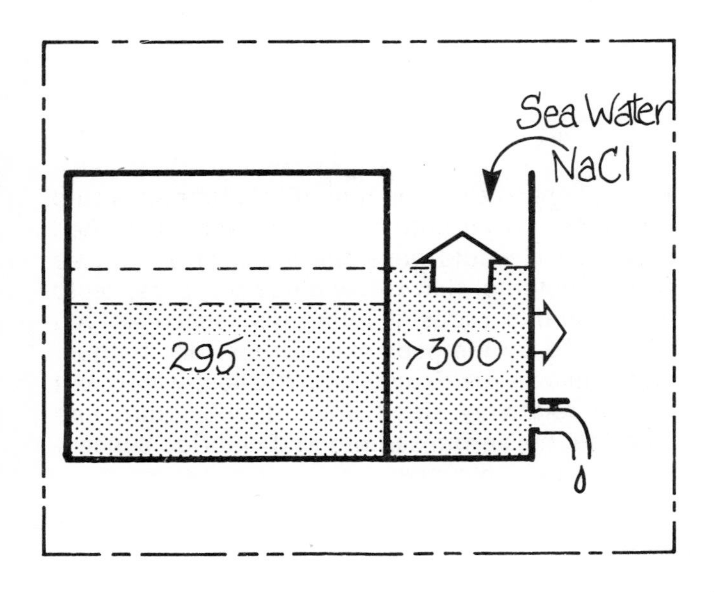

If he absorbs the salt and water that he ingests, it will stay in the E.C.F., increasing its volume and its osmolality. It cannot be excreted because the normal human kidney is unable (even in the most favorable circumstances) to excrete urine with a sodium concentration above about 300 millimoles per litre. This means that to excrete 100 millilitres of 450 millimoles per litre sodium chloride the kidney would have to excrete at least 150 millilitres of urine. In other words he could only drink sea water, and get away with it, so long as he had free access to additional fresh water.

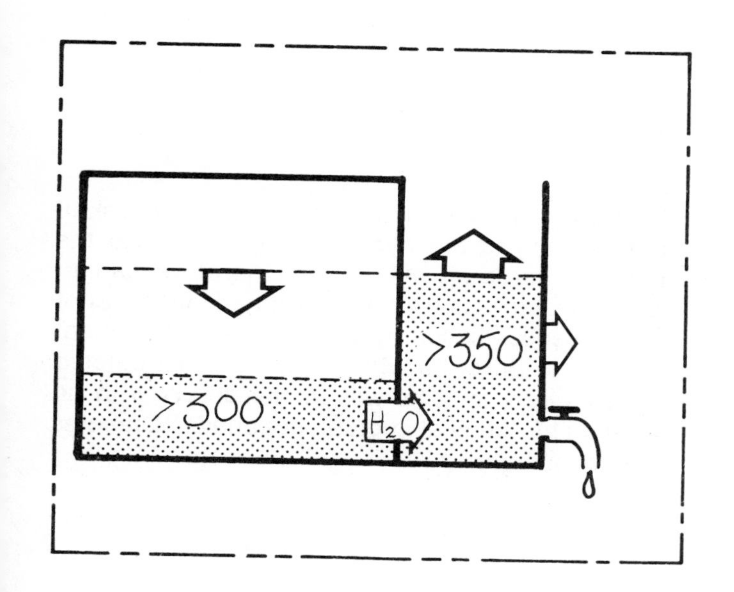

The E.C.F. will therefore become more hypertonic and its volume will tend to increase at the expense of the I.C.F. The osmotic gradient between the two compartments results in continued movement of water out of the cells thus reducing I.C.F. volume and increasing I.C.F. osmolality to match the osmolality of the E.C.F.

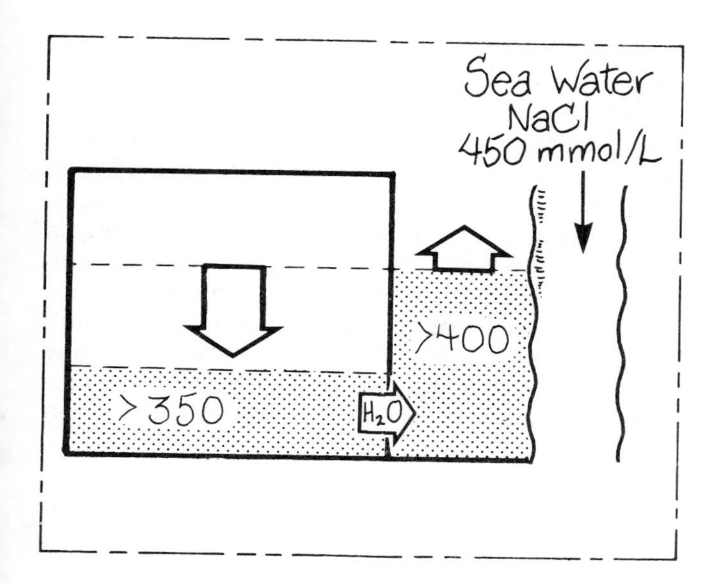

The degree to which the E.C.F. volume will in fact expand above normal depends upon additional factors; for example sea water contains ions such as magnesium which may induce diarrhea and vomiting, which will limit the degree to which the initially shrunken E.C.F. will expand.

Whether the hypertonic sodium of the sea water is absorbed in full, or exerts its effects via an intestine that can behave as if it is in communication with the E.C.F. makes little difference to its major effect which is progressive intracellular dehydration as the E.C.F. osmolality rises.

The ultimate disturbances of water distribution and volume regulation in the individual who drinks sea water are more dramatic and more rapid than those occurring in his friends who resist the drive to drink. The rapid cellular dehydration leads to rapid and severe disturbances of intracerebral function and an earlier departure!

These examples have been extreme, and for many of us theoretical, but they have been used to make a point. There are, however, more common pathological situations where the same mechanisms operate.

Patient with Diabetes Insipidus

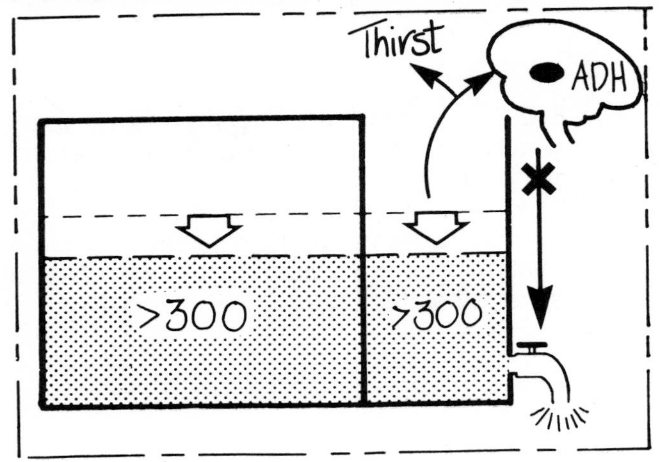

The patient with a fracture to the base of the skull and damage to the hypothalamus may lose the ability to produce ADH and therefore cannot control water loss from the kidney. If his level of consciousness becomes disturbed and he cannot respond to thirst he will become water depleted just as surely as if he was in the desert with no water. This clinical picture is called "Diabetes Insipidus."

An Unconcious Patient

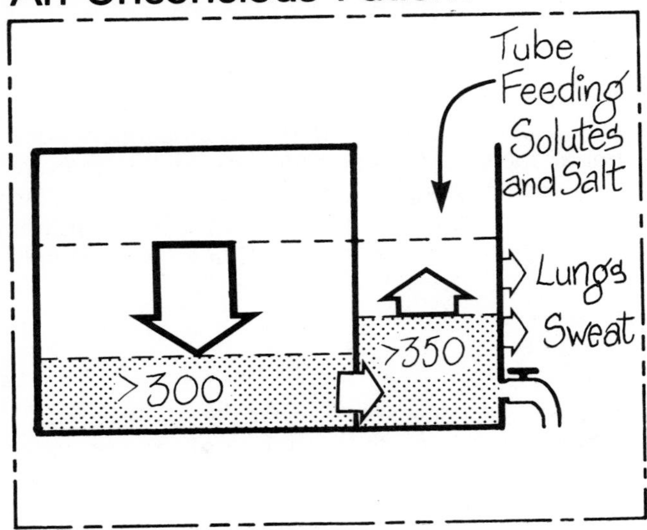

Another example is the patient who is unconscious and has a tracheostomy and a fever. Increased losses of water from the lungs and from increased sweating may be underestimated. Since he cannot complain of thirst (because he is unconscious), he may become deprived of water just as surely as our friend in the desert. His plight may not be recognized until the increased concentration of solutes (particularly sodium) is noted in his blood. If the problem of his inadequate water intake is compounded by an increase of the solute load by tube feeding with high calorie, low volume food concentrates he may closely approximate the conditions of our shipwrecked mariner who drinks hypertonic sea water.

Sodium, Water, and Volume

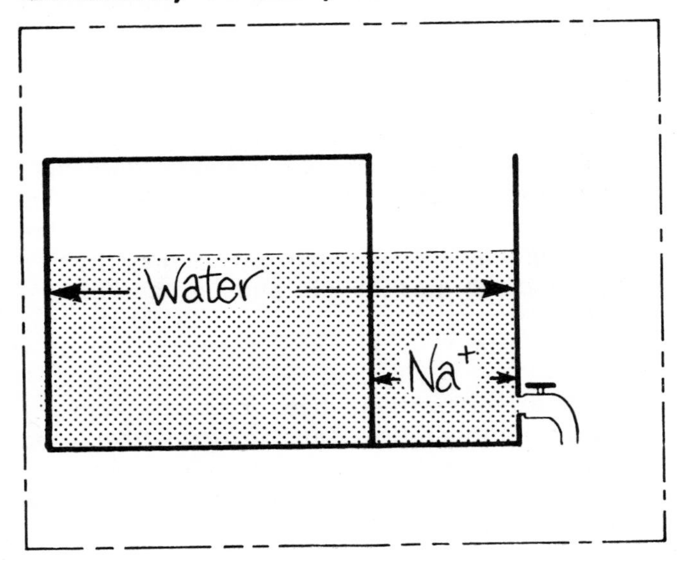

These situations demonstrate how close is the link between sodium and water in determining the volume of the intracellular and extracellular spaces.

In a healthy individual any excess of sodium that is ingested will result in a change in osmolality, thirst, subsequent drinking, and, finally, expansion of the E.C.F. As an example, we are all familiar with being thirsty after a salty meal, and it is no accident that whilst the salted peanuts are free, the thirst that they produce may be expensive!

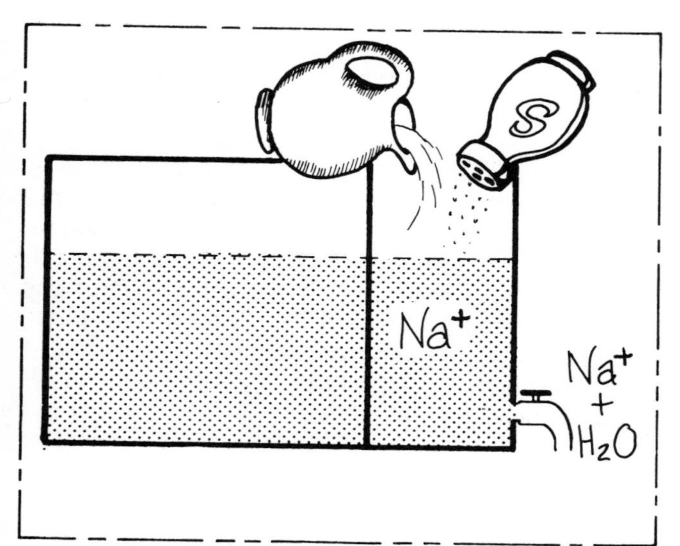

In a normal subject, this volume expansion is registered by volume receptors, resulting in reduced secretion of aldosterone and ADH which limit renal tubular reabsorption of filtered sodium and water until the excess has been excreted. In a *normal* subject with a healthy heart and kidneys it is almost impossible to overload the E.C.F. by eating salt and drinking water.

However if the heart is not normal, or if volume regulation in the E.C.F. is disturbed, this normal response will be incomplete.

Heart Failure

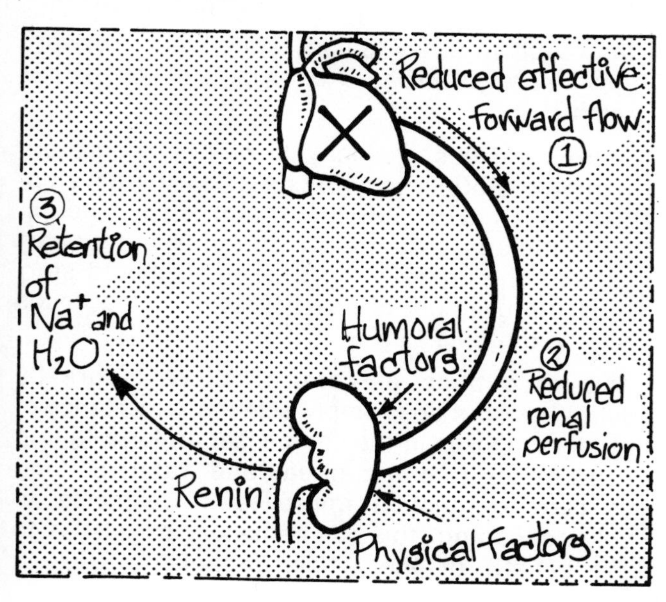

Let's look at someone with heart failure. The initial problem here is that the "effective forward blood flow" is reduced. The heart is not pumping adequately, and so the effective arterial volume and flow are reduced. This leads to underperfusion of the kidney and a series of responses which result in retention of sodium and water by the renal tubule. These responses include physical factors within the kidney and humoral factors acting on the renal tubule. (See Chapter 6)

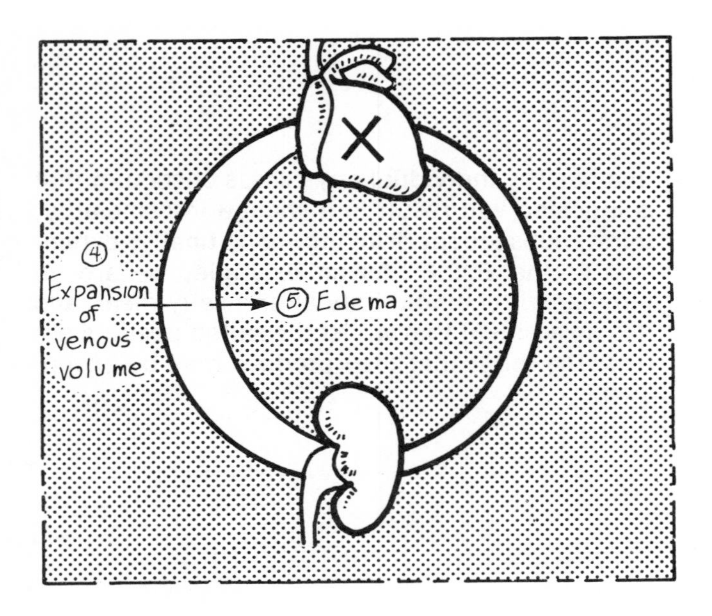

Since sodium is an extracellular ion, the additional volume remains in the E.C.F. If the heart is still failing as a pump, then the distribution of the retained volume will remain uneven. This will lead to the expansion of *interstitial* volume due to increased capillary hydrostatic pressure.

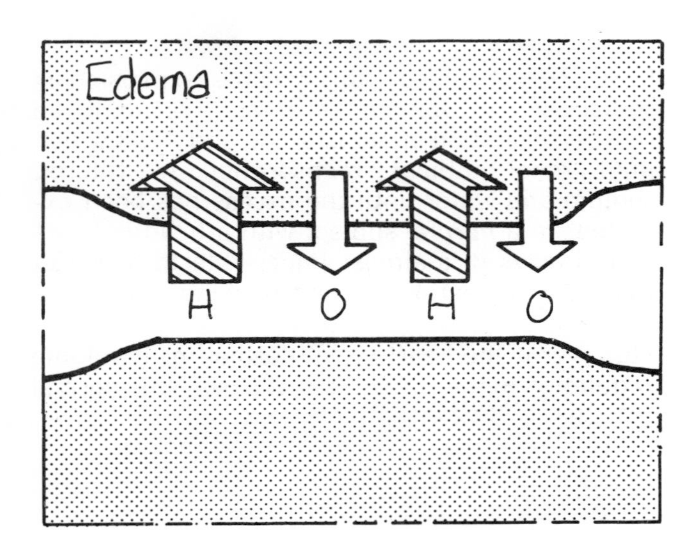

The hydrostatic pressure at the venous end of the capillaries rises, and the Starling hypothesis (relating to fluid movements in and out of the capillaries) tells us that volume will shift out of the plasma compartment into the interstitial space. So edema occurs, and continues to get worse if the heart cannot respond to increased venous filling by increasing output.

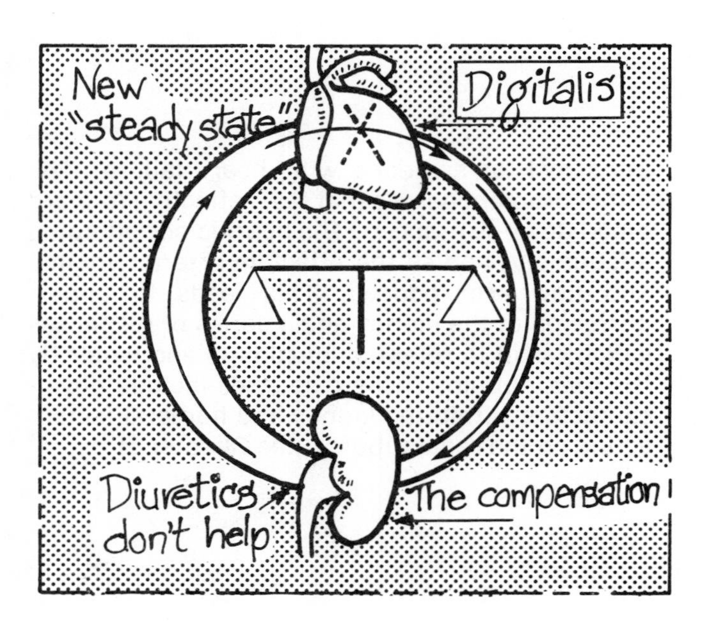

The expanded venous volume and increased venous return to the heart stretch the cardiac muscle fibers and this may result in improved cardiac output (Starling's Law of the heart). This can result in a new steady state where forward flow improves *because of* expansion of venous volume.

Diuretic drugs, which block compensatory volume retention, may in fact delay the development of a new steady state, and are not necessarily the appropriate first step in treating heart failure. The approach to treatment should start with improvement of heart function using drugs such as digitalis.

This description of cardiac failure is greatly simplified, but does emphasize the importance of renal responses in the compensation for a failing heart by the retention of salt and water. So long as cardiac function improves in response to volume loading, this compensatory process is to the benefit of the whole organism.

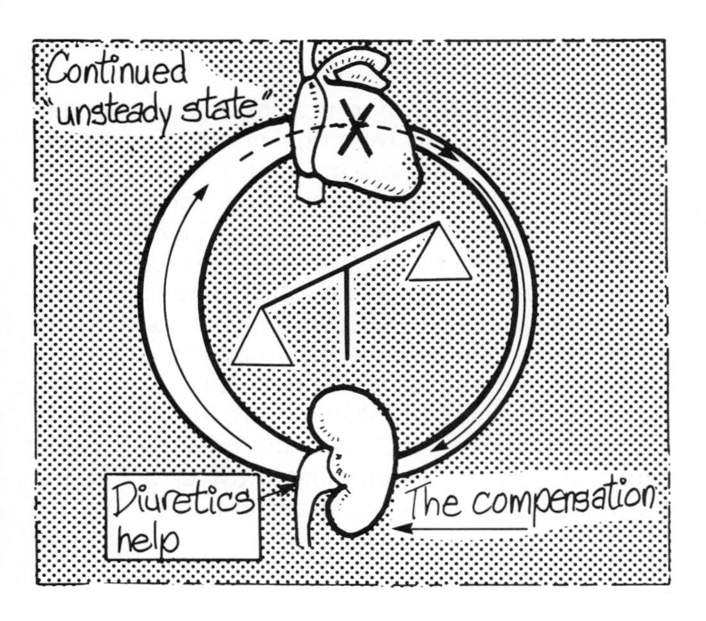

The improved cardiac output resulting from increased filling volume of the heart, and predicted by Starling's Law, occurs up to a point. Beyond that point, however, further increases of volume "behind the heart" do not produce further increase in output, and as the heart dilates, output may even fall. In this situation, the renal retention of salt and water compounds the initial problem of the failing heart, and far from being a compensation becomes an aggravating factor that perpetuates a deteriorating "unsteady state." Reduction of volume load with diuretics now becomes a priority of treatment (although the direct cardiotonic effects of digitalis remain important.)

Hypoalbuminemia

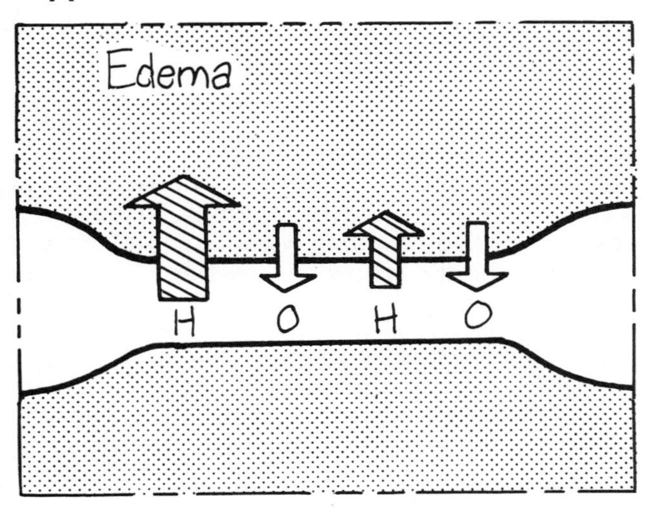

A similar situation exists when the serum albumin is reduced, such as by failure to synthesize it in liver disease. In this case, effective blood volume is reduced as fluid leaks out of the capillaries into the interstitial space because of a reduction of colloid osmotic pressure and *not* because of a failing heart and a rising hydrostatic pressure.

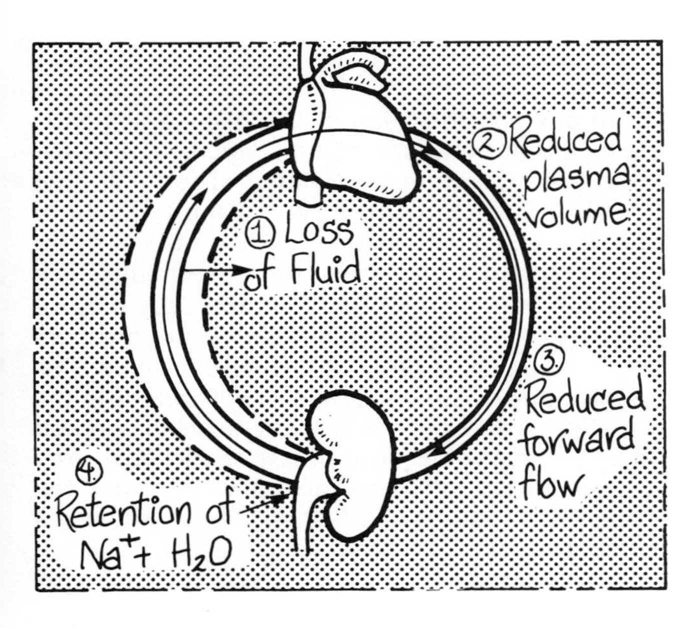

The kidney registers this as if it represented an overall reduction in plasma flow, and so the same changes that occurred in heart failure are set in motion. These result in retention of sodium and water, and the exaggeration of edema which continues to accumulate so long as the receptors in the kidney and elsewhere register a "volume deficit."

Thus salt and water are continually retained and continue to leak out into the interstitial space as long as the plasma albumin level remains inadequate.

Because the heart is healthy the *plasma* volume and flow are not "asymmetrical" as in heart failure, and the reduction in volume is on both the arterial and venous sides of the circulation.

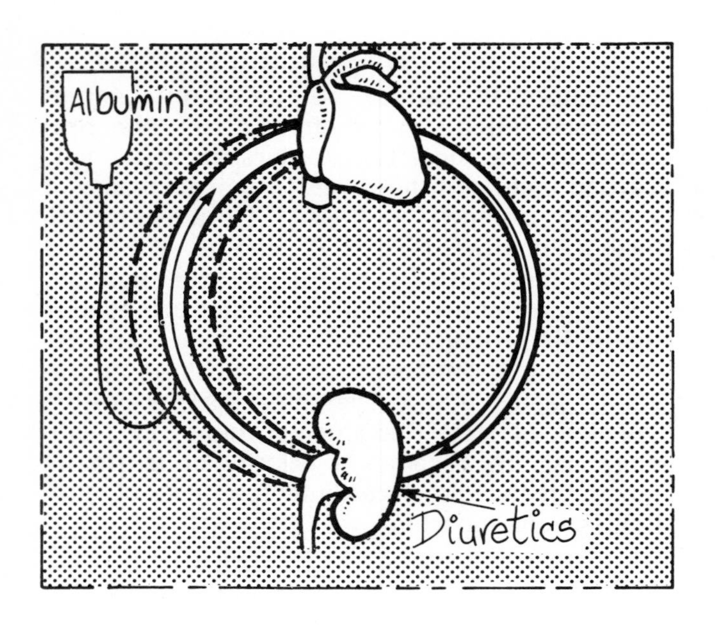

In this example, diuretics may improve the edema by blocking retention of salt and water, but once again they are blocking the response to a pathological stimulus and not correcting its cause. In fact, potent diuretics, by blocking volume retention, may actually *reduce* plasma volume and flow further, and thus make the situation worse by exaggerating the abnormality that started the volume retention at the outset. Infusing albumin is obviously the ''physiologically appropriate'' first line of treatment; just as in treating heart failure with digitalis the ''need'' for diuretics may be eliminated by correcting the primary cause of the volume retention.

Two Final Concepts

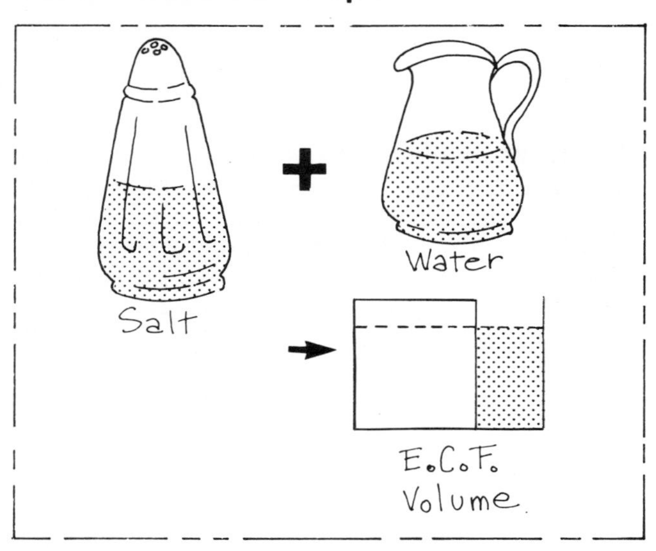

1) In these examples of disorders of body fluid regulation it should be clear to the reader that *volume* within the E.C.F. is tightly linked to the loss or gain of *sodium* which, by its osmotic effects, determines how much water will enter or leave the E.C.F. The regulation of *volume* and the regulation of *sodium* cannot be separated in attempts to understand the regulation of the E.C.F. and by inference the I.C.F. as well.

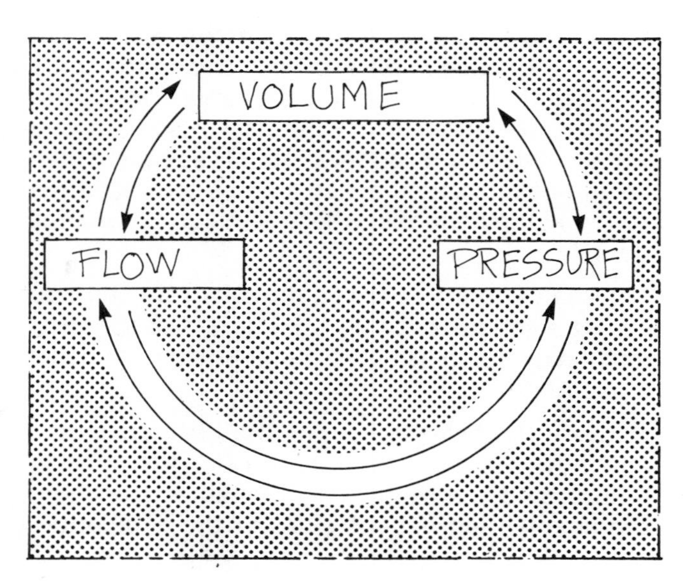

2) In any consideration of the plasma compartment it will be apparent that *volume, flow* and *pressure* are intimately related. Thus it is that the flow *into* the heart balances the flow *out* of the heart, whilst volume and pressure differ greatly between veins and arteries. Stimuli which result in *volume* retention by the kidney may be triggered by sensors that respond to *pressure* and *flow* (such as the juxtaglomerular apparatus in the kidney). Similarly, sensors of *volume* in the great veins and right side of the heart may result in an increase of arterial *pressure* and *flow*. Any change in one of these factors will modify the other two, and this is a fact of life in the plasma compartment.

Chapter 6

The Regulation Of Sodium

The major regulator of body sodium is the kidney. Since sodium is the major determinant of E.C.F. volume, the kidney is therefore the major regulator of E.C.F. volume.

Sodium Balance

Once dietary sodium has been absorbed, the only route of loss that can be regulated with precision is via the urine. Here we will look at this mechanism in more detail and consider the role of the kidney in states of sodium retention or loss.

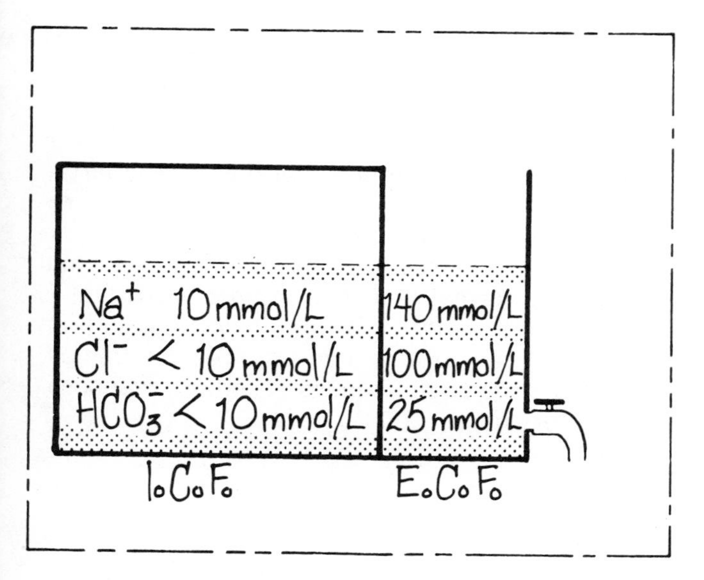

Sodium, and its accompanying anions, chloride and bicarbonate, are the major osmotically active particles in the E.C.F. None of them is present within the cells to any major extent.

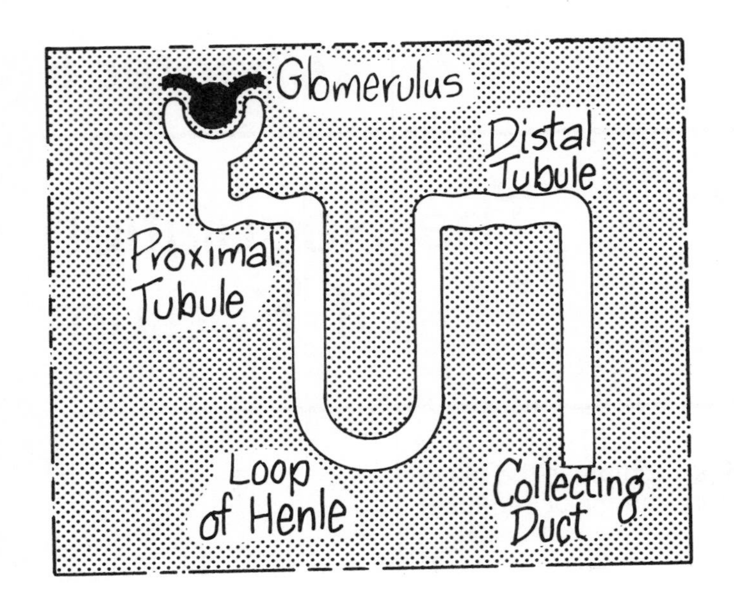

In looking at events in the kidney we will use a stylized diagram of the nephron. The important structures in the nephron are the glomerulus, the proximal tubule, the loop of Henle, the distal tubule, and the collecting duct.

Each kidney contains approximately a million of these units, in which structure and function are indivisible. Filtration at the glomerulus delivers fluid to the tubule, where its composition is modified by reabsorption, secretion or a combination of the two.

Glomerular Filtration

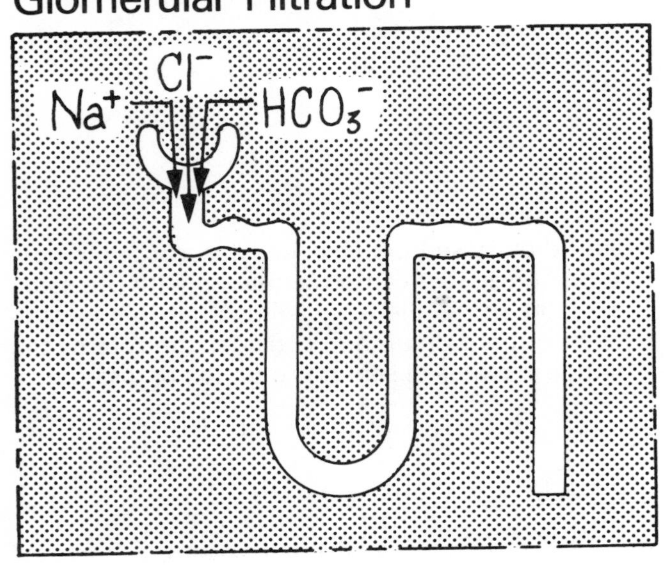

Because sodium is a small particle it is freely filtered at the glomerulus along with its accompanying anions, chloride and bicarbonate. In the absence of a barrier to filtration, and in the presence of relatively high serum concentrations, sodium, chloride, and bicarbonate will pass rapidly into the glomerular filtrate. Knowing the glomerular filtration rate, it is easy to calculate the filtered load which can then be compared to the amount appearing in the final urine.

	Amt. Filtered Per Day	Amt. Excreted Per Day	% Reabsorbed Per Day
Sodium	25,000 mmol	100 mmol	99·6%
Chloride	18,000 mmol	100 mmol	99·5%
Bicarbonate	5,000 mmol	Nil	100%
Potassium	700 mmol	50 mmol	93·0%
Water	180L	1L	99·4%

As is shown in this table, the amount of water and ions which are filtered per day is very large, but the efficiency with which they can be reabsorbed on their way down the tubule is remarkable. These figures are based upon a normal adult on a normal diet. If the same individual is put on a sodium-free diet the daily urine sodium losses will fall to only two or three millimoles, increasing the completeness of sodium reabsorption to very nearly 100%.

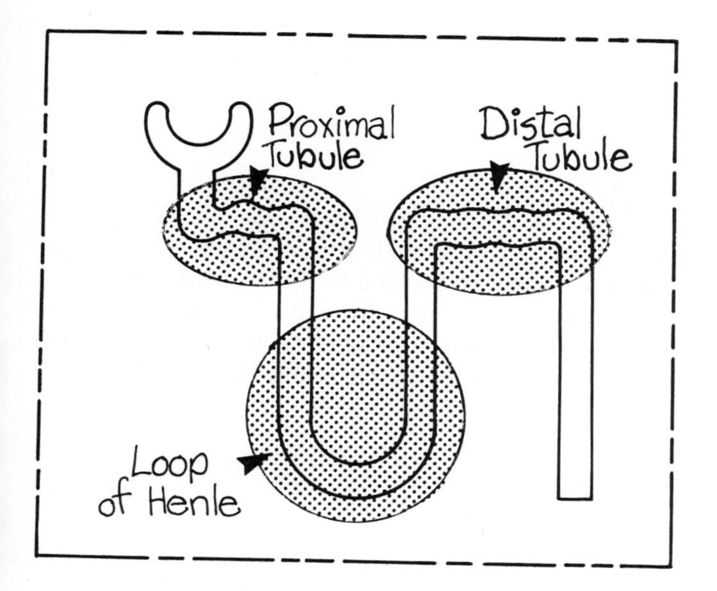

Obviously the kidney has a very well developed ability to recapture and conserve filtered sodium. Quantitatively most of this is reabsorbed in the proximal tubule but more distal sites of absorption are qualitatively just as important. Reabsorption at the three sites shown in this "stylized" nephron will be reviewed.

The Proximal Tubule

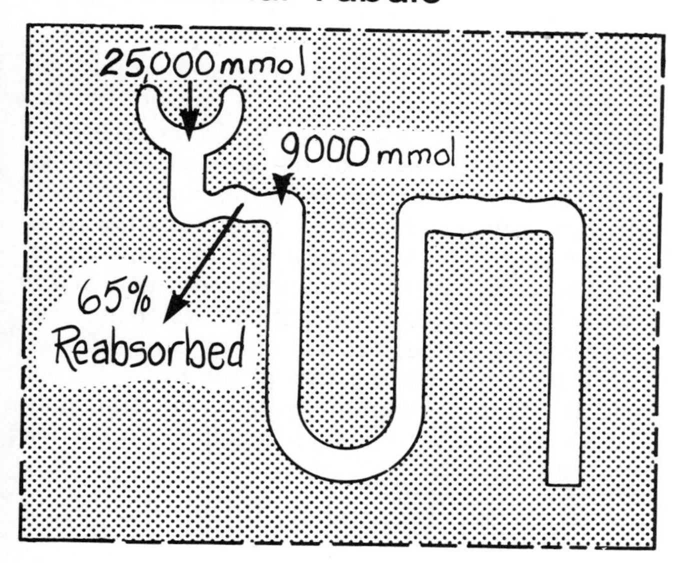

About 65% of the filtered load can be reabsorbed in this segment of the nephron, all of it by a process involving active transport of sodium from the tubular fluid, through the tubular cells and out into the peritubular space where it re-enters the E.C.F. Of the 25,000 millimoles filtered per day, only about 9,000 millimoles still remain by the time the tubular fluid reaches the beginning of the loop of Henle.

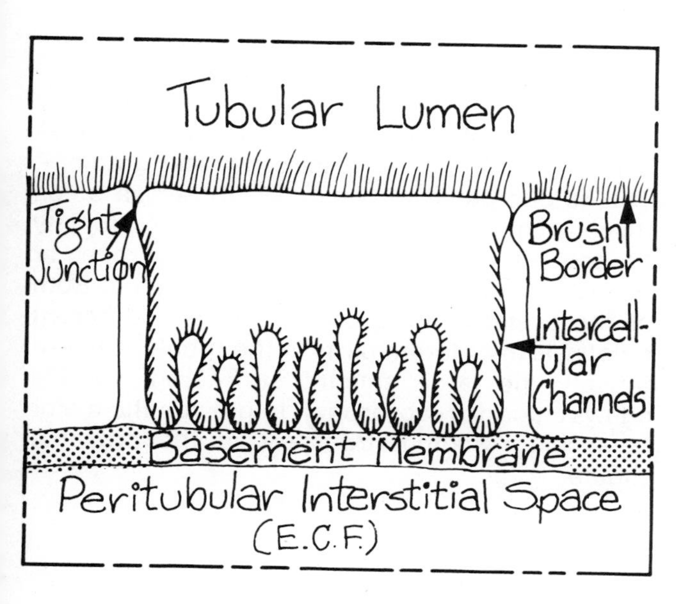

The cells concerned in this reabsorptive process line the proximal tubule and rest on a basement membrane which separates them from the peritubular interstitial space. The cells have many basal infoldings around which are many mitochondria. In between the cells are intercellular channels, but the communication between these channels and the lining of the tubule is blocked by a "tight" junction. The luminal border of the cell has a surface covered with projecting microvilli which increase its area. It looks like a brush border under the light microscope.

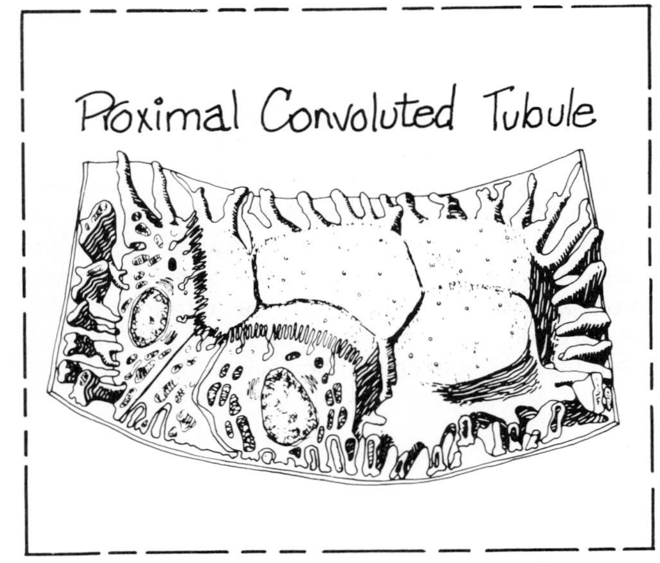

Redrawn from Rhodin J., Am. J. Med. 24661 (1958)

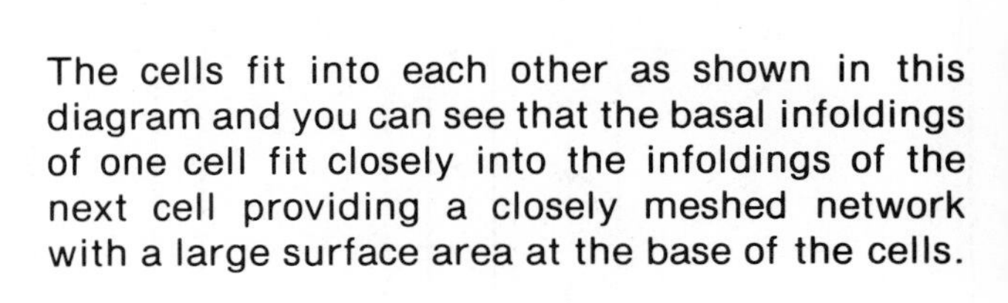
The cells fit into each other as shown in this diagram and you can see that the basal infoldings of one cell fit closely into the infoldings of the next cell providing a closely meshed network with a large surface area at the base of the cells.

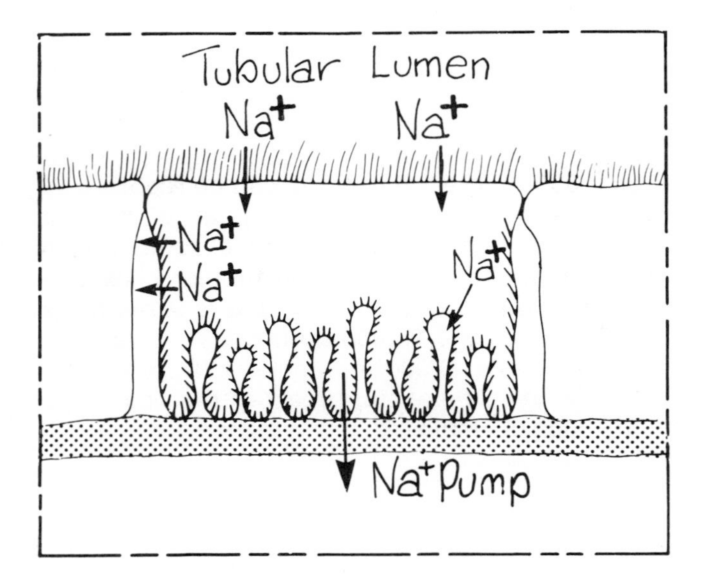

The presence of many mitochondria around the basal infoldings of the cell is shown by the shaded area in this drawing. Their presence suggests that the enzyme adenosine-triphosphatase (ATPase) is active in this area. There is good evidence that ATPase is directly linked to the provision of energy for a sodium pump which moves sodium from its low concentration *inside* the cell to its high concentration *outside* the cell. No pump is needed for sodium to *enter* the cell since the large surface area of the brush border and the high intraluminal sodium concentration favour the ready diffusion of sodium into the cell from the tubule.

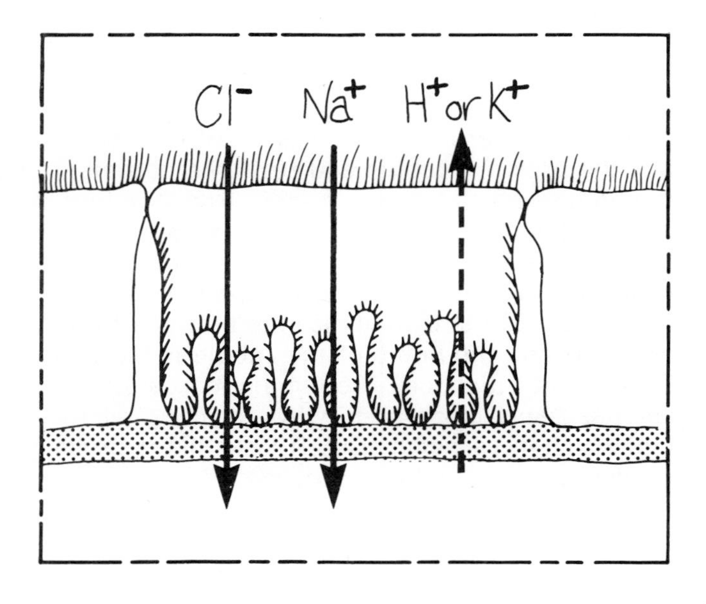

Sodium cannot of course be moved on its own without setting up an electrochemical gradient. This is dissipated in two ways. Either an anion (usually chloride) moves passively to accompany the sodium, or another cation (either hydrogen or potassium) moves in the opposite direction. Both of these mechanisms occur, but with different emphasis in different parts of the tubule. Quantitatively the large amount of filtered chloride allows chloride reabsorption to provide the major balance for sodium transport in the proximal tubule.

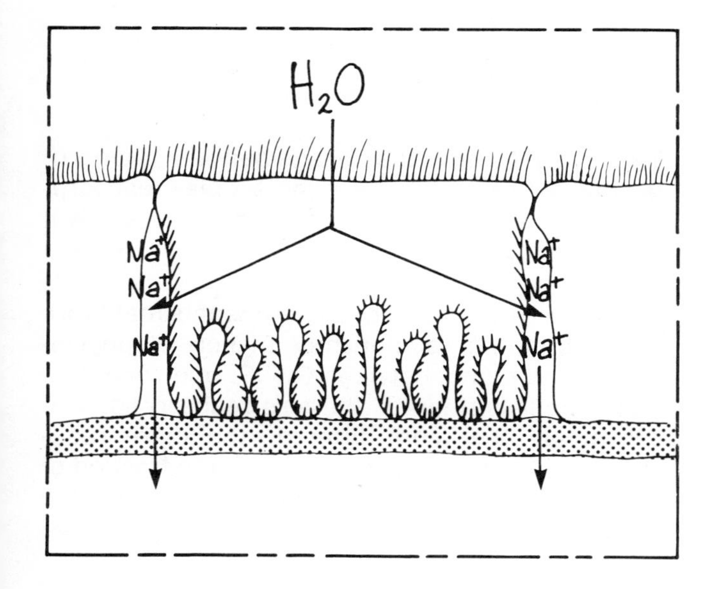

As sodium and chloride move from the cell into the peritubular space water moves with them as an inevitable osmotic event. Any osmotic gradient is thus also dissipated.

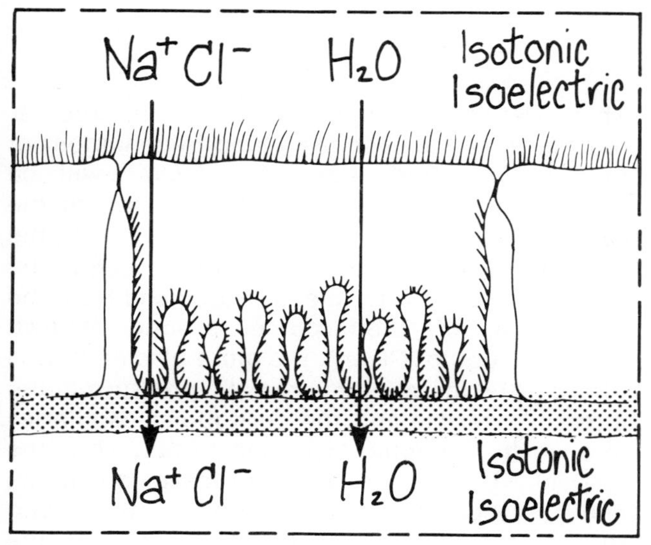

The result, in the proximal tubule, is the transport of isotonic sodium chloride from lumen to E.C.F. Thus the active transport of sodium results in the passive movement of chloride and water, and no measurable electrical or osmotic gradients exist across the wall of the proximal tubule.

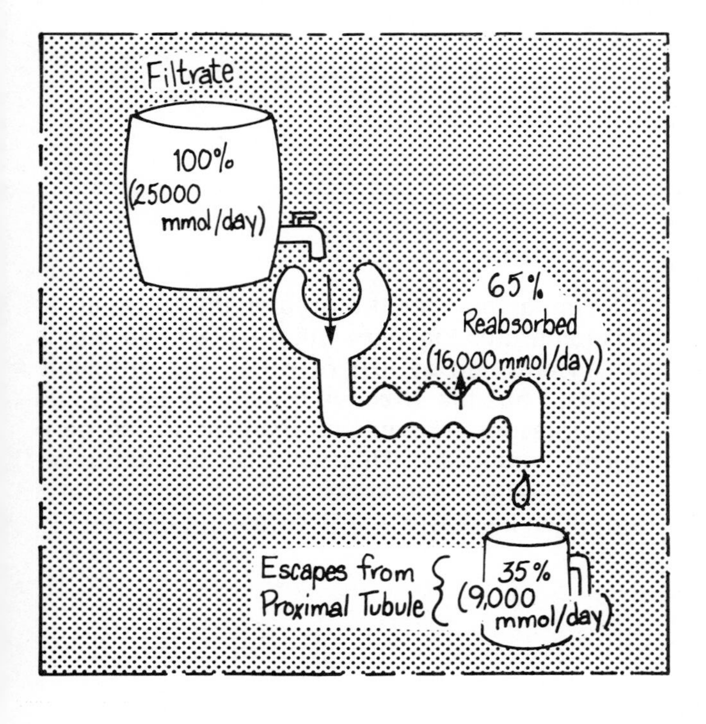

About 65% of the filtered sodium and 65% of the filtered water can be reabsorbed by the end of the proximal tubule and the luminal fluid remains *isotonic* as it enters the loop of Henle.

Only about 35% of the filtered load of sodium and water escape reabsorption to enter the loop of Henle.

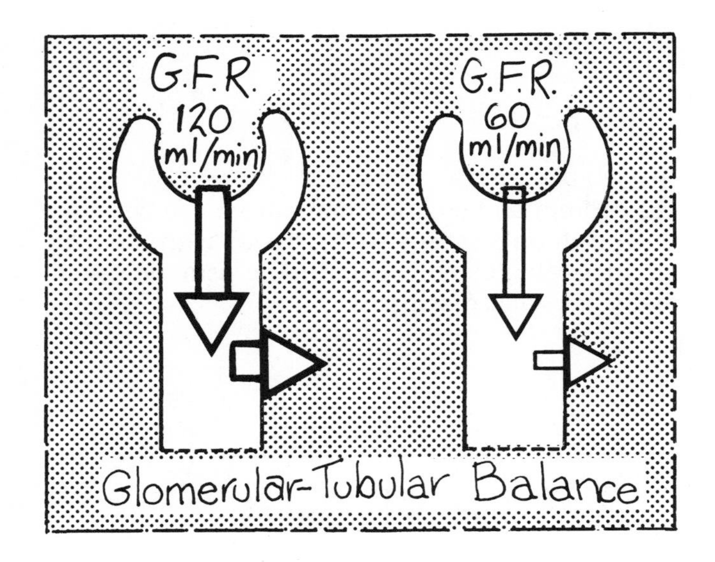

Two factors are known to exert a regulatory effect on sodium transport in the proximal tubule.

The first of these is the linkage between glomerular filtration rate (G.F.R.) and tubular reabsorption. Obviously if there is *no* filtration there will be *no* reabsorption.

What has been shown by experiment is that there is a continuing relationship, in normal physiological situations, between G.F.R. and reabsorption. The less filtered, the less reabsorbed, and conversely the more filtered, the more reabsorbed. Expressed in another way, the fraction of filtrate reabsorbed remains constant as the G.F.R. changes. This behaviour has been called *"glomerular tubular balance."*

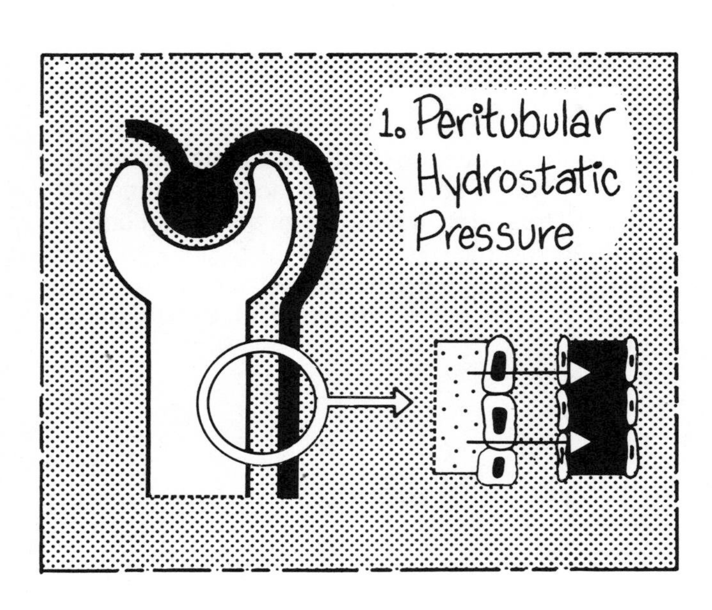

This relationship between G.F.R. and tubular reabsorption has been explained by a number of theories. Amongst these, two factors will be mentioned; they depend upon the fact that the capillary network carrying blood around the proximal tubule is, in fact, in direct continuity with the efferent arteriole, so that blood from the glomerular tuft then courses around its own proximal tubule. Clearly the more plasma that is filtered, the more the hydrostatic pressure in the efferent arteriole and the peritubular capillaries into which it empties will tend to fall. Thus the tendency for salt and water to leave the tubule and pass back into the peritubular space, and thus to the capillaries, will be greater if the G.F.R. is increased.

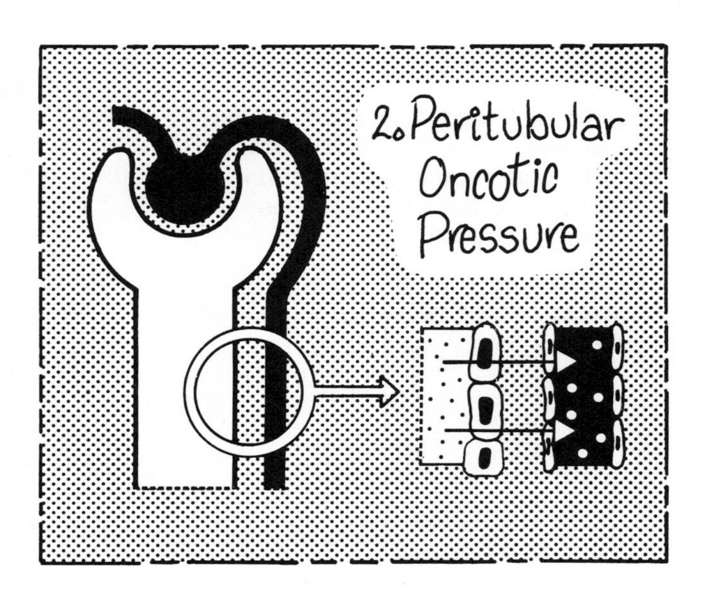

In a similar manner the greater the filtration at the glomerulus the more will the plasma proteins be concentrated by loss of water and diffusible solutes. Thus the effective osmotic pressure due to the plasma proteins will increase in the peritubular capillaries. This will also favour movement of sodium and water out of the tubule at a rate which increases with increasing filtration.

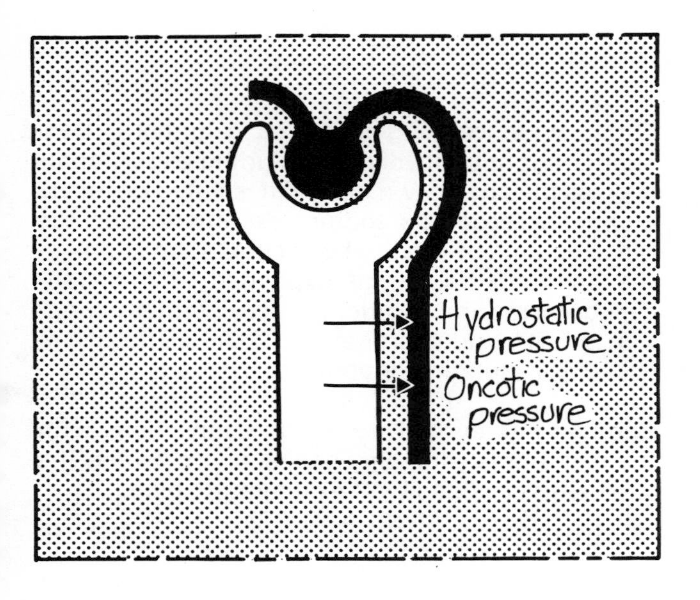

Thus two mechanisms, both relating to physical characteristics of peritubular fluid and dependent upon the renal vascular architecture, will favour a positive relationship between G.F.R. and proximal tubular reabsorption. The less the G.F.R., the less reabsorption is favoured; the more the G.F.R. the more reabsorption is likely to occur.

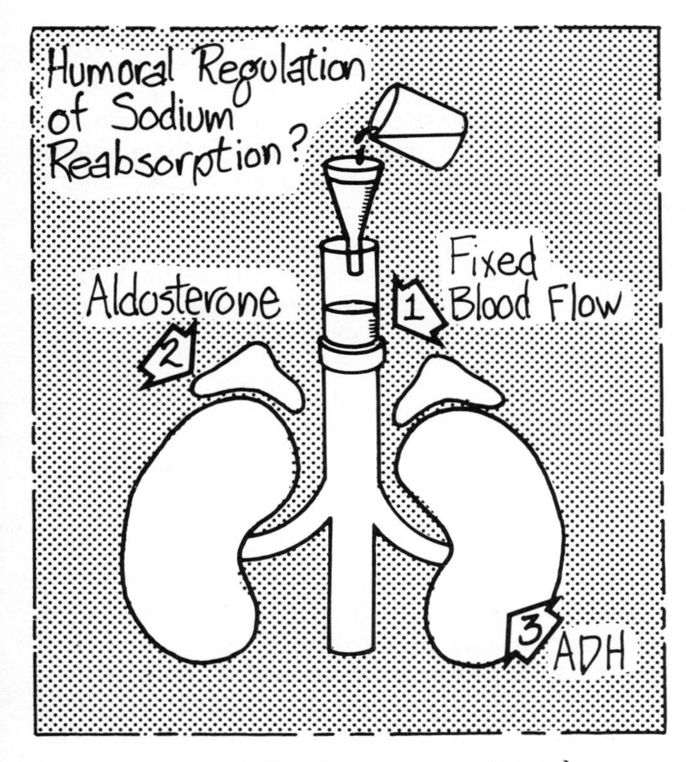

*de Wardener et al, Clinical Science 21, 249 (1961)

Other mechanisms by which proximal sodium reabsorption is regulated are still not fully explained, and in addition to the physical factors mentioned above, theories supposing a humoral regulator for the proximal tubule began with experiments like the one illustrated here, reported in 1961 by deWardener and his colleagues.* In an experimental animal they arranged the following conditions:

1) A fixed G.F.R. ensured by an aortic balloon to fix renal blood flow;
2) Large doses of minerallocorticoid to swamp the effects of endogenous aldosterone;
3) Large doses of ADH to swamp any effect this hormone might have.

Under these conditions, expansion of circulating blood volume caused a prompt increase in urinary sodium excretion, implying a change in tubular function independent of G.F.R. or known hormones.

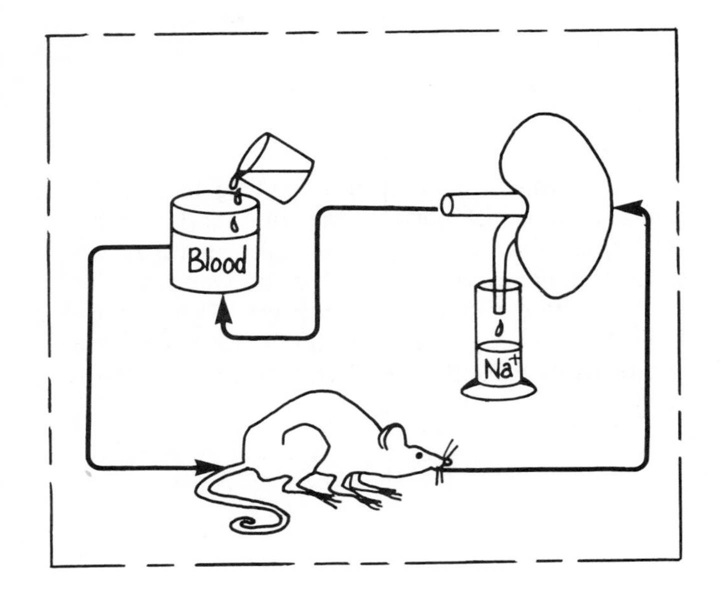

Other evidence suggested that a humoral messenger might be involved, including experiments such as the one shown here. An isolated perfused kidney increased its sodium excretion when the blood volume of a rat, whose circulation supplied the isolated kidney, was expanded. This expansion occurred in conditions where neither changing G.F.R., Aldosterone, nor ADH could be invoked as regulating agents.

The test kidney had no neural connections with the test animal, the only connection between the two being provided by the circulating blood.

Despite enthusiastic search no single hormonal agent has been found to account for these effects and ''third factor'' remains an elusive agent. Whether or not a distinct hormone is found, it is clear that more than a single mechanism is involved in the regulation of the large bulk of sodium reabsorption that takes place proximally.

The Loop of Henle

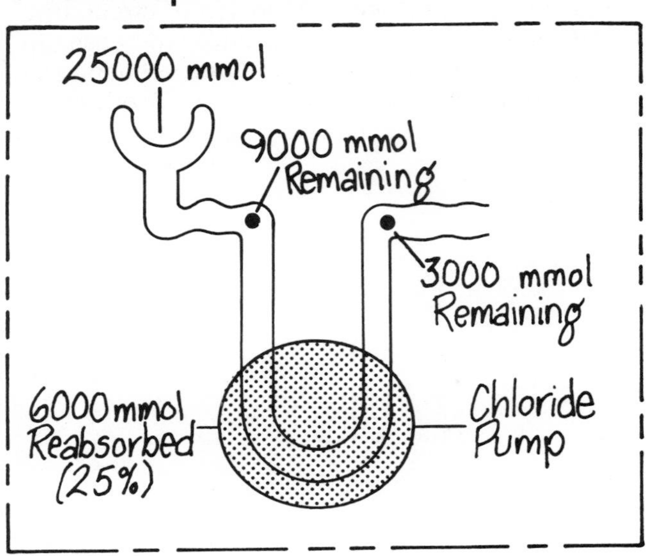

The second site in the tubule where sodium is reabsorbed is the loop of Henle. Only about 25% of the total load is reabsorbed here, but it is a critical 25% because on it depends the effectiveness of the countercurrent system for water conservation. Recent work has surprised physiologists by indicating that the mechanism here seems to be primarily a *chloride* pump with sodium as a passive accompanying cation.

This portion of sodium transport probably remains constant in most physiologic situations. It can, however, be influenced by the potent ''loop'' diuretic drugs which block ionic transport in the loop of Henle. Examples of such diuretics are furosemide and ethacrynic acid.

The Distal Tubule

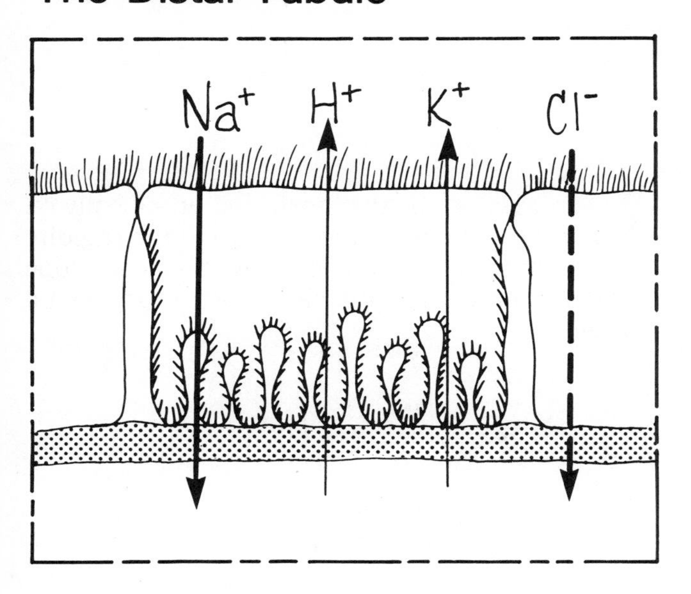

The third part of the tubule concerned with sodium transport is the distal segment. The mechanism here is similar to that in the proximal tubule with the important exception that active sodium reabsorption is more closely linked to exchange with hydrogen ions and potassium ions, and less to passive reabsorption of chloride ions.

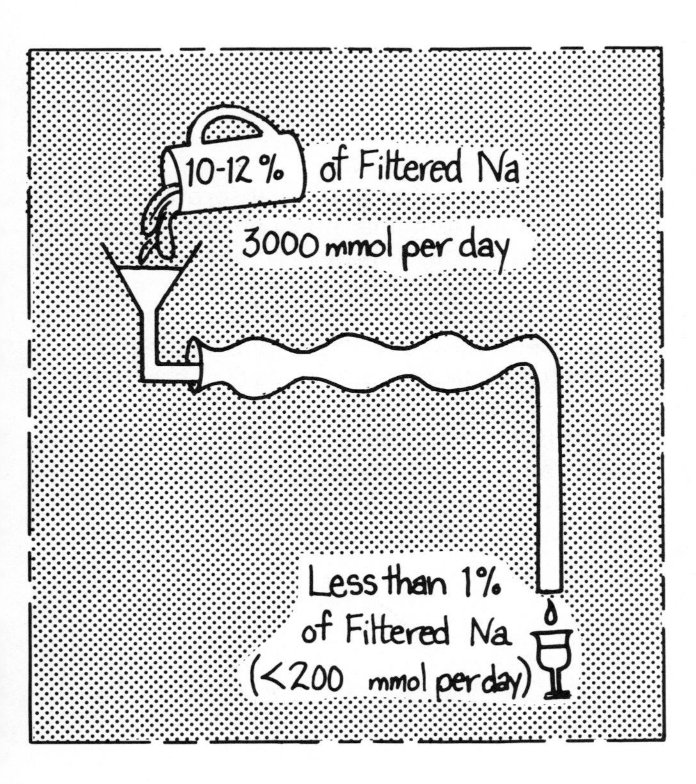

Only 10-12% of filtered sodium remains to enter the distal tubular segment (about 3,000 millimoles per day). This segment includes the cortical part of the collecting duct.

Although the total amount of sodium reabsorbed is much less than in the proximal segment, its efficiency can be much greater, since this is where the "fine regulation" of sodium recovery is determined. When sodium reabsorption is at its maximum, hardly any is allowed to escape from this segment of the tubule.

On a normal diet about 100-200 millimoles of sodium will be lost in the urine each day. This represents less than one percent of filtered sodium. Under conditions of zero sodium intake the excretion of sodium in the urine may fall to as low as 3 millimoles per day (0.01% of filtered load).

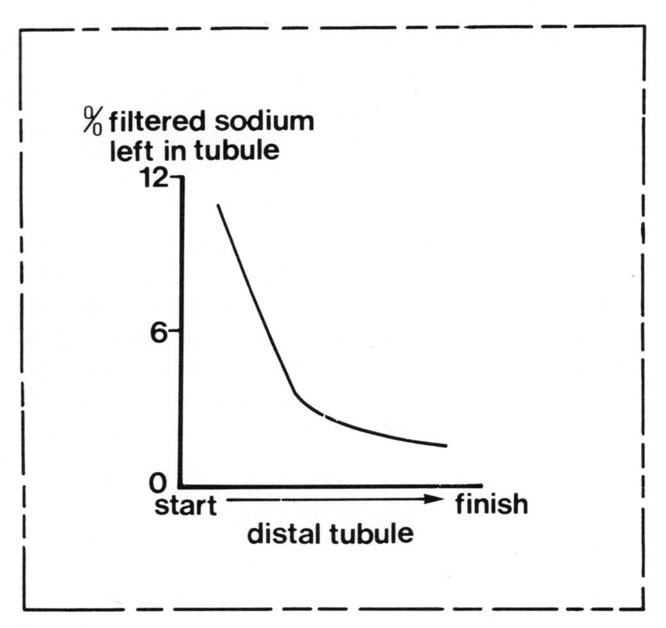

After Giebisch and Windhager Am. J. Med. 36, 643 (1964)

This process can be demonstrated graphically on the basis of micropuncture data, and the rapidity of sodium transport is shown by the steep slope for disappearance in the early parts of the distal tubule.

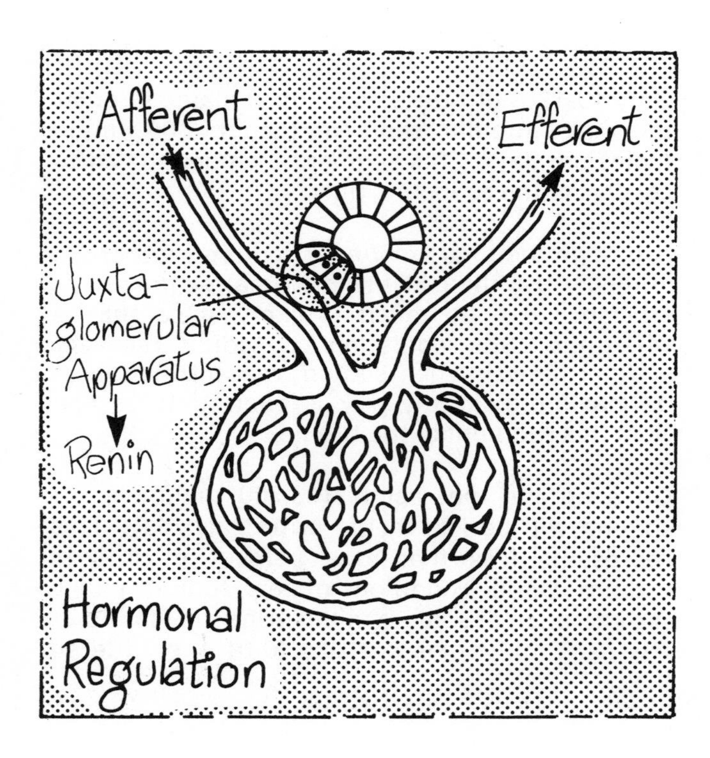

The regulation of sodium reabsorption here is carried out by hormonal mechanisms whose final messenger is the steroid hormone *aldosterone.* The first step in this regulation is the release of a protein called *renin* from granules in the wall of the afferent arteriole where it becomes closely related to the distal tubule of its own nephron.

The cells in the tubule wall at this point are closely applied to the arteriolar wall and are called the ''macula densa.'' This area is collectively called the ''juxtaglomerular apparatus.''

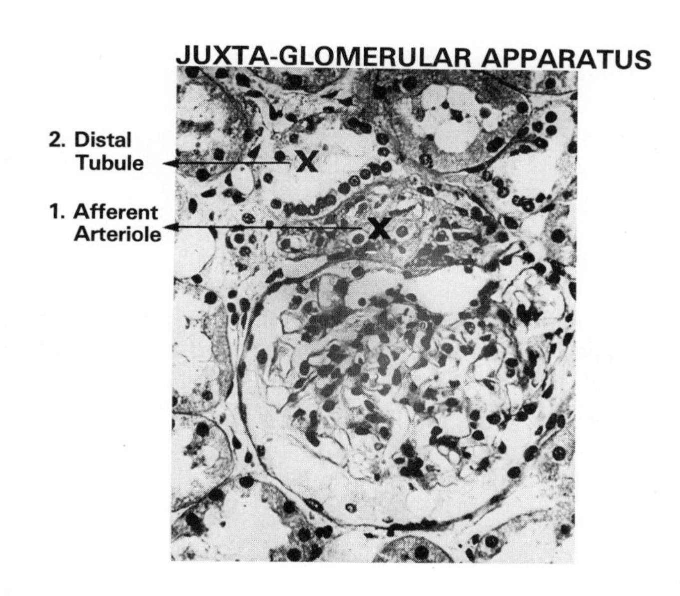

In a microscopic section cut through this region the granulated cells in the afferent arteriole (1) can be seen next to the regularly organized cells of the distal tubule (2) which are called the "macula densa."

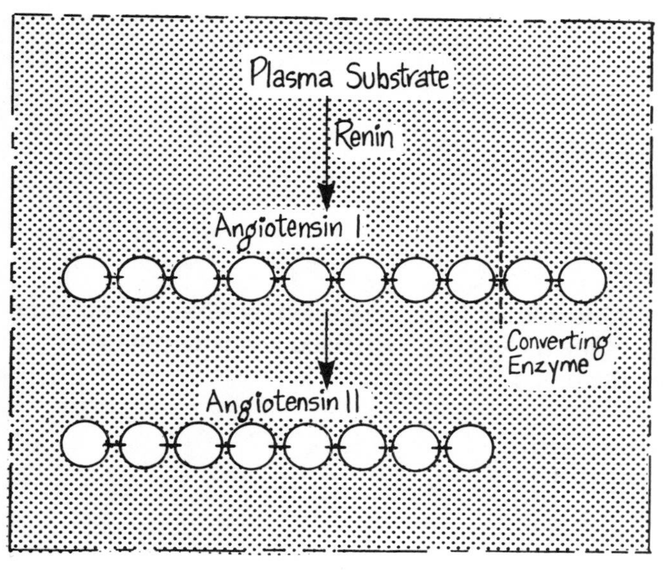

Renin is released from this area in response to changes of pressure and/or flow in the afferent arteriole and can also be released by β-adrenergic stimulation in the renal nerves. Once released it acts on a substrate in the plasma to produce *angiotensin I* which in turn is converted to *angiotensin II* by an enzyme in the plasma (converting enzyme).

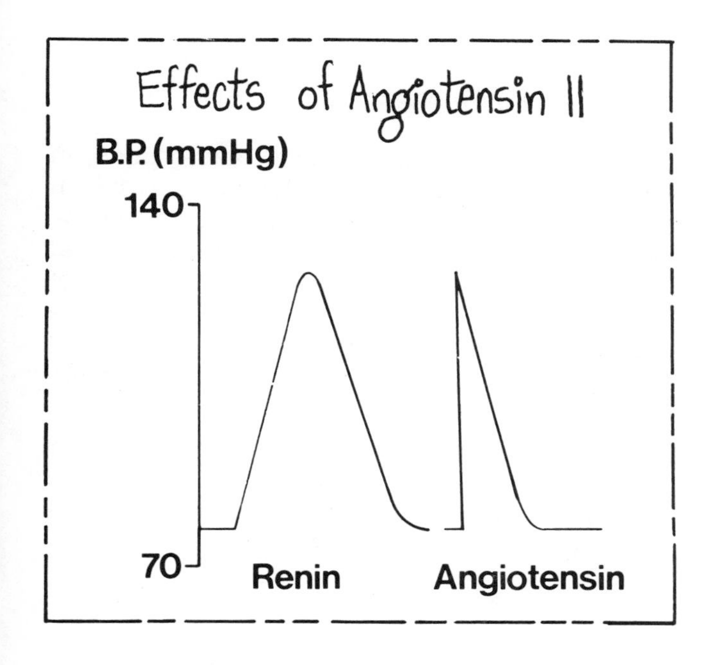

Angiotensin II has two actions. One is to raise the blood pressure by increasing peripheral resistance. In an animal model, the effects of infusing angiotensin II are potent and immediate. The effects of infusing renin are delayed because the chemical reactions needed to produce angiotensin II take a finite amount of time.

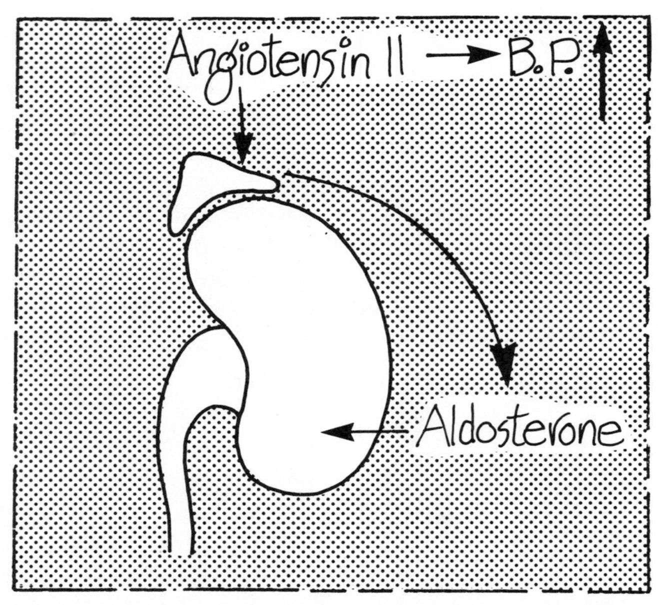

The second effect of angiotensin II is to cause the release of aldosterone from the adrenal gland, which then acts upon sodium reabsorption in the distal tubule and cortical collecting duct.

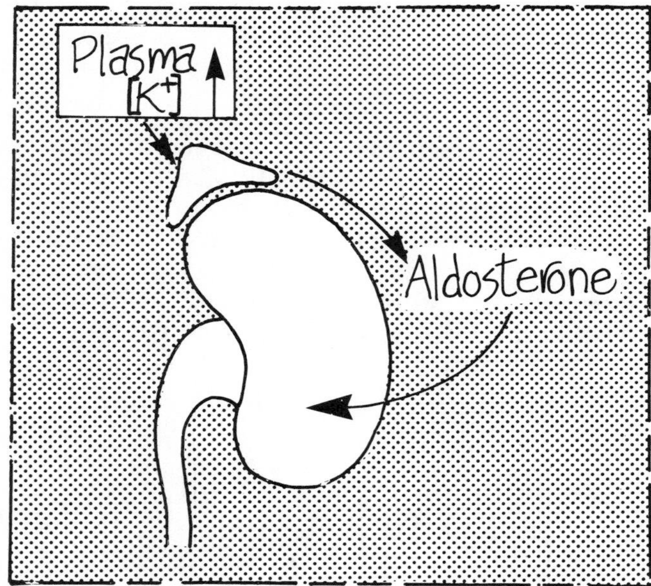

Aldosterone is also released by *direct* stimulation of the adrenal by increased levels of potassium in the blood.

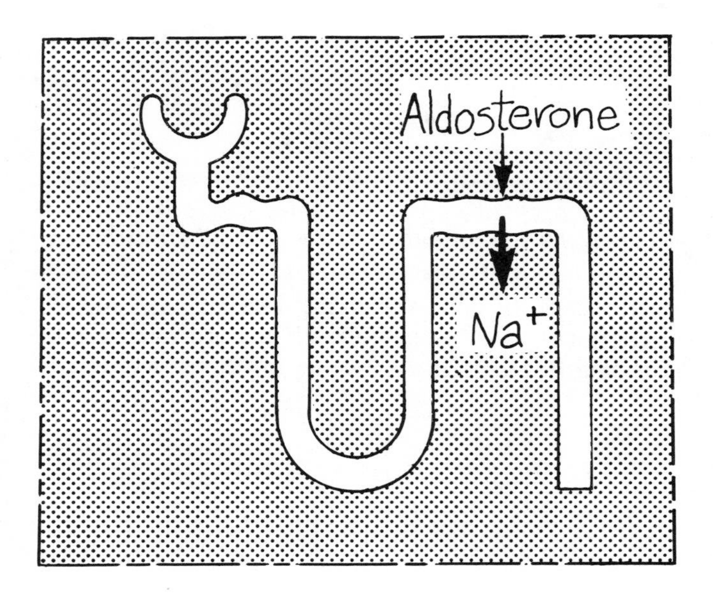

Aldosterone activates sodium reabsorption in exchange for potassium. It probably acts by increasing the activity of the sodium pump in the basal area of the cell. Some people think that increased permeability of the luminal border of the tubular cell may also occur. Whatever the molecular mechanism the result is a striking increase of sodium reabsorption and potassium loss.

Aldosterone also acts upon sodium transport in the colon, the sweat glands and the salivary glands, but these actions are much less impressive than its action on the renal tubule.

Summary

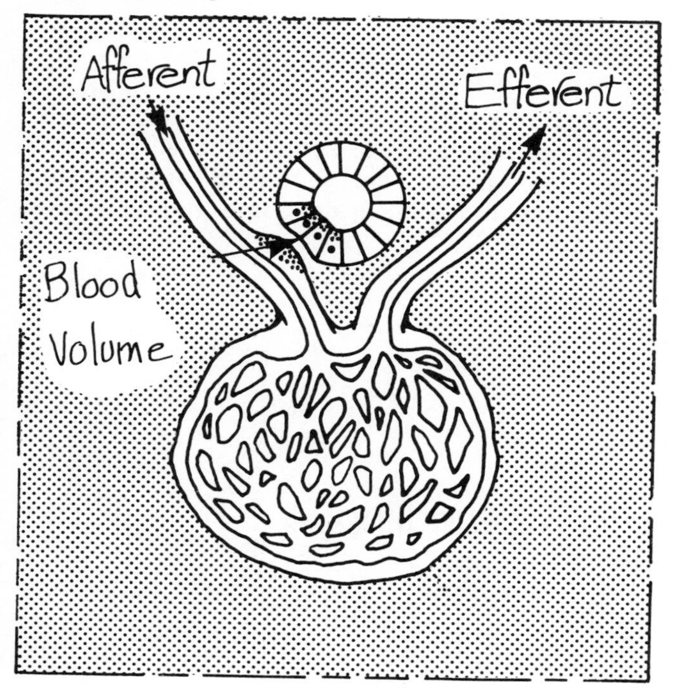

In summary changes of blood volume and blood pressure can be sensed at a renal level resulting in the release of renin and the initiation of a cascade of events that cause the retention of salt and water and a correction of blood volume.

In some abnormal situations (e.g. stenosis of a renal artery) an inappropriate secretion of renin may occur with the generation of large amounts of angiotensin and aldosterone.

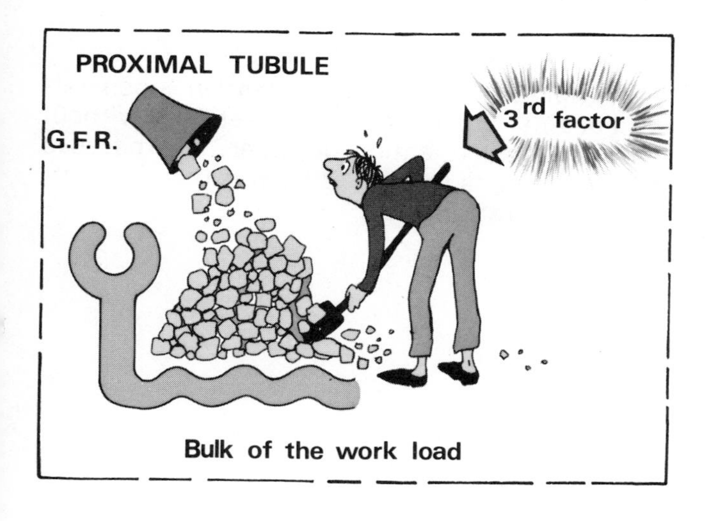

Most of the filtered sodium is reabsorbed in the *proximal tubule.* The degree of this reabsorption is related to the G.F.R., the peritubular physical environments, and "third factor", although this last factor is not fully understood. However large the amount of sodium reabsorbed proximally, a significant fraction is left to enter the loop of Henle and the distal tubule.

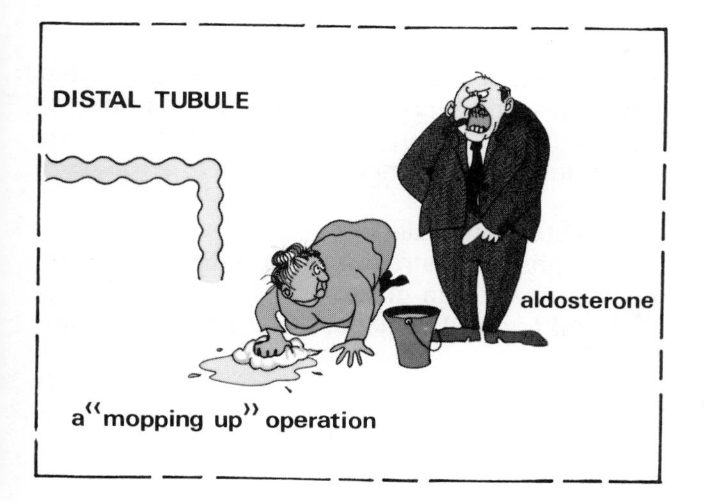

In the distal tubule the *amount* of sodium reabsorbed is not nearly so large, but the *completeness* of its reabsorption in times of stress may be extreme.

The major regulator of sodium reabsorption at this site is aldosterone.

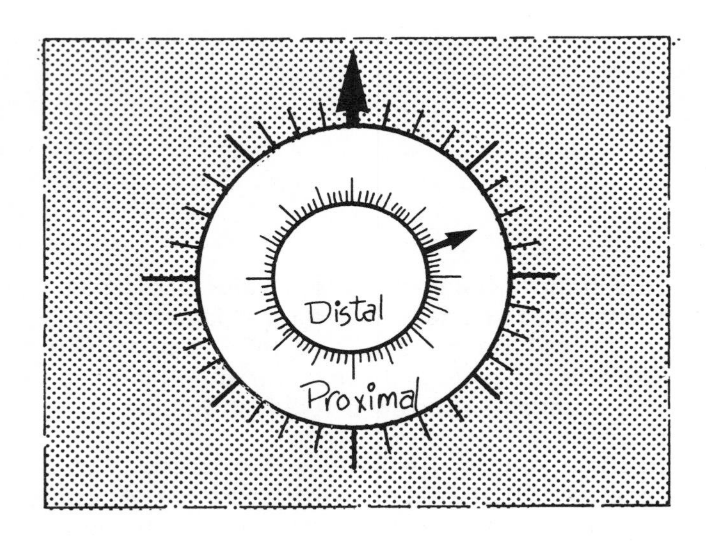

Proximal regulation of sodium recovery may be considered as the ''coarse tuning'' of the system while the more accurate ''fine tuning'' takes place distally.

Clinical Examples Of Disordered Sodium Handling

Sodium Retention

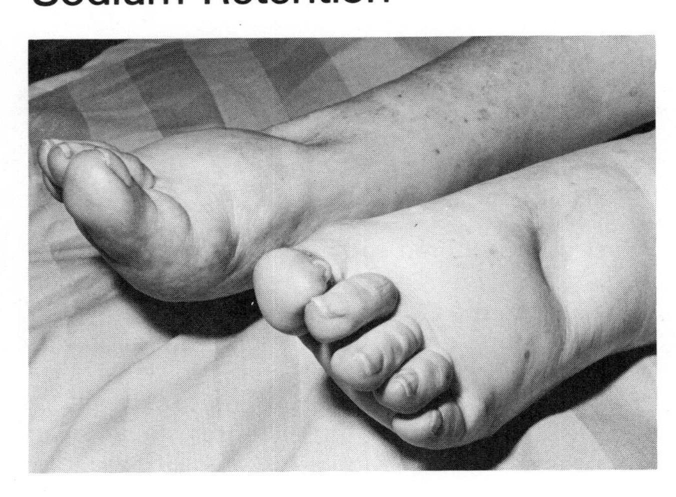

The cardinal sign of sodium *retention* is generalized edema. Sodium cannot be retained without the inevitable retention of water, but because sodium remains in the E.C.F., it is this compartment that expands.

Edema

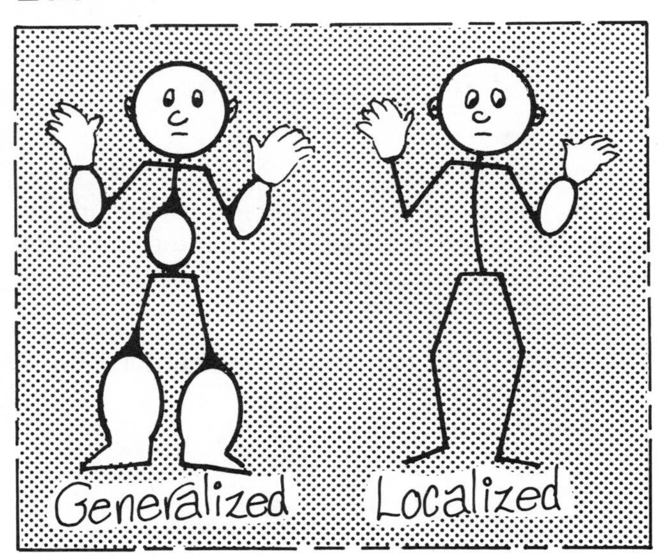

Edema can, of course, be localized, and in such a case does not represent a primary disorder of sodium metabolism. *Localized* edema implies one of the following mechanisms:

1) Venous obstruction
2) Lymphatic obstruction (in which case the edema may be non-pitting)
3) Capillary wall damage, usually due to inflammation.

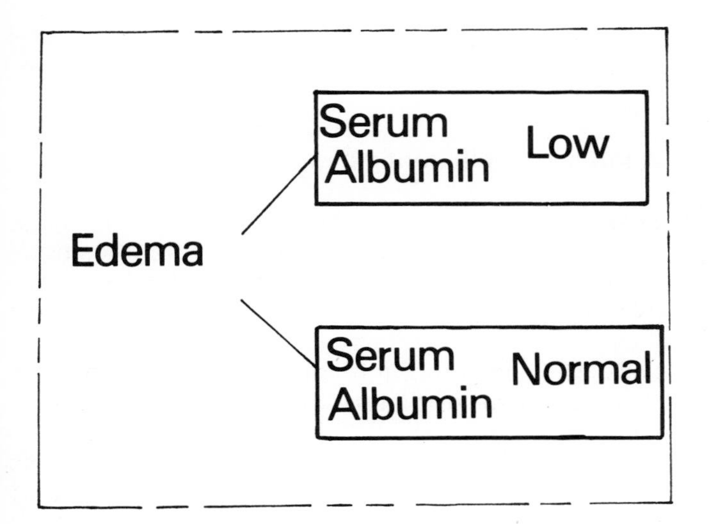

Generalized edema implies an overall change in the osmotic or hydrostatic events in the capillary bed throughout the body and may be associated with a normal or a reduced serum albumin level.

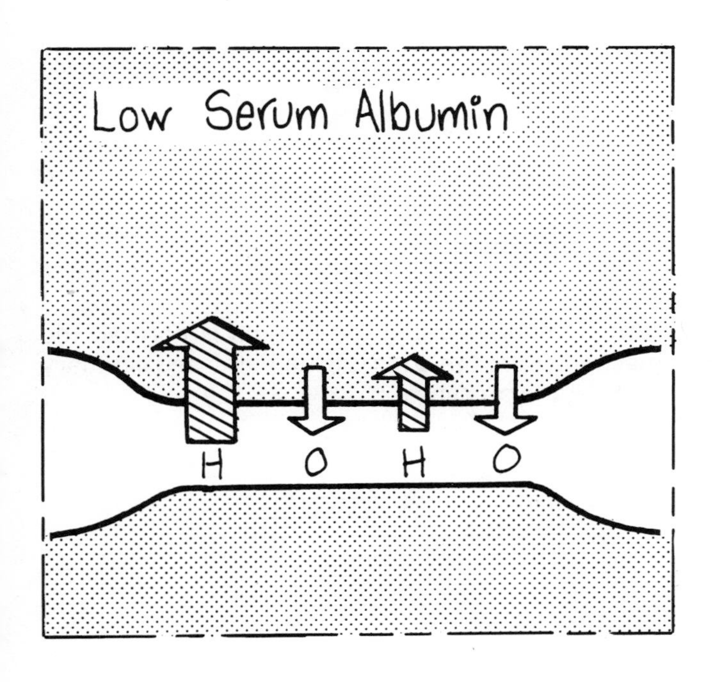

If the albumin level is reduced the mechanism of the edema is a reduced oncotic pressure in the capillaries and a consequent leak of fluid out of the plasma space into the interstitial space, with a resulting diminution of plasma volume.

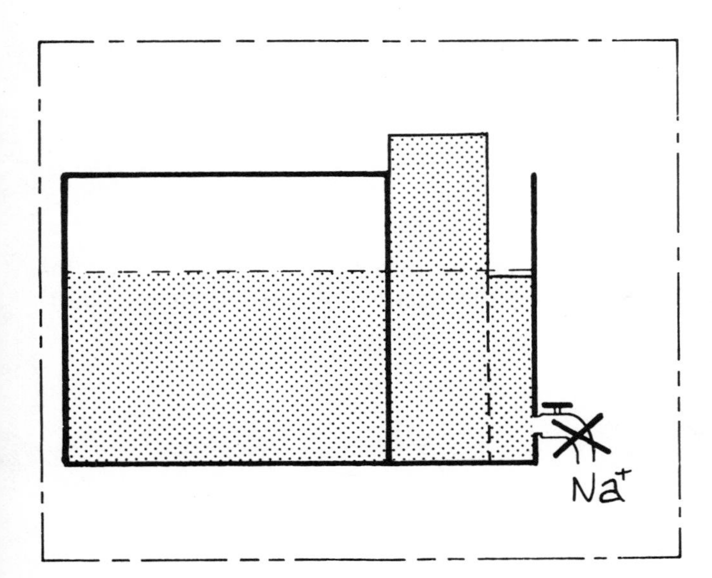

Since the plasma volume is relatively reduced the neck veins are not engorged and the blood pressure is not elevated. Since effective renal blood flow is also relatively reduced, sodium retention will occur due to direct and indirect (hormone dependent) renal mechanisms. Because of the low plasma oncotic pressure, however, the retained sodium (and water) continues to move out into the interstitial space increasing interstitial edema without plasma volume expansion.

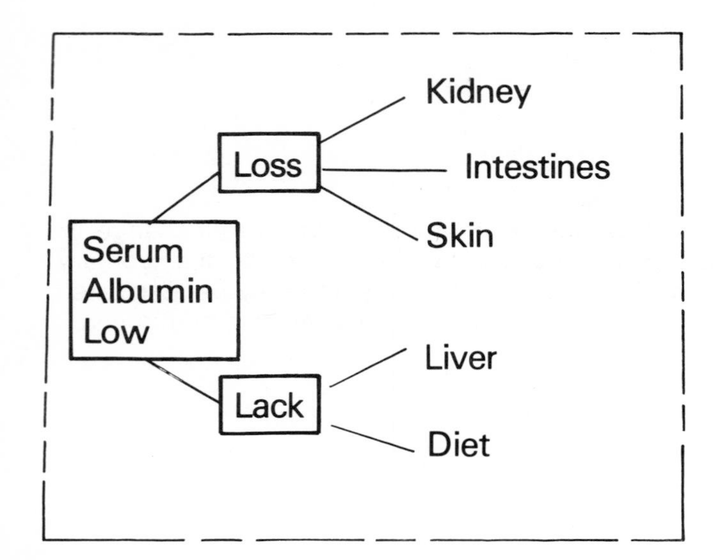

A low serum albumin may be due to loss or lack.

Loss can occur through the glomeruli (nephrotic syndrome), or through the gut (protein-losing enteropathy), or (rarely) through the skin (burns; exudative skin disease).

Lack can occur through failure of manufacture (liver disease), or dietary inadequacy (kwashiorkor; malabsorption).

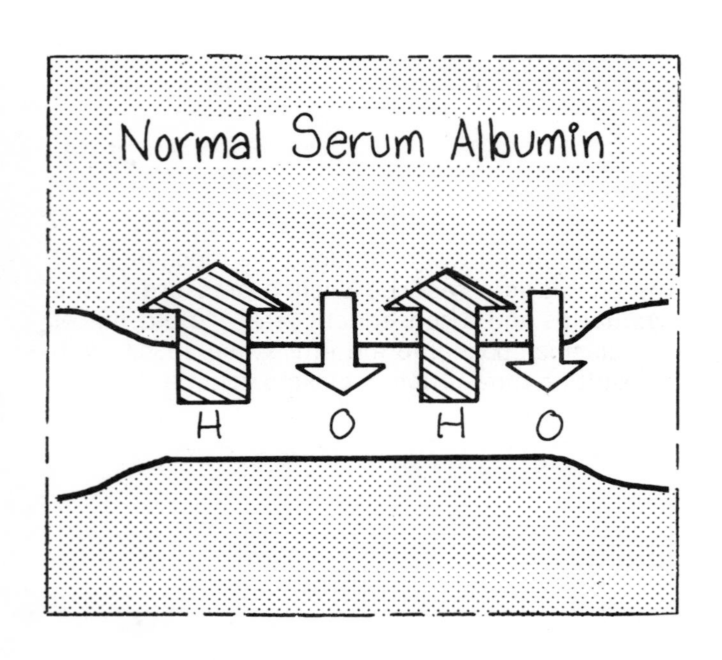

If the albumin level in the plasma is normal, the mechanism of the edema is an increased capillary hydrostatic pressure.

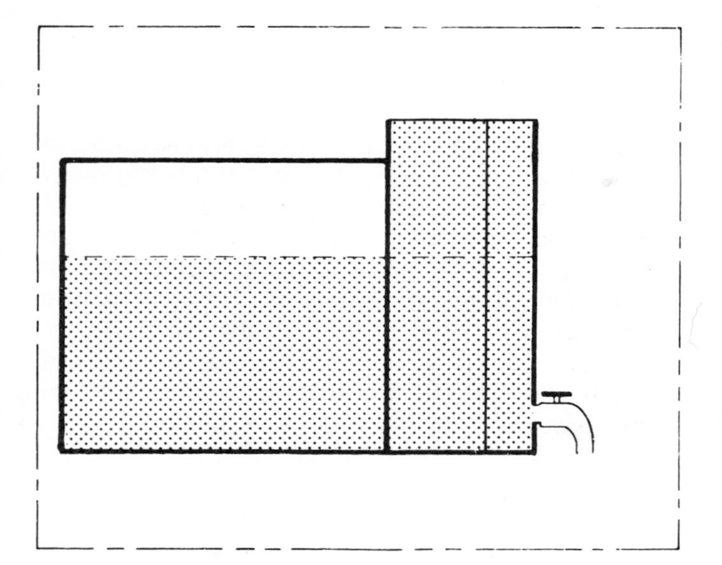

Total plasma *and* interstitial volume will be increased and the neck veins will be visible, indicating a raised venous pressure.

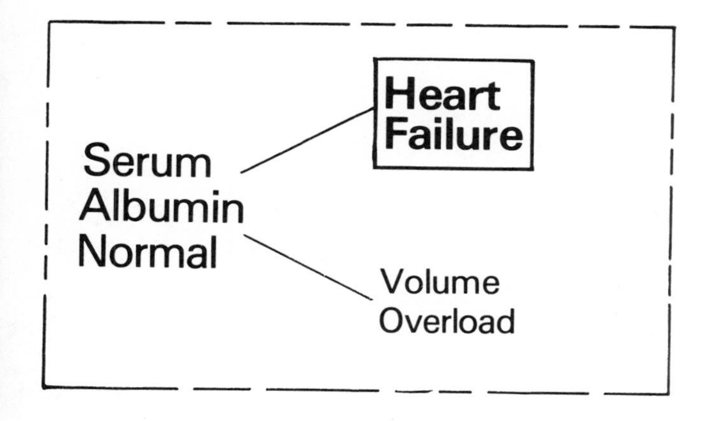

One mechanism is cardiac failure. Here effective forward flow is reduced, due to a defective pump, and thus venous engorgement occurs, but effective renal blood flow falls. This sets in motion all the mechanisms in the kidney that will retain sodium (and water) and thus continue to increase the edema. The blood pressure is often low because of reduced cardiac output.

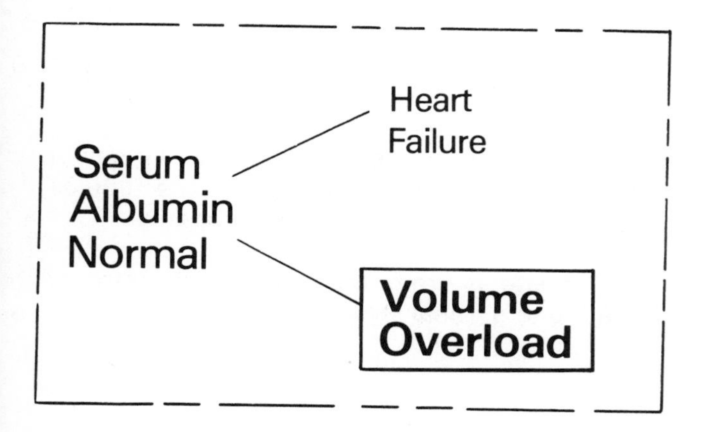

Another mechanism is volume overload where sodium and water are retained in the presence of a healthy heart. This usually is caused by primary renal disease such as acute glomerulonephritis.

It is also seen in acute or advanced chronic renal failure where sodium loading has been allowed to occur.

In this situation the venous pressure *and* the arterial pressure are increased.

Sodium Loss

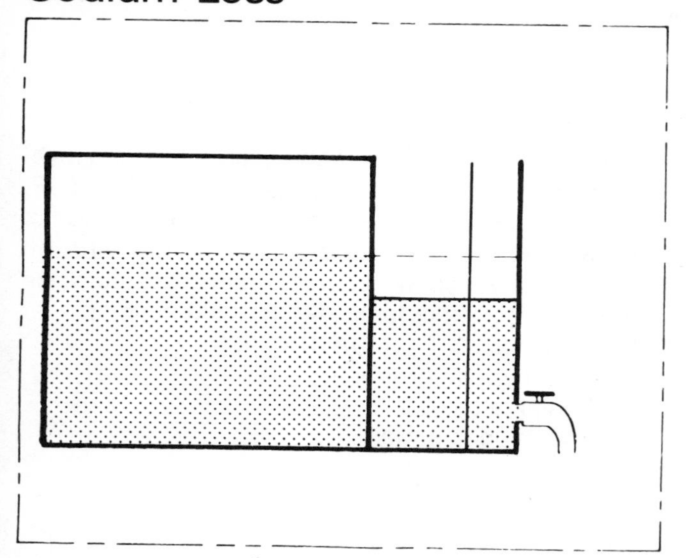

The cardinal sign of sodium *loss* is volume depletion.

The plasma volume will be reduced as indicated by low blood pressure and rapid pulse.

The interstitial volume will be diminished as indicated by lack of edema, reduced skin turgor, sunken eyes, and dry mucosal surfaces.

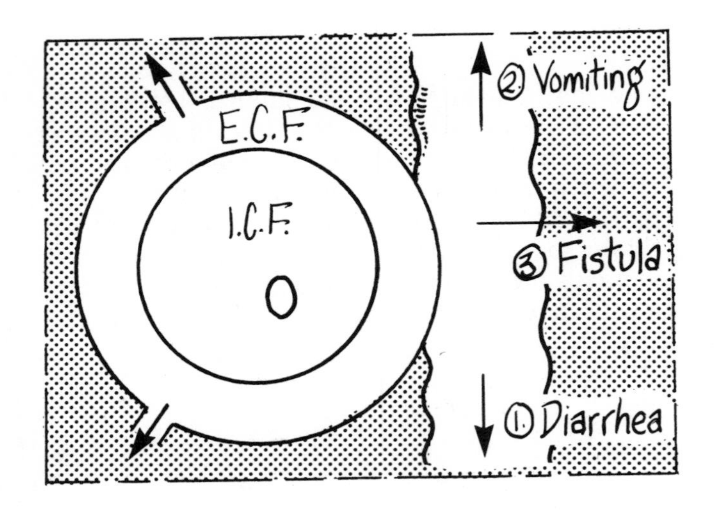

The common site for sodium (and water) loss is from the intestine due to (1) diarrhea; (2) vomiting; (3) external fistula.

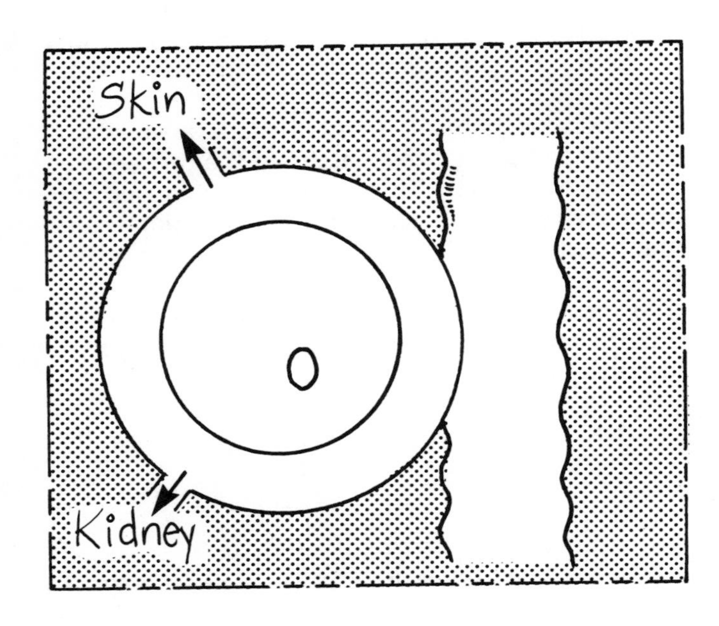

A second site for loss is from the kidney in diseases that predominantly affect the tubules, e.g., pyelonephritis; obstructive uropathy.

A third site is from the skin in uncontrolled sweating, e.g., heat stroke.

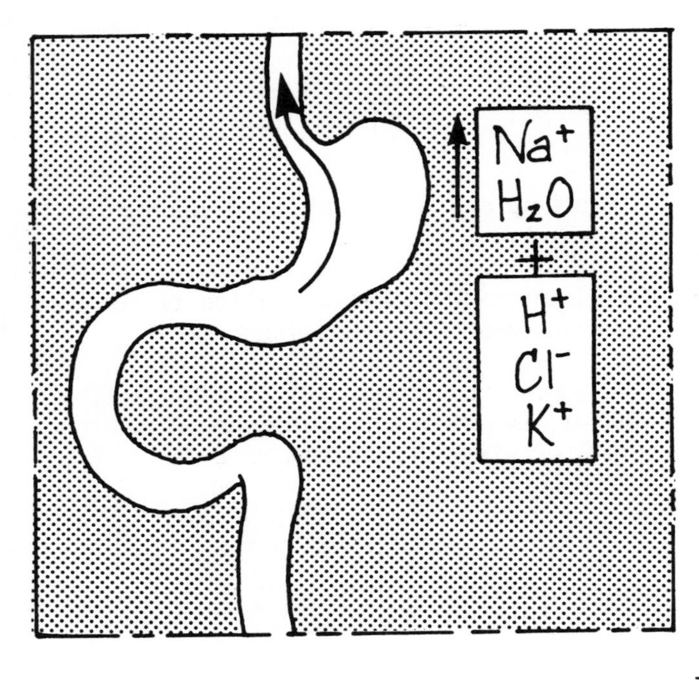

Losses from the intestine can, for practical purposes, be considered as losses of E.C.F. volume, and are usually approximately isotonic. Above the pylorus they are isotonic but with an excess of hydrogen, chloride, and potassium. So the patient with pylonic stenosis will develop

1) Isotonic E.C.F. volume depletion
2) With hypochloremia
3) And alkalosis
4) And hypokalemia

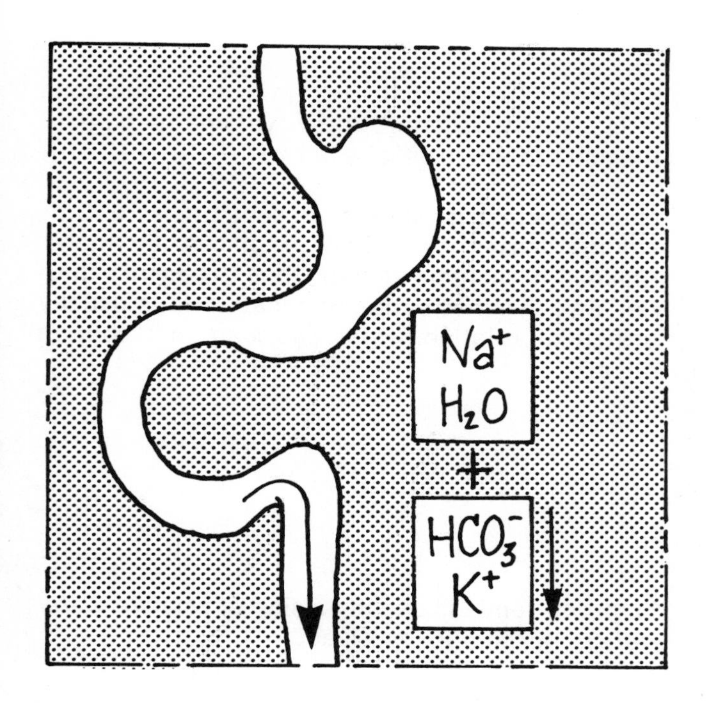

Losses from below the pylorus are of E.C.F., but with an excess of bicarbonate and potassium. The patient with cholera will therefore develop

1) Isotonic E.C.F. volume depletion
2) With Metabolic acidosis
3) And Hypokalemia

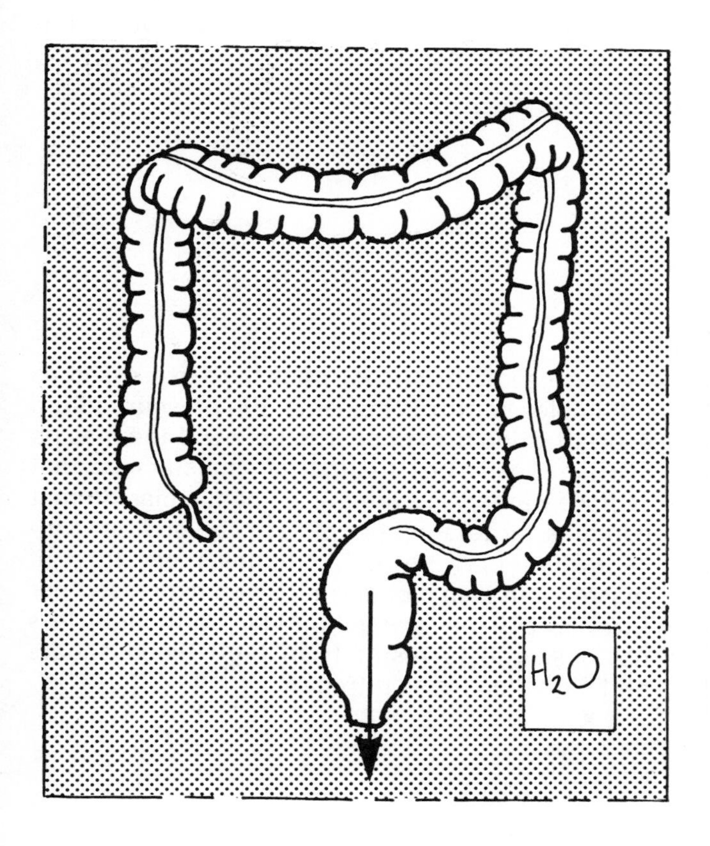

Only with diarrhea from the *large* intestine are the losses hypotonic, with losses of water being greater than of solutes.

Hyponatremia

Usually
Water Overload

Not
Sodium Deficit

Since sodium is usually lost in isotonic solutions it is usually recognized by *volume depletion.* Plasma sodium is usually *normal.*

Hyponatremia is seldom present unless E.C.F. volume losses have been replaced by water (instead of isotonic saline), thus producing dilution and hyponatremia. Hyponatremia is usually a sign of relative *water overload,* not sodium deficit.

Hypernatremia

Usually
Water Deficit

Not
Sodium Overload

Hypernatremia is usually a sign of relative *water deficit, not* sodium overload.

Since these chemical events are largely due to mismanagement of the handling of water rather than sodium they will be discussed in more depth in the next chapter.

'Factitious Hyponatremia'

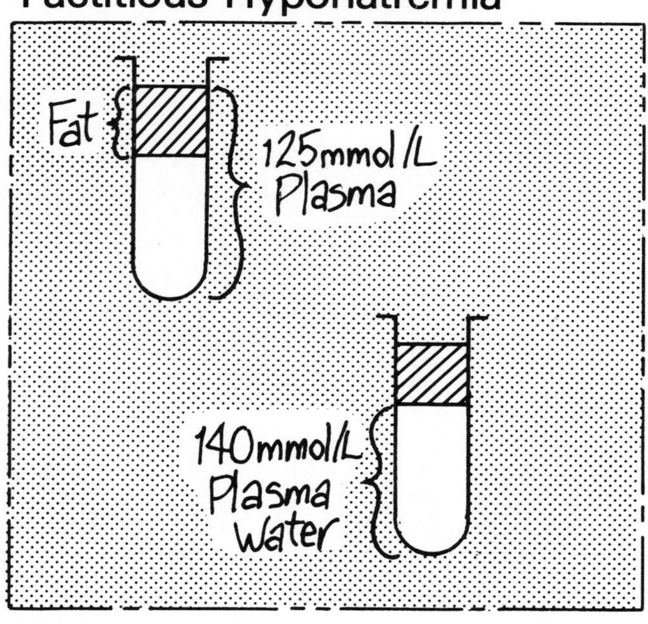

Because sodium concentrations are expressed as millimoles per litre of plasma, the *plasma* sodium concentration may appear to be low if the plasma contains a large volume of substances that replace water. An example is lipemic plasma that contains large amounts of fat. The sodium concentration may be 125 millimoles per litre of plasma, but is in fact normal when expressed per litre of plasma *water.* Similar, although less obvious, effects result from high blood sugars and high blood urea nitrogen.

Chapter 7

The Regulation of Water

Under normal circumstances we all maintain a water ''balance'' so that our intake and our losses are equal. Sources of loss include moistening the expired air, sweating, and of course the urine. Because of losses from sites other than the kidney, our urine output is always less than our total water intake.

When we are deprived of water, or when our losses by routes other than the urine increase, the urine becomes smaller in volume and more concentrated. We can all recognize this by the smaller volume and more intense colour of the urine we pass in hot weather when our losses via perspiration increase.

When our fluid intake is high or our unmeasured losses are small, our urine output will increase, and its colour will fade.

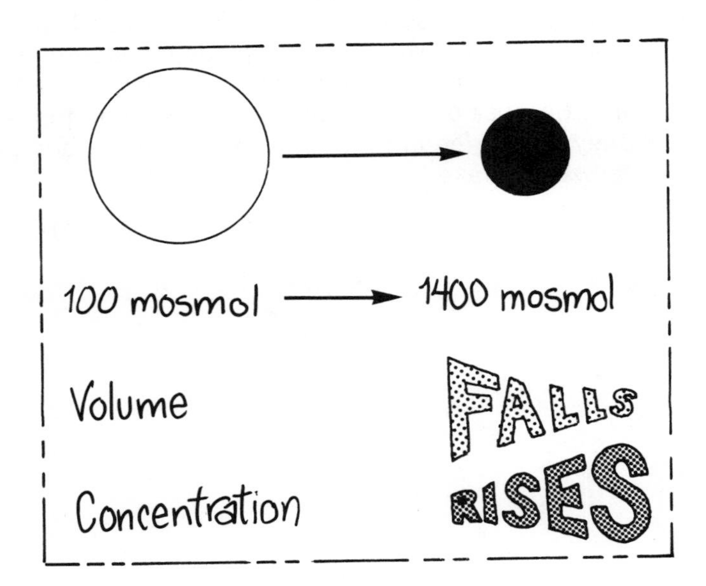

The ability to vary the volume and concentration of the urine in accordance with our needs to lose or retain water is an illustration of the precise role of the kidney as a regulator of normal function.

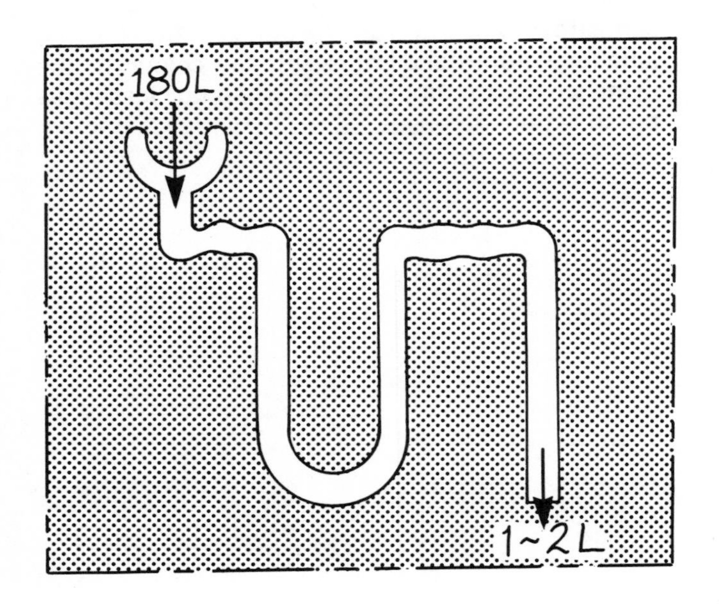

In the previous chapter it was noted that the total volume of glomerular filtrate each day, for an adult, is about 180 litres, but the final urine output is a litre or two, indicating that about 99% of filtered water is reabsorbed.

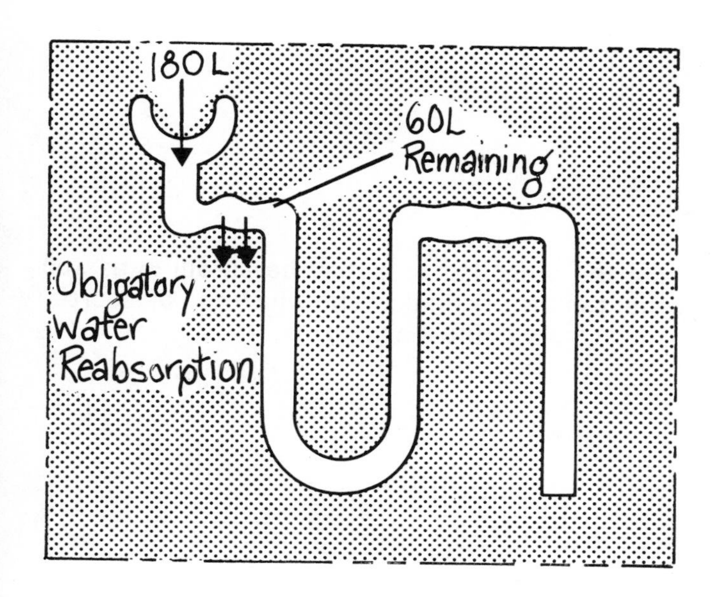

Most of this volume is reabsorbed in the proximal tubule as a passive osmotic event, secondary to the active movement of sodium and chloride out of the lumen. This water reabsorption is *obligatory* and will only vary to the extent that sodium reabsorption varies. Thus up to 120 litres may be reabsorbed in this way, but it is all ''captive'' water, tied to the osmotic effects of active ionic transport.

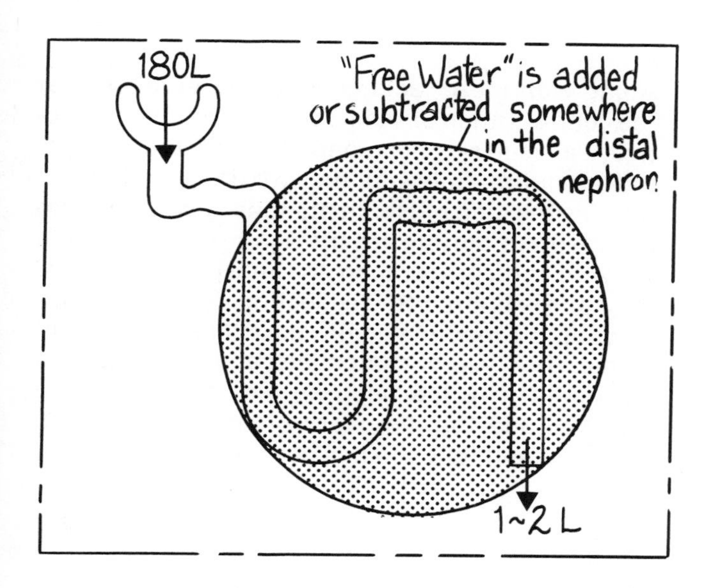

Losing or retaining water independently of these osmotic inevitabilities implies the reabsorption or non-reabsorption of water *free* of such osmotic demands. Thus the term ''*Free Water*'' has been coined for water that is added to the urine (thus diluting it) or subtracted from the urine (thus concentrating it) without any initial movement of solutes. This all takes place beyond the end of the proximal tubule.

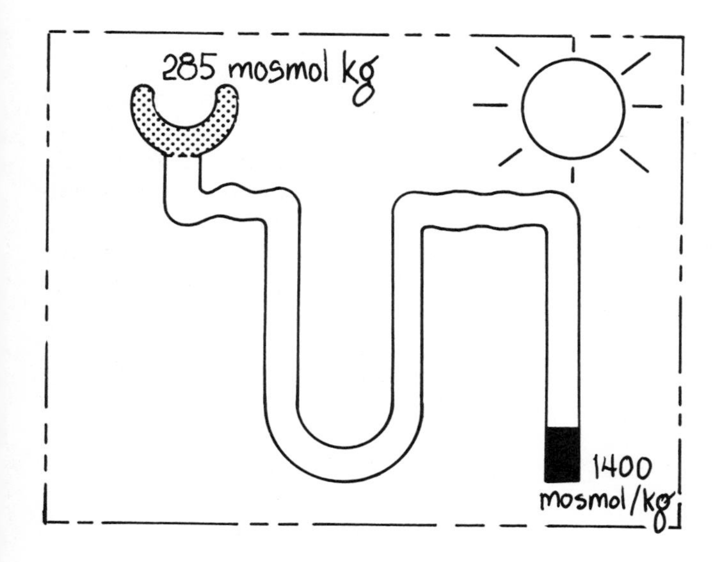

The mechanism for urinary concentration and dilution is a remarkable example of the link between structure and function and begins with the observation that whilst the glomerular filtrate is *always* isotonic with plasma, the urine may be as concentrated as 1400 milliosmoles per kilogram.

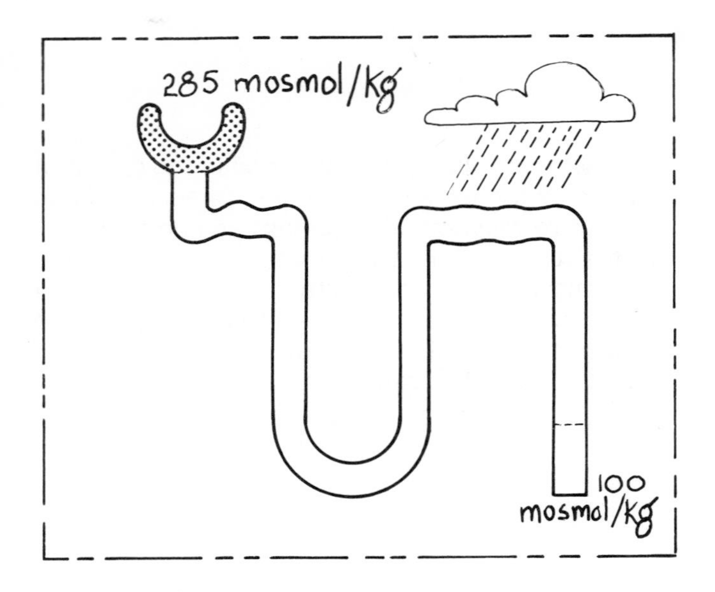

On the other hand, it may be as dilute as 100 milliosmoles per kilogram which is, of course, much more dilute than plasma.

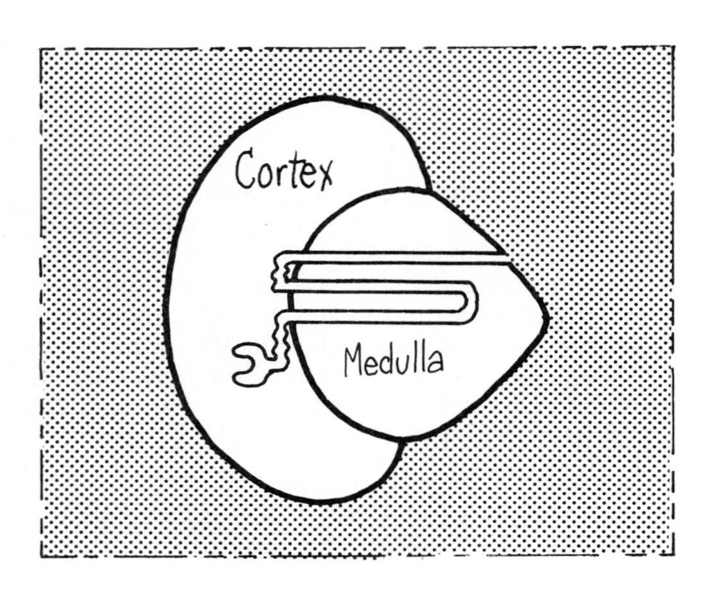

Along with these facts goes the observation that the configuration of the nephron is highly distinctive, particularly those nephrons arising close to the medulla. Between the proximal and distal tubules the loop of Henle plunges down deep into the medulla only to return again (like a hairpin loop) to the cortex; then the most distal part of the entire nephron (the collecting duct) runs back through the medulla again, parallel to the loop of Henle. For years anatomists and physiologists wondered why the nephron was constructed in this way and the discovery of the link between structure and function has been an exciting story.

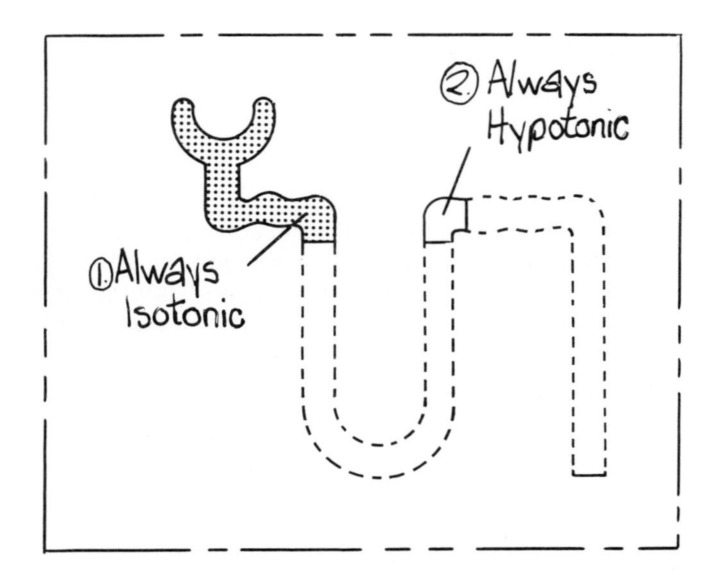

Two important facts were found in animals by micropuncture. These were

(1) That the filtered urine remained isotonic with plasma to the end of the proximal tubule, whatever the final urine concentration.

(2) That the urine was *always* hypotonic at the end of the loop of Henle—whatever the final urine concentration.

Clearly something was happening in the loop of Henle to *dilute* the urine.

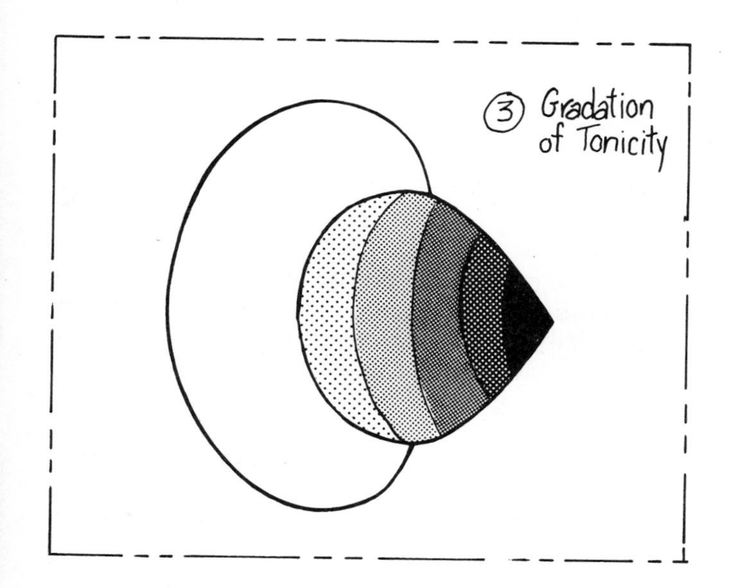

A third observation was that the tonicity of the tissues in the medulla showed a gradation, being concentrated at the tip and isotonic at the cortico-medullary junction. The extent of this gradient was related to the extent to which the animal being studied was able to concentrate the urine; a long medullary papilla and long loops of Henle correlated with a big osmotic gradient and an ability to produce highly concentrated urine (for example in desert rodents).

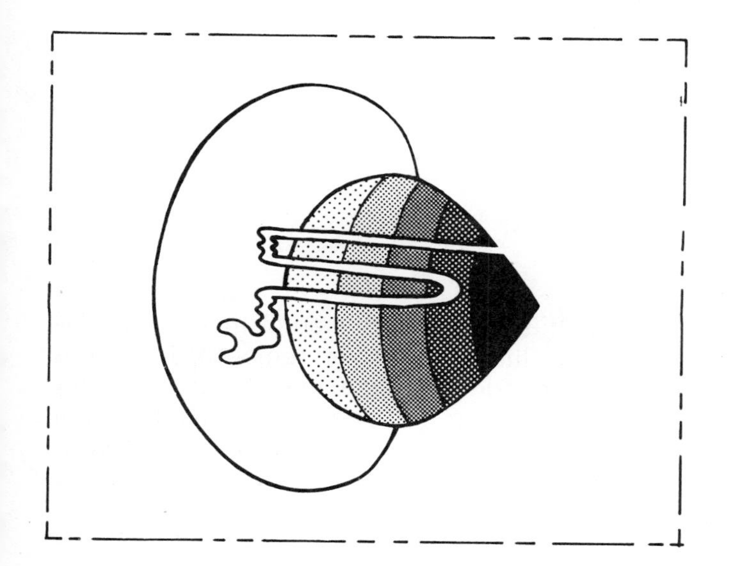

The anatomical arrangement of the tubule appeared to be related in more than a casual way to the osmotic gradients within the medulla, but the relationship to the way in which the kidney regulated water balance was still not clear.

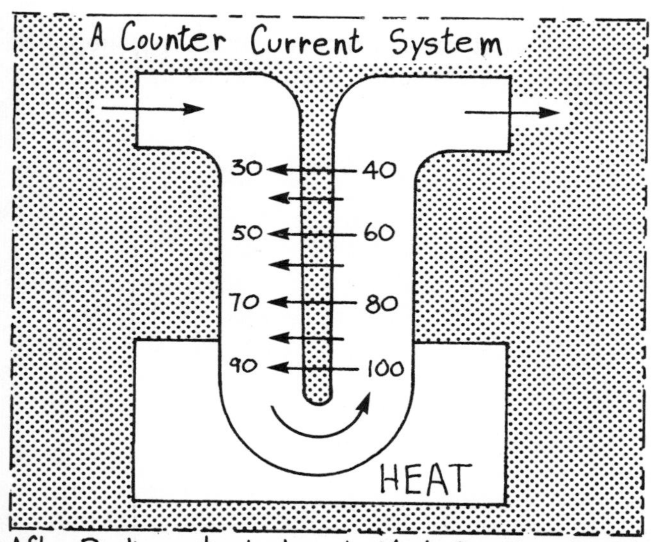

After Berliner et al Am. J. Med 24, 730 (1958)

Some investigators, however, recognized that this arrangement looked as if it might function like a ''counter current'' exchange system familiar to engineers. In a heat exchange system, for example, heat can be conserved by placing the incoming cold water close to the outgoing hot water, thus trapping heat that would otherwise be wasted. This kind of hypothesis has turned out to be correct.

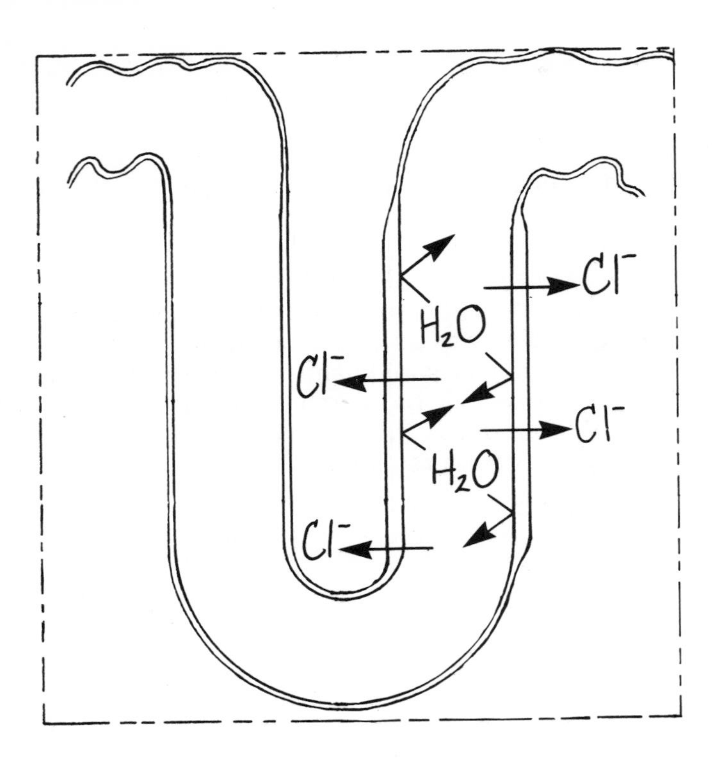

It has now been shown that the thick ascending limb of the loop of Henle has remarkable properties.

First, it is capable of pumping ions out into the interstitium. It is now thought that the transport is primarily an active *chloride* pump with sodium following passively, the exact reverse of what was first thought.

Second, the wall of the thick ascending limb, whilst allowing solutes to be pumped out, does not allow water to follow the osmotic gradient created by the movement of solutes.

A Countercurrent Multiplier

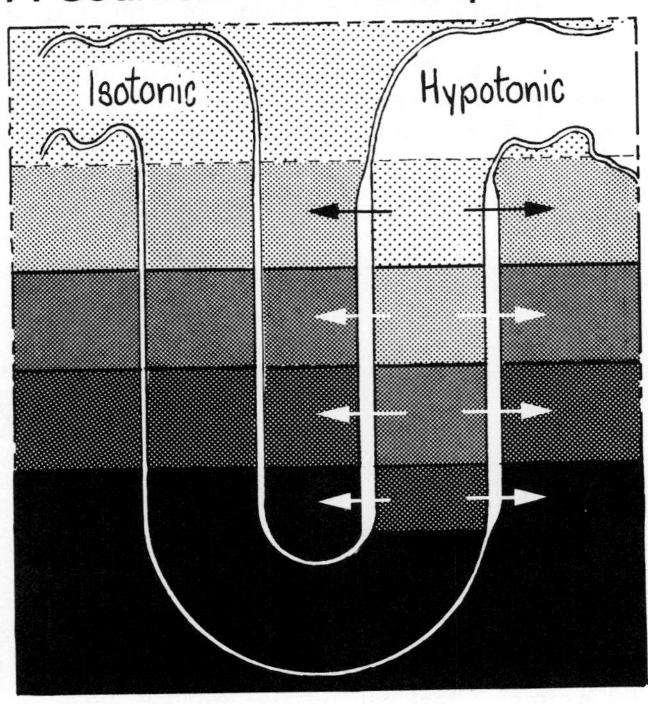

The result is that at each step of the way up the ascending limb, solute is removed from the tubule and added to the interstitium, leaving water behind. As this process is repeated over and over again in many small steps it can be seen that the urine becomes progressively more dilute. At any given level the interstitial fluid is relatively more concentrated than the urine in the ascending limb. Thus, by the end of the loop of Henle the urine has become hypotonic. As a result of small concentration steps, multiplied many times over, a gradient of solute concentration has been built up throughout the medulla with the greatest concentration at its tip.

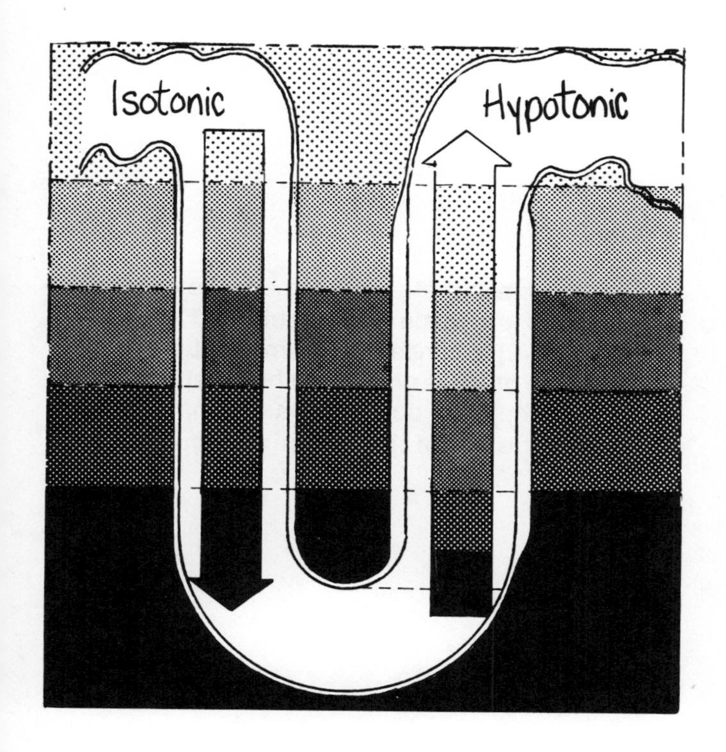

Because the *descending* limb of the loop of Henle is freely permeable to water it will share all the concentration steps of the interstitium around it: Water moves passively across its wall to equalize the osmolalities.

This process of concentration of the interstitium and dilution in the ascending limb of the tubule by a series of very tiny steps will go on until an equilibrium is reached. This will depend upon the rate of flow of the urine, the length of the loop, and presumably the number of pump sites. The ''hairpin'' configuration of the loop is, of course, essential in order to ''trap'' and maintain the concentration gradient.

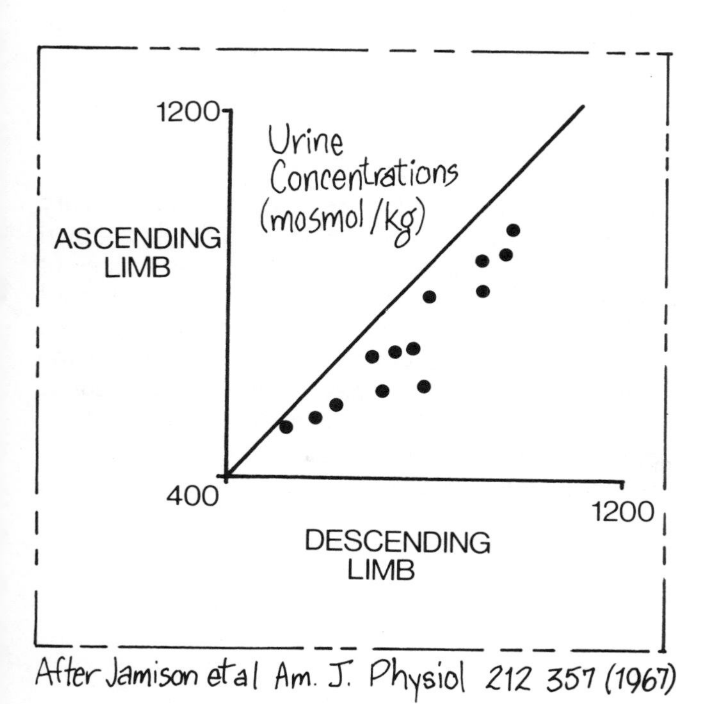

After Jamison et al Am. J. Physiol 212 357 (1967)

Support for this theory has been shown by a number of experiments, such as the one shown here. As the theory predicted, micropuncture studies in animals showed that at all levels of the loop of Henle the fluid in the ascending limb is *always* less concentrated than the fluid in the descending limb.

Recent studies have added much more refinement to our understanding of the way in which this system functions. It is more complex than has been outlined here but this forms a reasonable basis for understanding the concepts involved. More detailed approaches are given in the references at the end of the book.

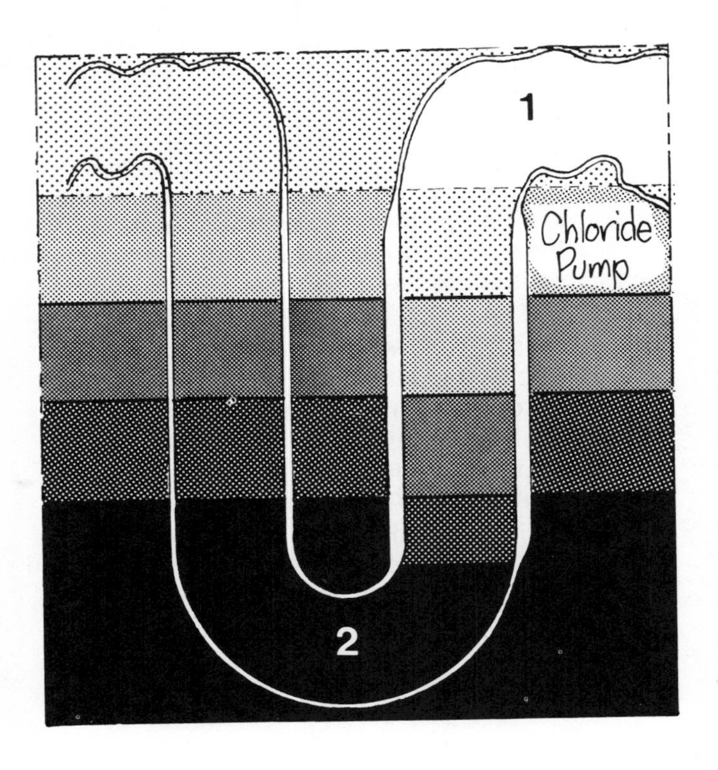

So far it is clear that, at all times, the properties of the loop of Henle will result in (1) a dilute urine, and (2) a concentrated medullary tip. The mechanism of urinary *dilution* has therefore been explained, but how do we explain urinary *concentration* by the end of the collecting tubule?

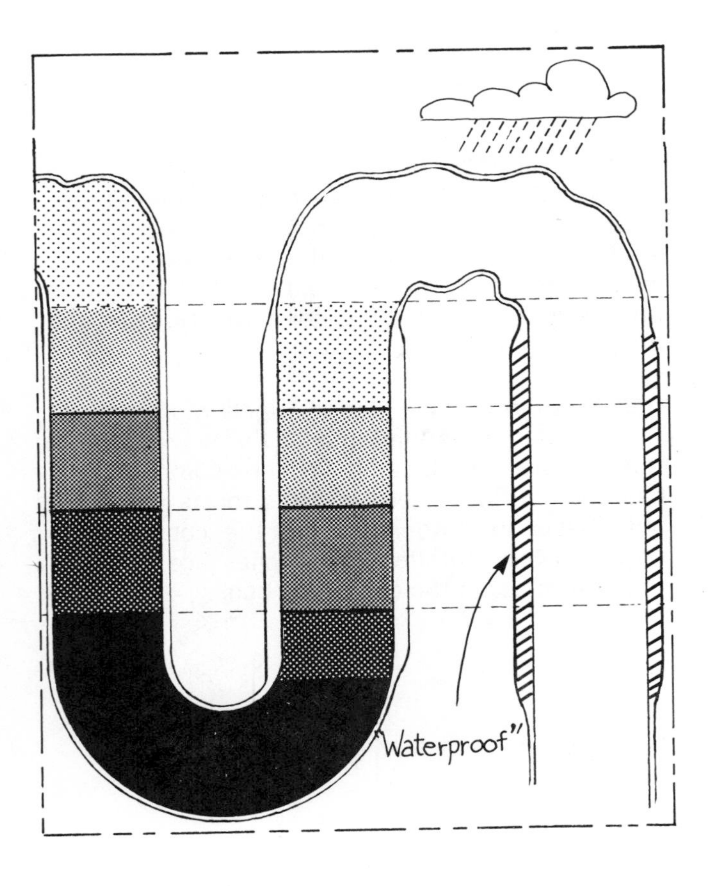

Another anatomical fact now becomes very important: the collecting duct plunges down through the medulla, amongst the loops of Henle, and through the concentration gradient of the medullary interstitium. Under conditions where water conservation is not needed the urine remains dilute until it leaves the kidney because the collecting duct is effectively ''waterproof'', and is thus not affected by the osmotic gradients through which it passes.

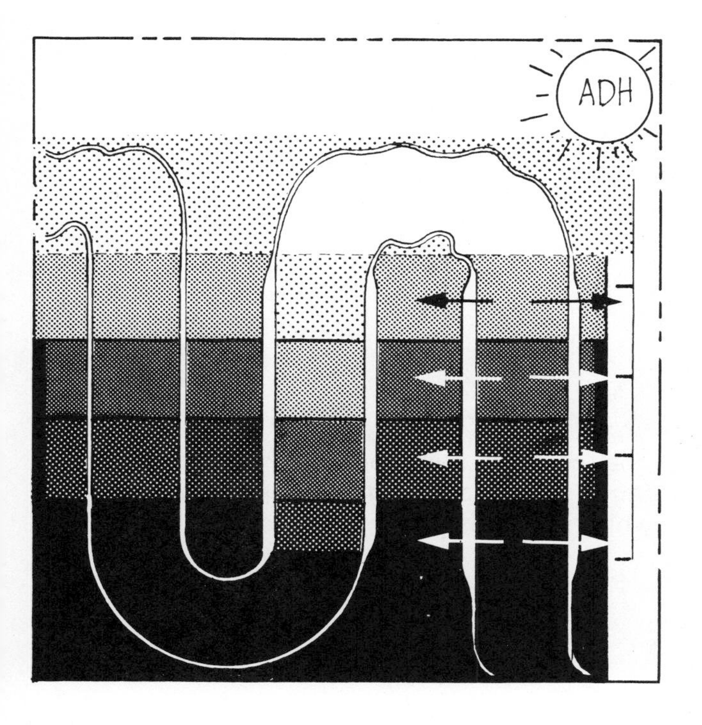

When, however, water conservation is required, antidiuretic hormone (ADH) is released from the hypothalamus and renders the wall of the collecting duct permeable to water. As a result, the osmotic effects of the interstitium cause the removal of water from the collecting duct, thus concentrating the final urine. As well as rendering the collecting ducts permeable to water, ADH may also increase the activity of the ionic pump in the ascending limb of the loop. This will also favor distal water recovery by increasing the osmotic gradients but has not been proven to occur in man.

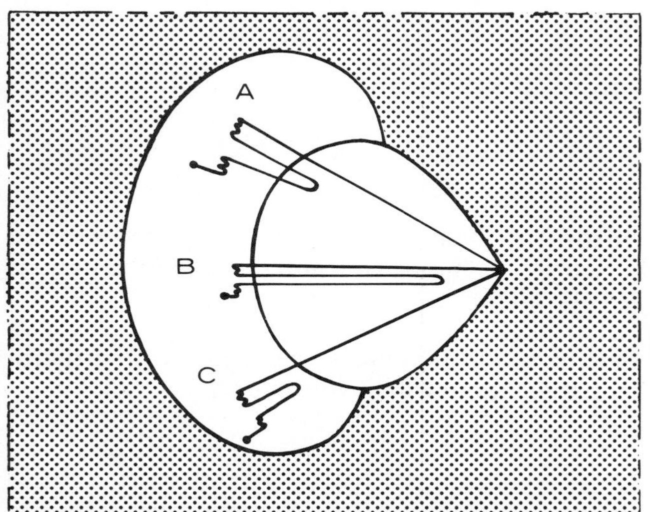

Two final notes must be made. The first is that, in man, only some nephrons (B) have long loops of Henle and it is only these that are responsible for setting up the conditions whereby medullary gradients become established and urine concentration may occur. The loss of these nephrons (by disease of the medulla, for example) will upset the concentrating ability of the kidney as a whole.

A Countercurrent Exchanger

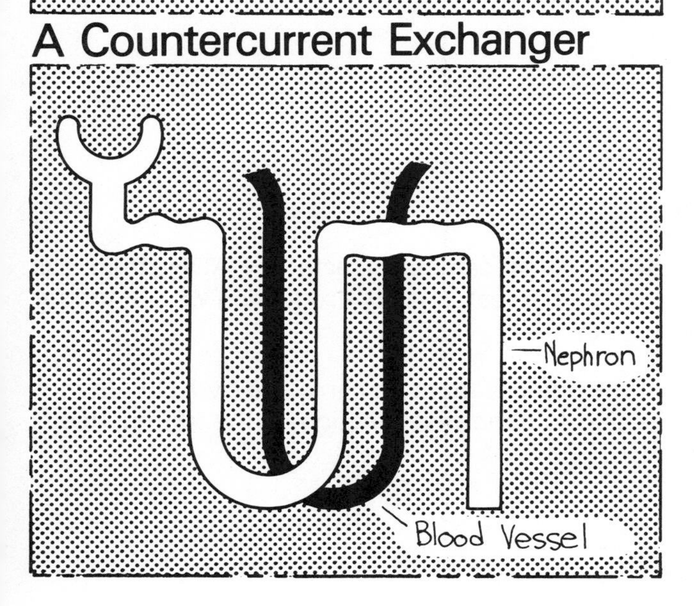

The second point is that the blood vessels coursing through the medulla also have a ''hairpin'' configuration. This is essential to maintain the concentration gradients set up by the ''countercurrent multiplier'' of the loop of Henle. The loops of the blood vessels act as a *countercurrent exchanger* and maintain the gradients rather than washing them out as would occur if they went straight through the medulla.

Clinical Examples of Disordered Water Handling

Water Depletion

The cardinal symptom of *water depletion* is thirst. The cardinal sign is a high plasma sodium, which indicates a high E.C.F. osmolality.

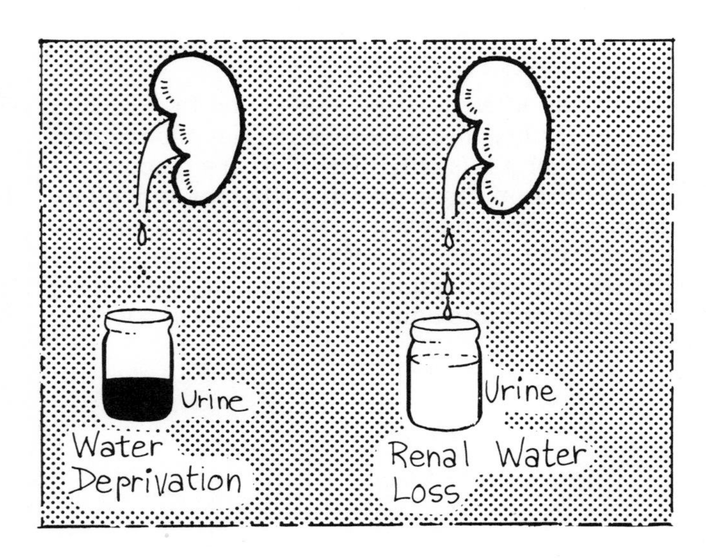

If the depletion is due to *water deprivation* the urine will be highly concentrated, and the patient will be oliguric.

If it is due to *renal loss,* the urine will be dilute and the patient will have polyuria.

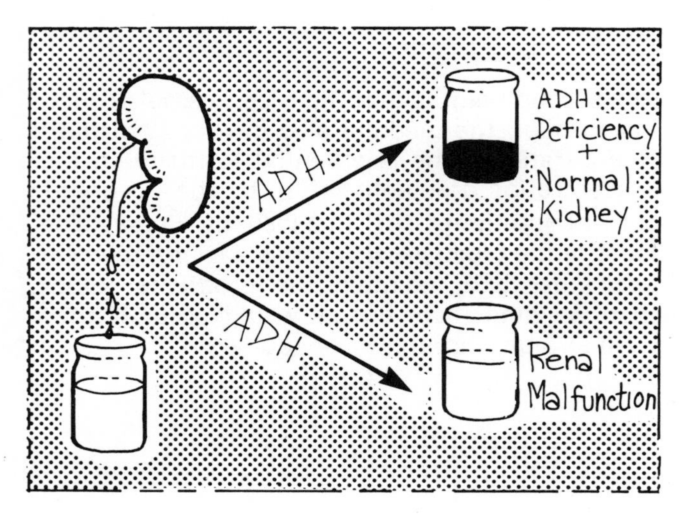

If the defect can be corrected by ADH, then it is due to a defect in ADH production with a normal kidney.

If the defect cannot be corrected by ADH, then it is due to abnormal function of the kidney itself.

ADH Dependent Water Loss

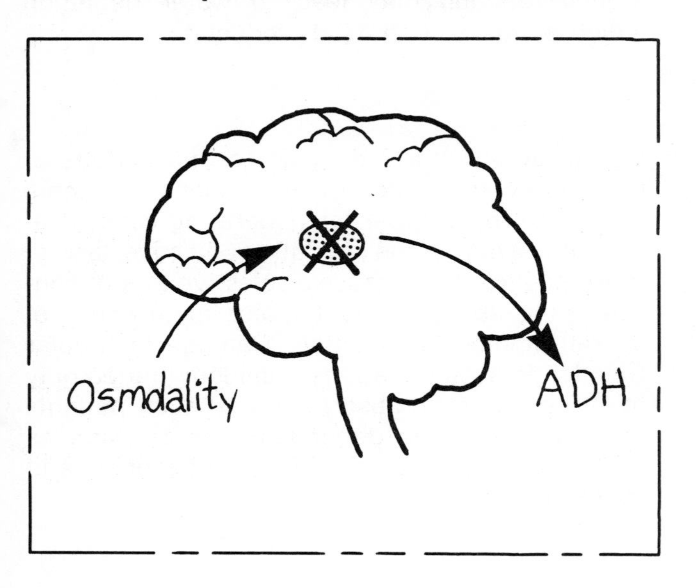

If the defect is ADH dependent, then it implies a healthy kidney but some defect in the production or release of ADH from the hypothalamus. This condition is characterized by profound polyuria and thirst and is called diabetes insipidus; it has already been referred to in Chapter 5.

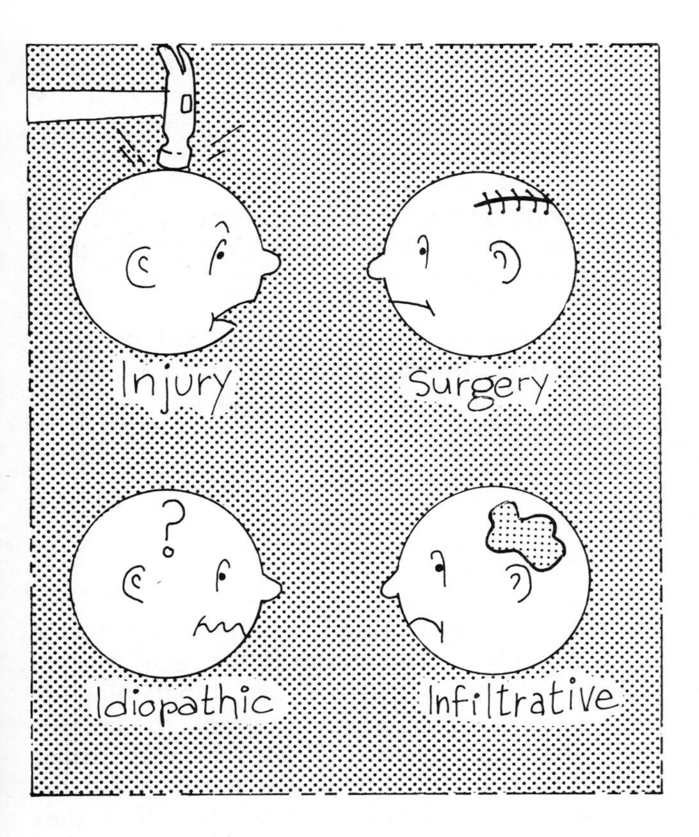

This may be caused by trauma, with fracture of the base of the skull, complications of cranial surgery, (particularly pituitary gland surgery), and by infiltrative processes such as basal meningitis or neoplasm. Many cases, however, appear with no obvious cause and have to be classified as ''idiopathic.''

Abnormal Kidney Function

There are three mechanisms by which disordered kidney function may lead to water depletion which is unrelated to ADH production.

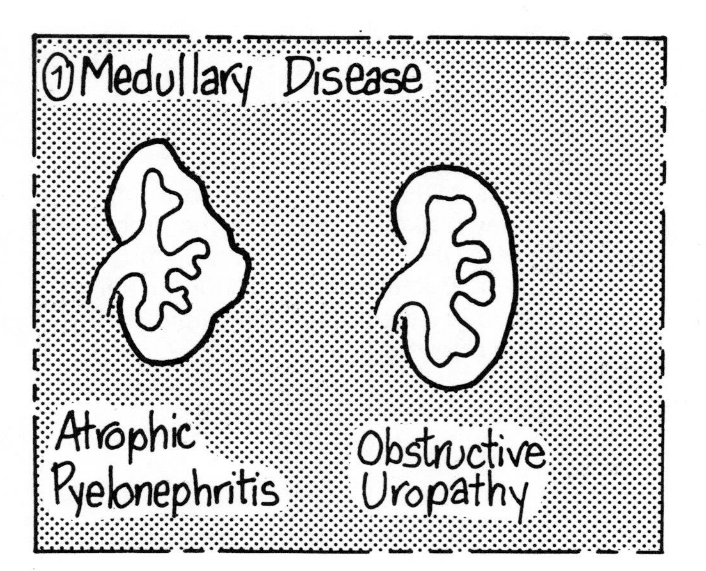

One is by structural damage to the medulla of the kidney involving the long loops of Henle which are responsible for generating concentration gradients in the medulla. Selective loss of these nephrons can cause a selective loss of concentrating ability without major disturbance of overall glomerular function. Damage to tubular function in such situations is seldom limited only to ''free water'' reabsorption and such patients usually have solute dependent loss of water as well. Examples are atrophic pyelonephritis and obstructive uropathy.

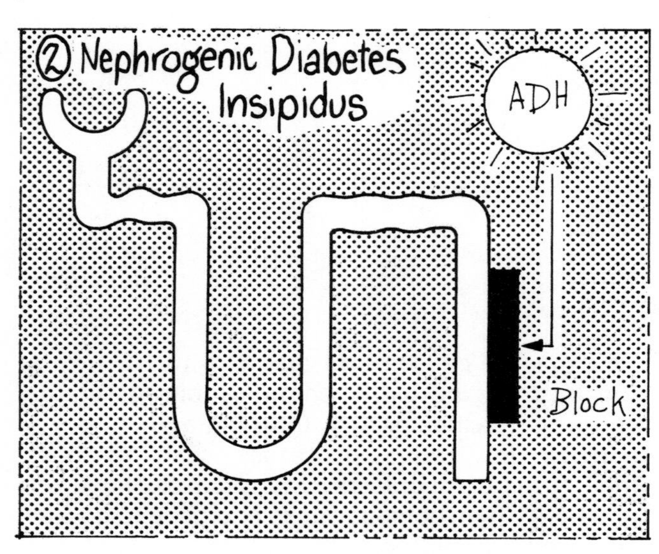

Another mechanism is of renal non-response to ADH due to a functional abnormality of the tubule. This is usually called ''*nephrogenic diabetes insipidus.*''

This condition may be a spontaneously occurring problem in a male infant whose immature kidney fails to respond to normal ADH.

It may, however, be a side effect of certain drugs, particularly lithium carbonate which is widely used in manic-depressive illness. Lithium appears to block the tubular actions of ADH.

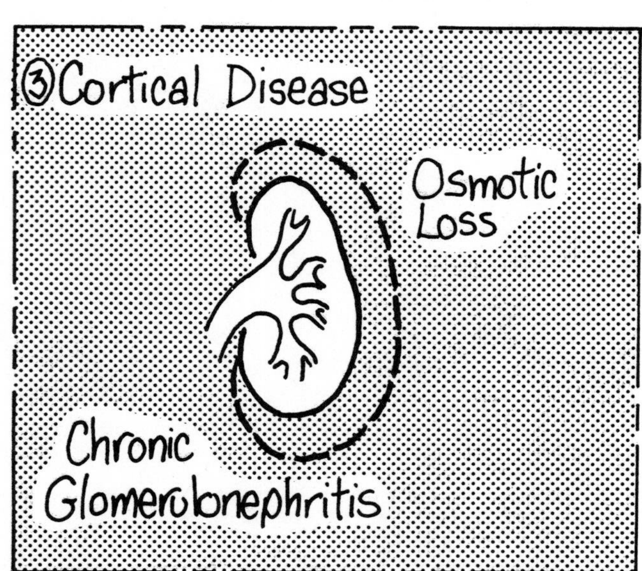

A third mechanism is the osmotic effect due to high levels of filtered solutes in the relatively few remaining nephrons that are functioning in advancing cortical disease such as chronic glomerulonephritis. This means that those nephrons that still survive are handling a much higher load of solutes, such as urea, than they were when all nephrons were functioning; such solutes provide an osmotic load which will increase tubular flow and ''wash out'' the medullary concentration gradient. This explains the polyuria of most patients with chronic renal failure.

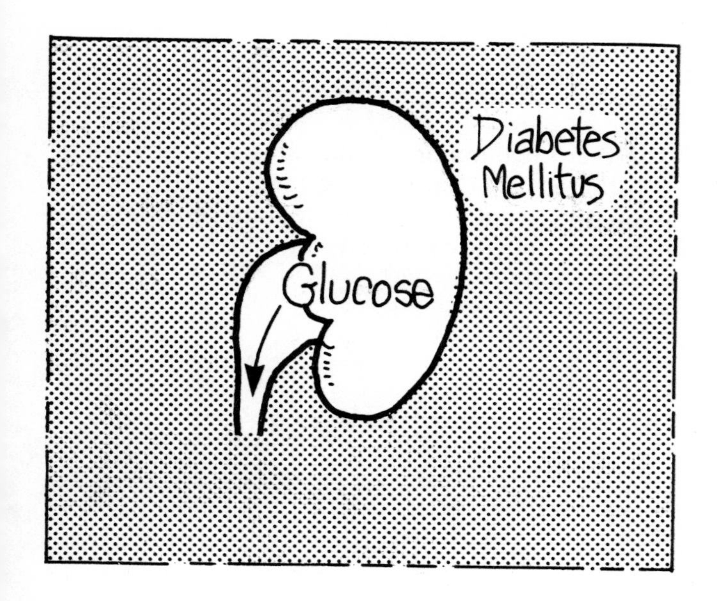

A similar mechanism operates in uncontrolled diabetes mellitus where sugar is the osmotic solute in an otherwise normal kidney. Because of the high blood sugar and the consequent "overflow" into the glomerular filtrate, the ability of the tubule to reabsorb glucose is overwhelmed. Glucose left within the tubule provides an effective osmotic load.

Water Retention

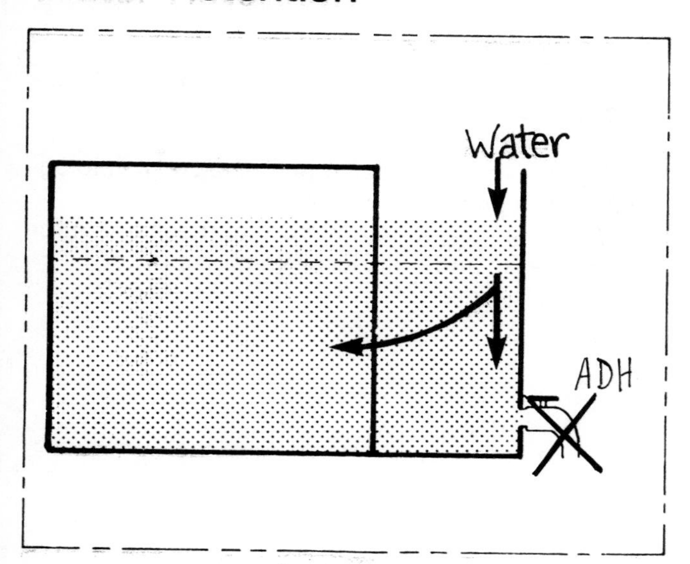

Water *retention* occurs whenever water intake exceeds the ability of the body to excrete the excess load. Since a water load is evenly distributed throughout body water, it will cause proportionately similar expansion of E.C.F. and I.C.F. compartments. It will also dilute the solute concentrations in the body fluids. Hyponatremia is therefore an invariable finding which indicates an excess of water and *not* a deficit of sodium. ADH production will normally be shut off by such events.

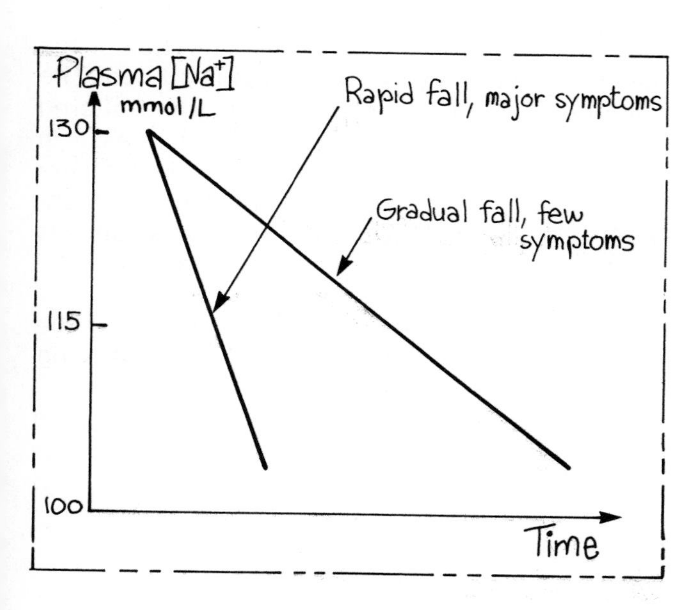

Retention of large volumes of water over a short period of time causes a rapid fall in serum sodium and the development of major neurological symptoms due to osmotic shifts of water into the I.C.F., most notable in the intracranial structures.

Gradual retention of water is, however, a more common clinical situation and does not produce dramatic cerebral signs. A gradual fall in plasma sodium with the gradual development of headache and confusion are seen, but when slowly developing, a plasma sodium as low as 100 millimoles per litre may be surprisingly asymptomatic.

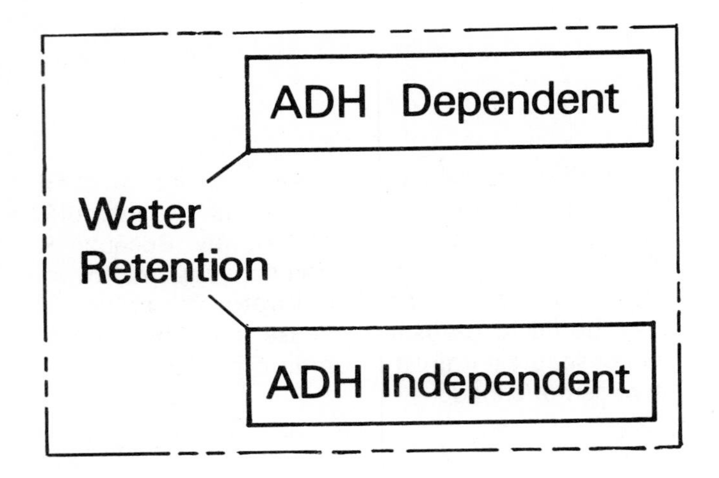

There are two major mechanisms whereby water is retained: (1) Due to the effects of ADH acting upon the kidney; (2) Due to failure of an abnormal kidney to excrete a water load, regardless of ADH activity.

ADH Dependent Water Retention

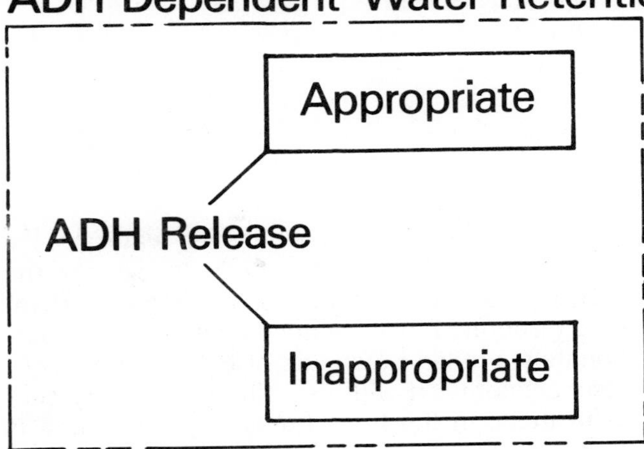

ADH dependent water retention may be divided into two categories:

1) Those which depend upon known physiological stimuli and are therefore "appropriate."
2) Those unrelated to any normal mechanism for ADH release and therefore "inappropriate."

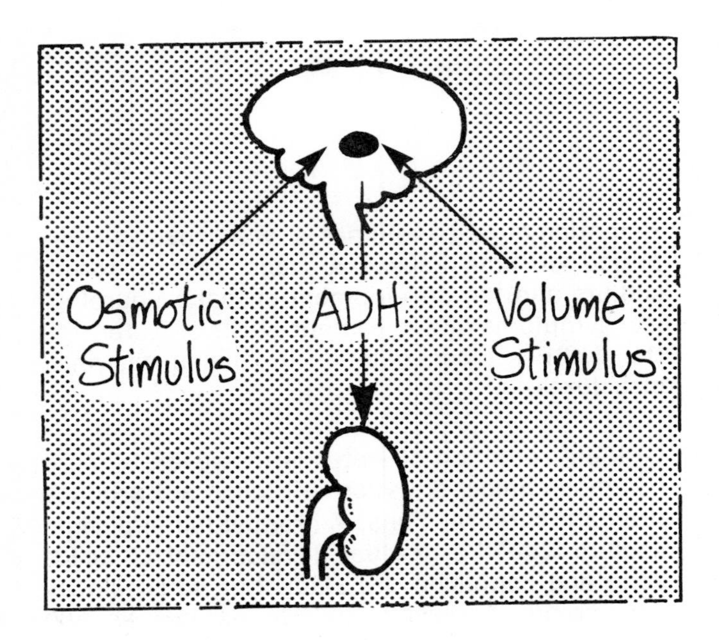

It is important to recall that there are two stimuli to ADH release. The first is an increase in *osmolality* and normally results in dilution of the urine appropriately as the plasma is diluted. The *second* is a stimulus resulting from altered plasma *volume* and may cause ADH release to override "osmotic appropriateness." This occurs when the volume stress is persistent, and unrelieved by sodium conserving mechanisms which retain water in company with salt.

ADH released in this context will result in the retention of "free water," without accompanying solutes, by the kidney.

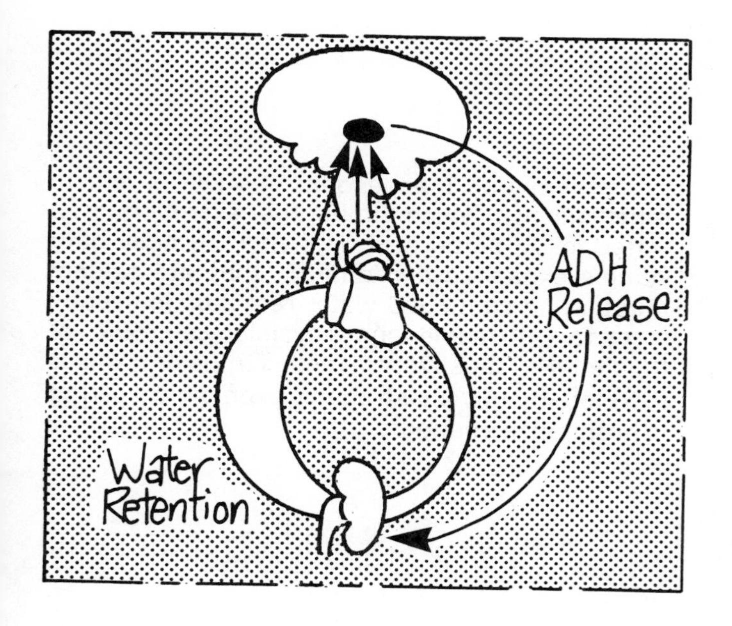

This is seen in heart failure or persistent hypoalbuminemia where all the sodium retaining mechanisms are in play, but still the effective forward flow is inadequate in spite of obvious edema.

In these situations afferent stimuli reach the hypothalamus from a number of receptors which monitor "volume" in some way and probably include sites in the aortic arch, the right and left side of the heart and the great veins.

The resulting ADH release causes water retention (and hyponatremia) in patients who are often already being treated with potent diuretics and considered to have become resistant to them. Because it has a physiological explanation such ADH production is "appropriate".

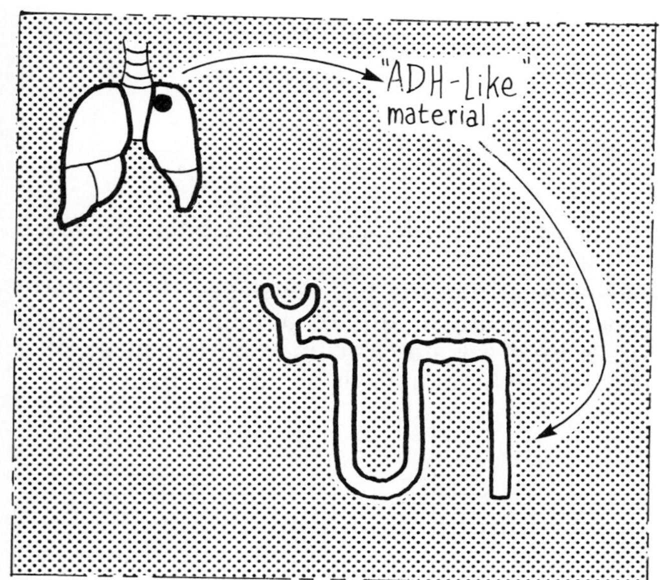

Sometimes, however, water is retained due to ADH release which has no physiological explanation, that is by production of ADH-like material independent of any osmotic or volume stimulus. The source of this substance is often a neoplasm, most commonly a carcinoma of the lung. This results in the syndrome of *"inappropriate ADH secretion."*

The laboratory indicators are a low plasma sodium with an inappropriately concentrated urine. Because free water retention causes expansion of all body compartments, including the E.C.F., the kidney tends to release sodium into the urine since the volume expansion inhibits its sodium retaining mechanisms. Thus, edema does not occur and volume expansion is not clinically obvious.

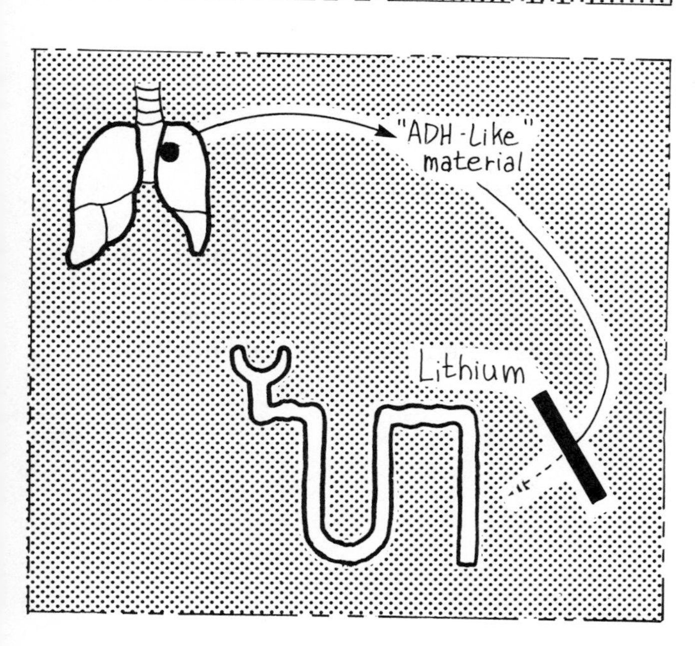

This syndrome can be treated by restriction of water intake or by attempting to block the action of ADH on the renal collecting ducts, with drugs such as lithium carbonate or demeclocycline.

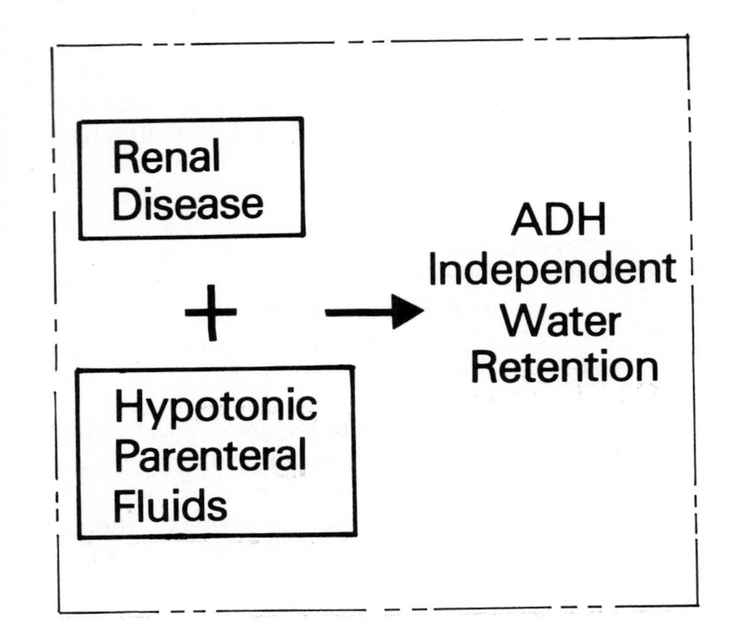

Water retention independent of ADH is due to renal disease which limits the excretion of a water load. Since such patients become both volume expanded and diluted, they are seldom thirsty, and continued water retention is usually secondary to some non-physiological route of fluid intake (usually intravenous infusion); this is therefore mainly an iatrogenic problem occurring in hospitalized patients.

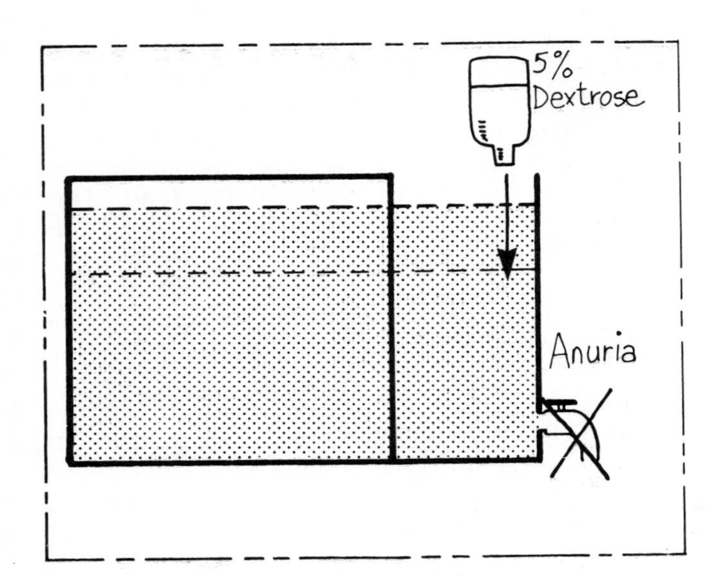

Such a situation can arise when a patient receives continued loading with intravenous dextrose and water after oliguria has developed. In the early phase of acute oliguric renal failure the inability to excrete a water load may be unnoticed at a time when fluid intake is being encouraged.

HYPONATREMIA

usually means

WATER RETENTION

Retention of "free water" in all these situations will cause *hyponatremia*, and as mentioned in the previous chapter a low serum sodium is usually due to water retention and not sodium loss.

Chapter 8

The Regulation of Hydrogen Ions

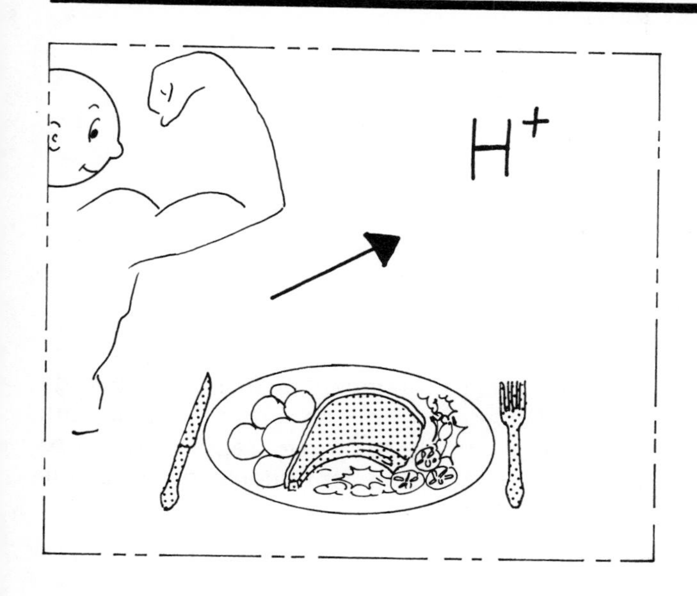

On a normal North American protein-rich diet and with normal metabolic activity, the body is constantly producing protons which threaten the stability of the hydrogen ion concentration upon which the cells rely if they are to function normally. In health most are removed by oxidation; if not, they have to be dealt with in other ways, which is the topic of the present chapter.

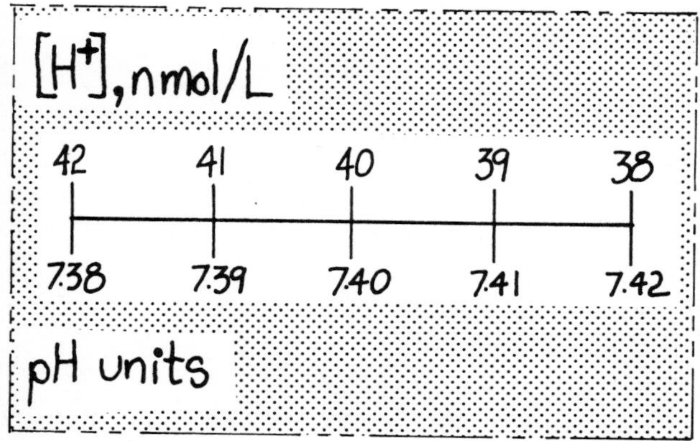

Remember that a *proton* is a hydrogen atom that has given up an electron and is the same as a *hydrogen ion;* an *acid* is any substance that can give up a proton, whilst a *base* can receive a proton.

Normal E.C.F. hydrogen ion concentration is about 40 nanomoles per litre (40 x 10^{-9} moles per litre) which is a pH of 7.4. In the events of everyday life the variation of extracellular pH is very narrow; around this normal range one nanomole of hydrogen ions per litre equates with 0.01 pH unit (a convenient coincidence).

[H+], n mol/L
125
20
6.9
7.7
pH units

In abnormal situations, much wider ranges may be seen, and this diagram indicates the kind of maximal aberrations beyond which the limits of survival are usually passed. In practice a pH of 6.9 or 7.7 will only be seen in profound pathological situations.

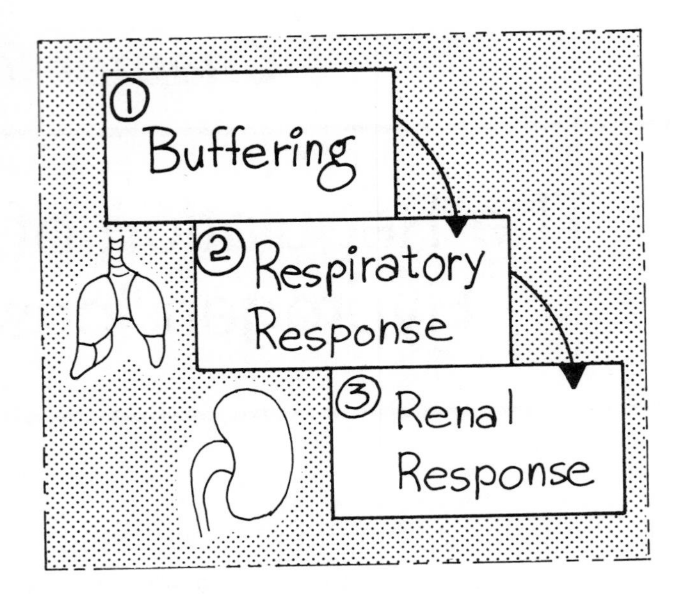

Three mechanisms are available to regulate changes in the pH of the E.C.F. and they operate in the sequence shown here. Buffering is more or less immediate, the respiratory response is delayed and the renal response is the slowest to develop.

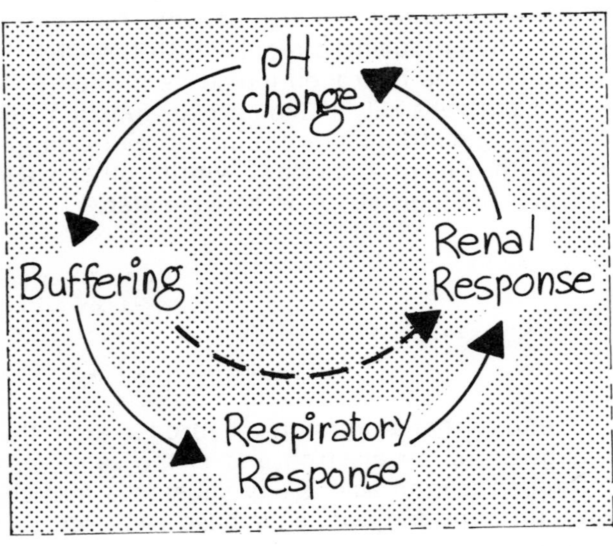

In physiological situations, however, where large and rapid changes of pH do not occur, a smooth coordination of these three regulating steps provides a precise and continuous control of the pH of the E.C.F.

Buffering

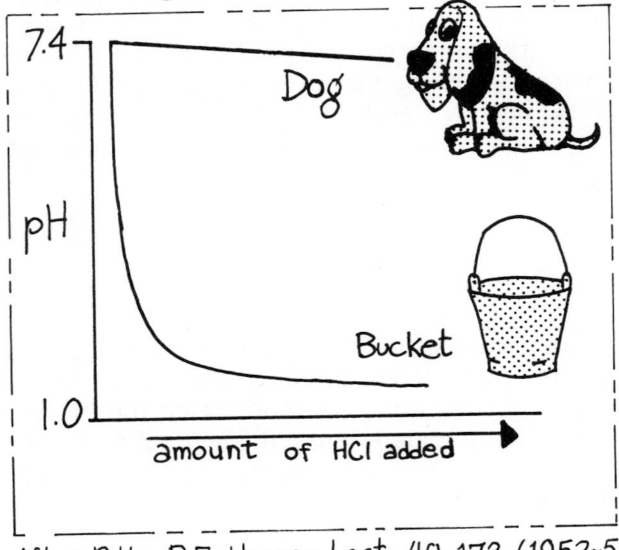

After Pitts R.F. Harvey Lect. 48 172 (1952-53)

The concept of "buffering" is well illustrated by this old experiment. If hydrochloric acid is added to a bucket of water, the pH drops rapidly. If the same amount of hydrochloric acid is infused into a dog with the same total body water as the bucket, little fall in pH occurs. The hydrogen ion load has been "buffered".

Strong Acid

$$HCl \rightleftarrows H^+ + Cl^-$$

Weak Acid

$$H_2PO_4^- \rightleftarrows H^+ + HPO_4^{2-}$$

Buffering is achieved by the binding of hydrogen ions in *weak* acids. Because they do not completely dissociate, they limit the amount of free hydrogen ion available. Hydrochloric acid is a *strong* acid and almost completely dissociates to produce a high concentration of hydrogen ions. Dihydrogen phosphate is a *weak* acid and only partially dissociates to give less free hydrogen ion than an equimolar amount of hydrochloric acid.

The "Buffer Equation"

$$[H^+] = K_A \frac{[H^+ \text{ donor}]}{[H^+ \text{ acceptor}]} \quad (1)$$

can be re-written as:

$$pH = pK_A + \log \frac{[H^+ \text{acceptor}]}{[H^+ \text{donor}]} \quad (2)$$

The relationship between the two sides of a "buffer reaction" can be expressed by the general equation of Henderson (1) which is based on the law of mass action.

However the use of actual hydrogen ion concentrations has not gained general clinical acceptance and so Henderson's equation has been modified, by a simple mathematical trick, to express hydrogen ion concentration as pH. In this form it carries the names of Henderson and Hasselbalch (2).

The two equations say the same thing, but in different ways, and to avoid confusion we will stay with pH as the unit of hydrogen ion concentration.

Buffer Reactions:

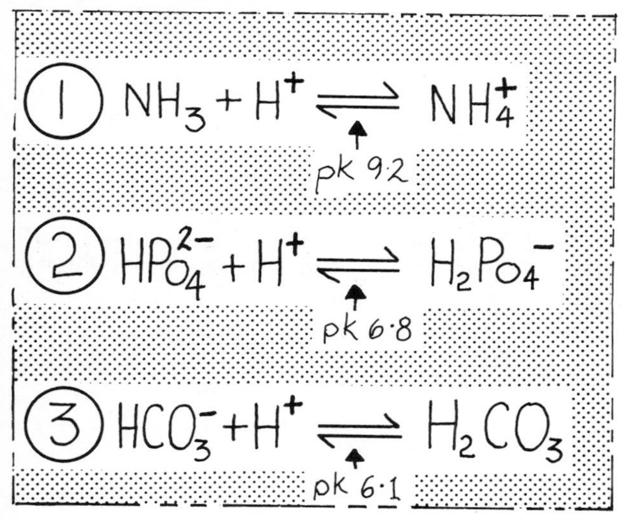

The pK_A for any buffer is a constant which depends upon the buffer being considered. It is the pH at which the reaction is evenly balanced between the two components of a buffer pair so that each is present in equal concentration.

$$pH = 6.8 + \log\frac{[HPO_4^{2-}]}{[H_2PO_4^{-}]}$$

For phosphate, as an example, the reaction can be represented as shown here. In any buffer reaction the relationship between pH and the ''buffer pair'' represented on each side of the reaction is mathematically predictable. If the pH is known, then the ionic concentrations of the reactants can be calculated.

Intracellular Buffers

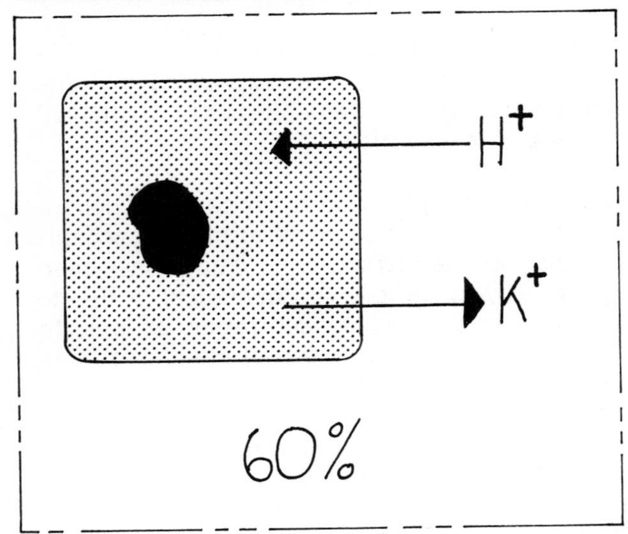

Sixty percent or more of buffering occurs inside the cells. To enter the cells the hydrogen ions exchange with potassium ions, which explains why a sudden acid load may be associated with an increase in the potassium concentration in the E.C.F. In states of long-standing systemic acidosis hydrogen ions may enter bony tissue in exchange for calcium ions and this can eventually lead to demineralisation of the bone (osteomalacia).

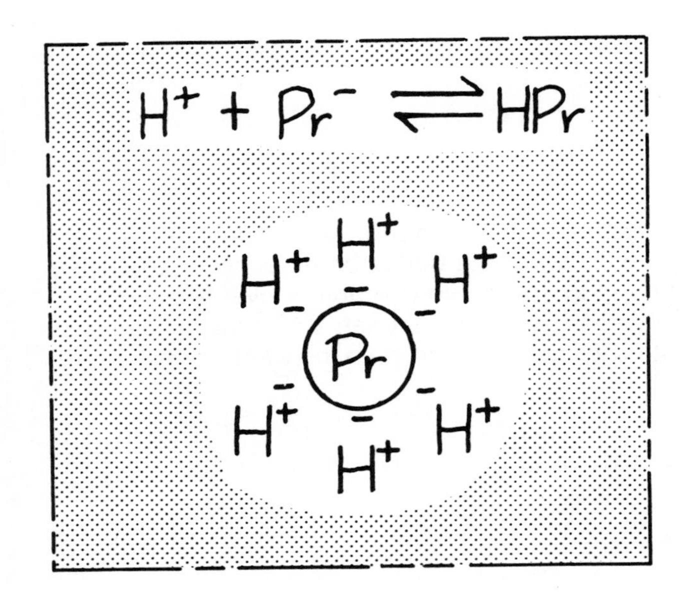

Intracellular buffers are mainly proteins and include hemoglobin in the red cells. Although designated in the equation as monovalent, proteins often function as polyanions, with a number of sites where cations may bind.

Extracellular Buffers

$$H^+ + HPO_4^{2-} \rightleftharpoons H_2PO_4^-$$

Extracellular buffers such as phospate are only effective to a limited degree. They rapidly become saturated as the end product accumulates and brings the reaction to equilibrium.

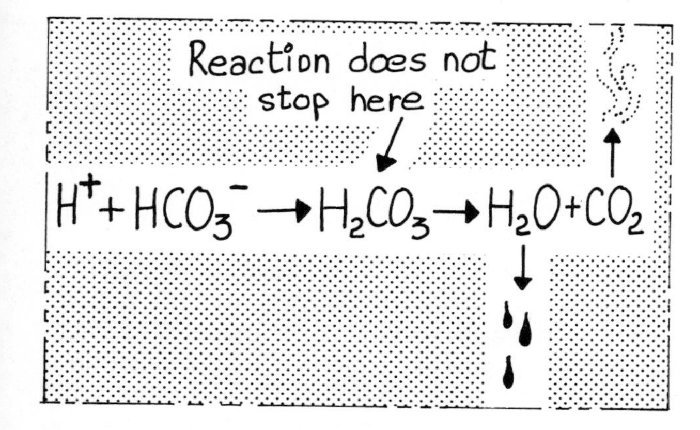

However, the reaction which converts bicarbonate to carbonic acid does not become saturated, and for this reason the value of bicarbonate as a buffer in the E.C.F. is unique. The end products of the reaction can be dissipated: Carbon dioxide via the lungs and water into the general water pool in the body.

$$pH = 6.1 + \log \frac{[HCO_3^-]}{[H_2CO_3]}$$

The relationship between pH and the action of bicarbonate as a buffer can be mathematically predicted by the Henderson-Hasselbalch equation. Clearly if any two of the variables are known the third can be calculated, and many nomograms exist whereby such calculations are rendered simple.

OR

$$pH = 6.1 + \log \frac{[HCO_3^-]}{0.03\,P_{CO_2}}$$

The equation can be rewritten in a number of ways which accept the fact that carbonic acid is in equilibrium with the dissolved carbon dioxide in the body fluids. Carbon dioxide is expressed in terms of its partial pressure, using the term P_{CO_2}.

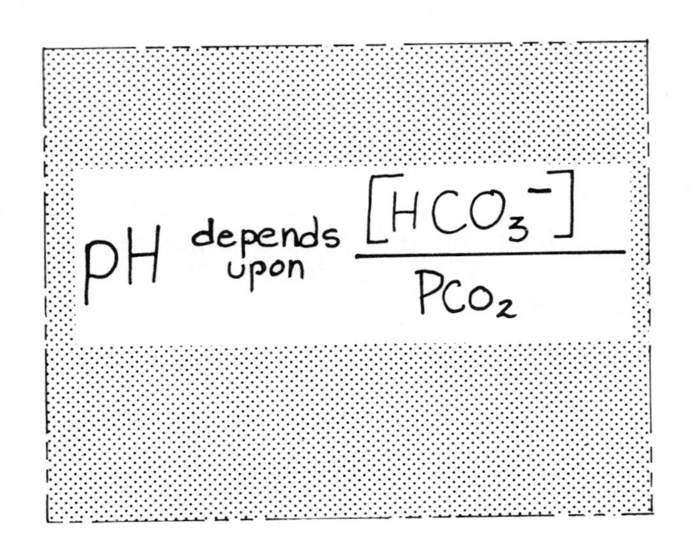

In conceptual form, which simplifies things yet further, the relationship can be rewritten in this way, which will prove useful in the clinical situations to be discussed later.

It is important to recognize that this relationship means that the pH will vary as the bicarbonate and P_{CO_2} change; in other words, if other things remain constant:
1) Removing bicarbonate, raising the P_{CO_2} *or* adding free hydrogen ions will all have the *same* effect - a fall of pH.
2) Adding bicarbonate, lowering P_{CO_2} or removing free hydrogen ions will all produce the *same* effect - a rise of pH.

The Respiratory Response

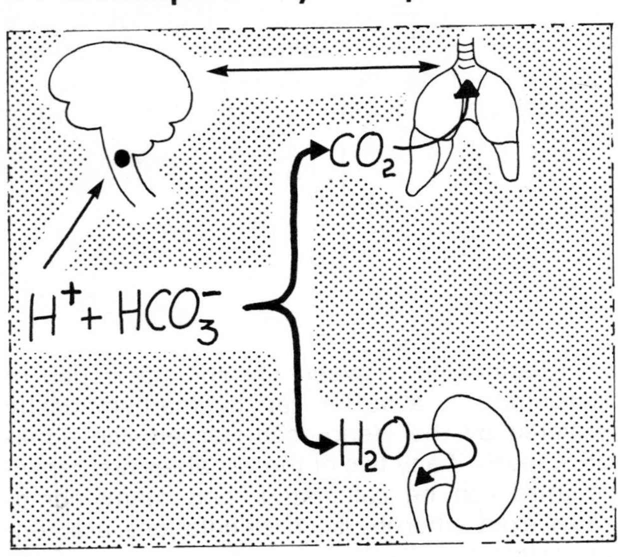

It is now apparent what constitutes the respiratory response: in effect, hydrogen ions can be turned into water and carbon dioxide. The reaction can keep moving from left to right, without any saturation due to buildup of the end products of the reaction, because carbon dioxide can be rapidly excreted through the lungs.

The stimulus to increased ventilation is probably an increase of hydrogen ion concentration sensed by cells in the brain stem. Provided the lungs are healthy, they can increase ventilation to excrete a large amount of carbon dioxide.

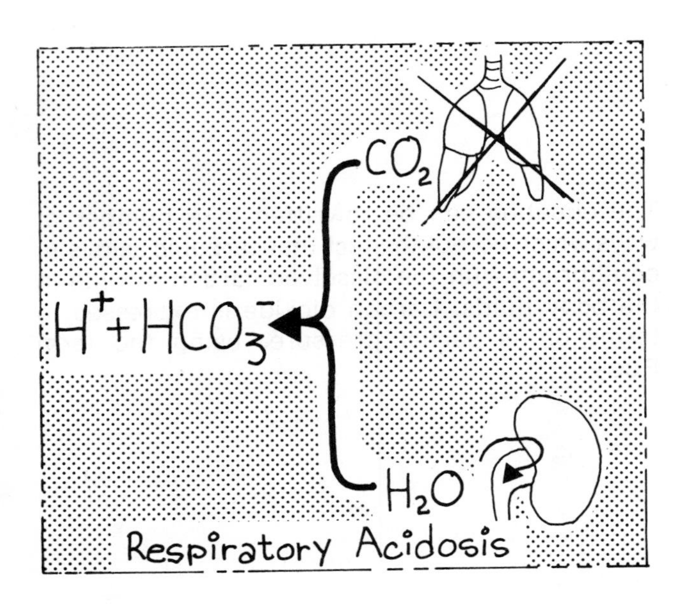

If the lungs are unable to excrete carbon dioxide, then of course, the E.C.F. is unable to handle hydrogen ions and a respiratory acidosis results, because the reaction cannot move from left to right as carbon dioxide accumulates.

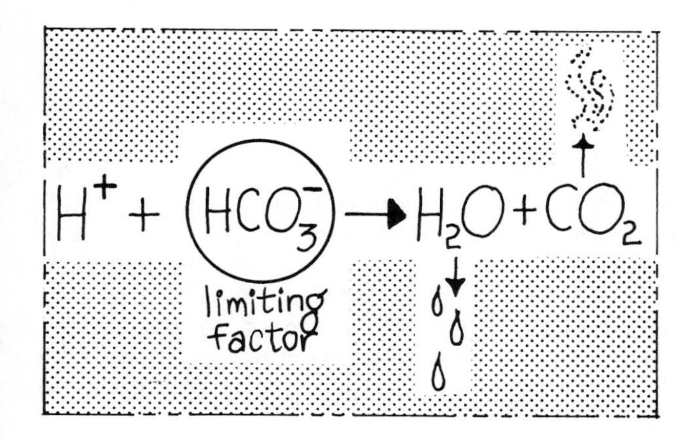

The limit to this reaction in the presence of normal lungs is therefore *not* a buildup of the end products of the reaction (water or carbon dioxide) but a "running down" of available bicarbonate to fuel the continued buffering of hydrogen ion.

Renal Response

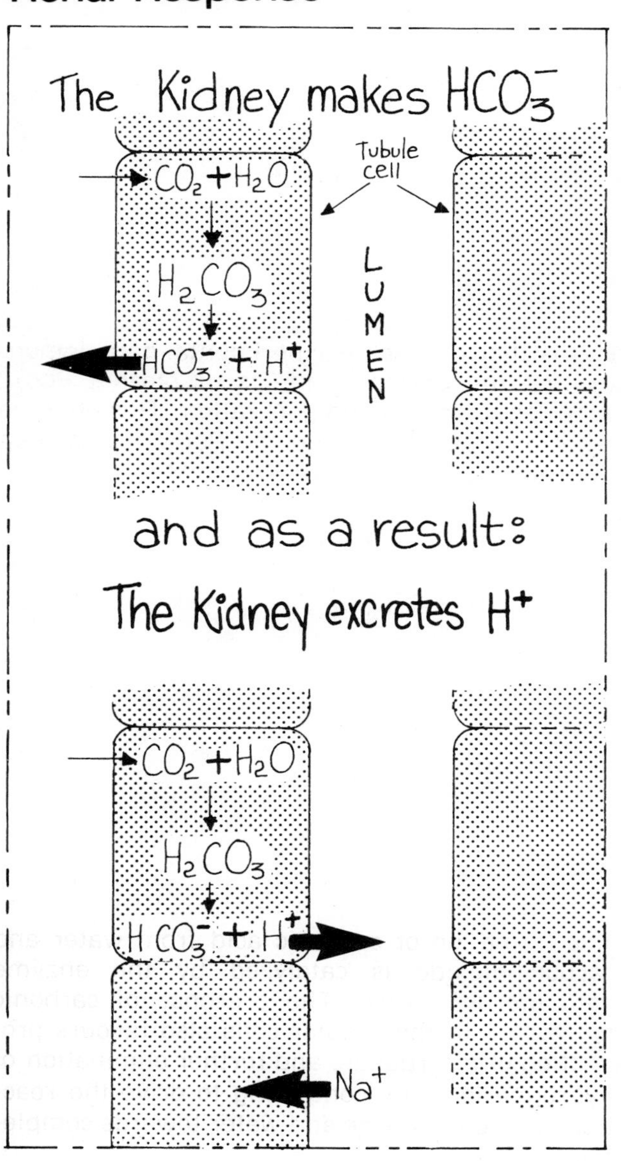

This is where the renal response comes in. The kidney has the capability of regenerating the bicarbonate supply and in effect reverses the processes involved in the respiratory response. Carbon dioxide diffuses into the cells of the renal tubule and leads to the production of bicarbonate which returns to the E.C.F. to replenish the E.C.F. bicarbonate.

The beauty of this mechanism is that for every bicarbonate ion that is returned to the E.C.F. one hydrogen ion is ejected into the tubular lumen, thus providing an additional mechanism for removing hydrogen ion from the E.C.F. Since there cannot be a uni-directional flow of positively charged ions, the movement of hydrogen ions out of the cell is associated with the movement of sodium into the cell by active transport. The transport mechanisms for hydrogen and sodium ions are probably separate but in effect there is an exchange of one for the other.

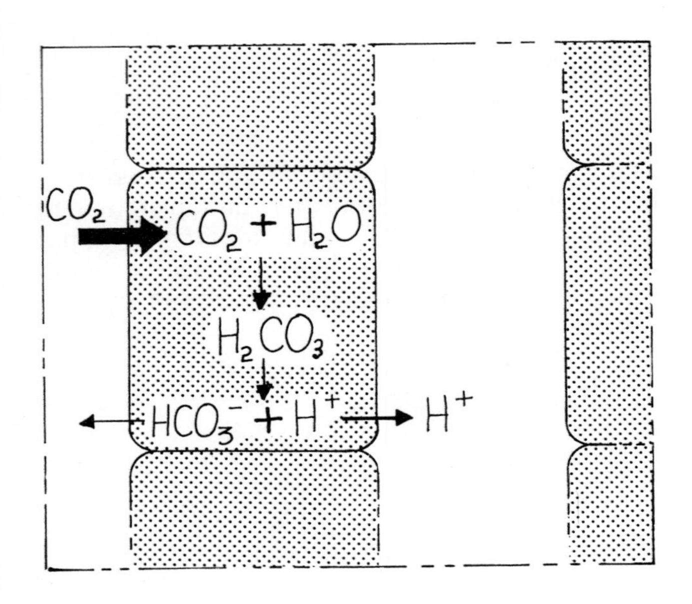

Bicarbonate production, and thus hydrogen ion excretion, will be increased by a rising partial pressure of carbon dioxide which drives the reaction between carbon dioxide and water to produce carbonic acid and then on to bicarbonate and hydrogen ion. This occurs in respiratory failure where the renal response is to excrete hydrogen ions and regenerate bicarbonate to stabilize the pH.

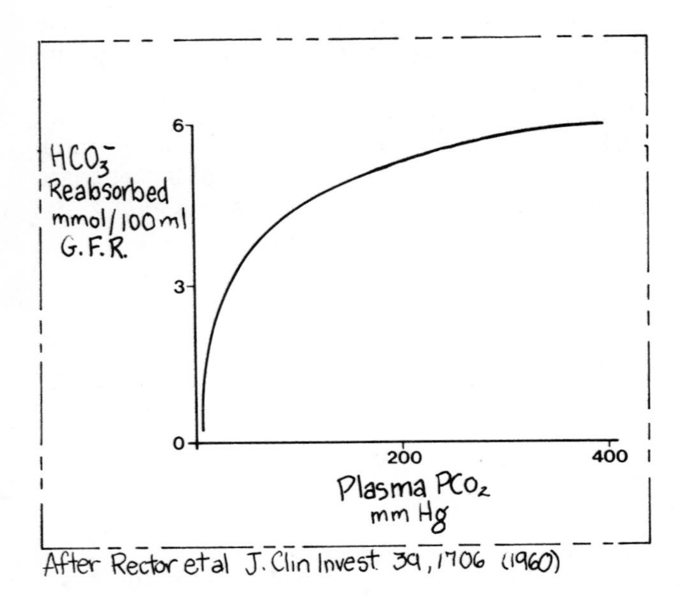

After Rector et al J. Clin Invest 39, 1706 (1960)

This is shown by some experimental data demonstrating the dramatic effects of plasma P_{CO_2} upon renal bicarbonate generation. This effect is especially marked around the P_{CO_2} levels seen clinically (between 40 and 100 mmHg).

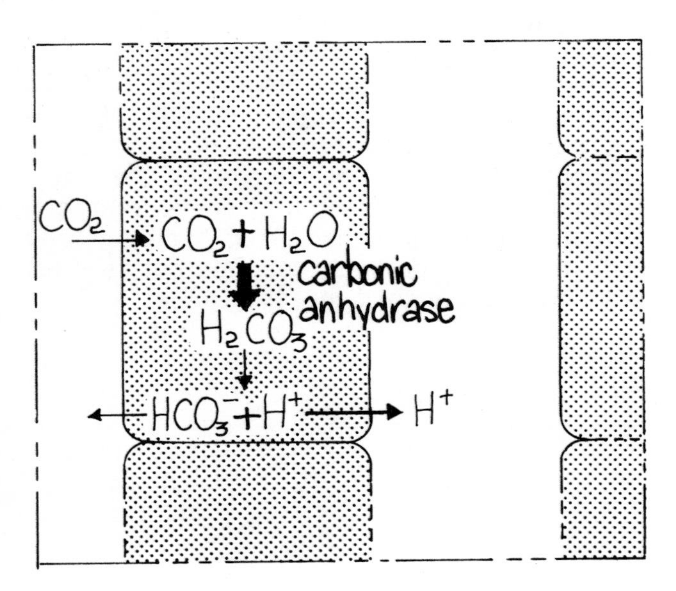

The formation of carbonic acid from water and carbon dioxide is catalyzed by the enzyme *carbonic anhydrase.* The presence of carbonic anhydrase in the tubular cells also favours production of bicarbonate and further elimination of hydrogen ions, by continuing to drive the reaction of carbon dioxide and water towards completion.

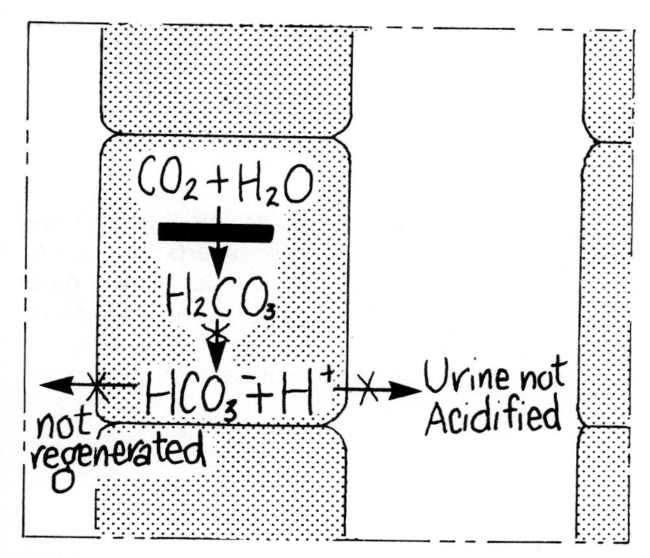

Blocking carbonic anhydrase, with a drug such as acetazolamide, will rapidly block both bicarbonate regeneration and hydrogen ion production by the renal tubule. As a result, filtered bicarbonate will escape in the urine and the buffering capacity in the plasma will fall as the bicarbonate falls. In short, the urine becomes alkaline and the plasma pH falls, a condition called ''renal tubular acidosis''.

Buffering of Renal Tubular Hydrogen Ions

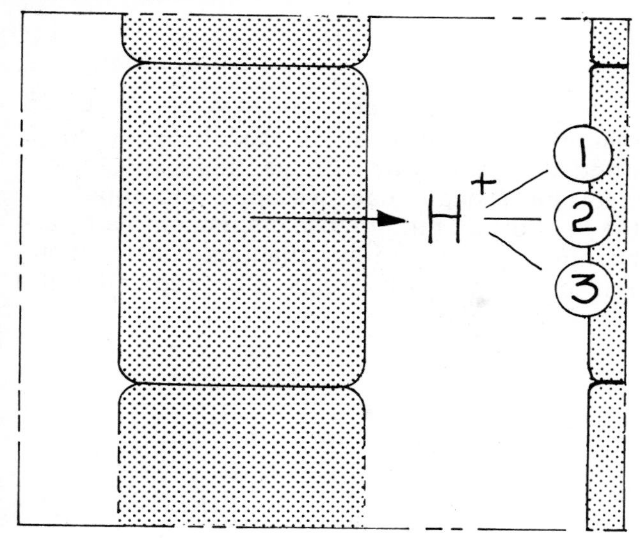

Once the hydrogen ion has been secreted into the renal tubule it is handled in a way that prevents a fall of pH in the lumen which would otherwise stop further hydrogen ion secretion, since in the *proximal* tubule the cells cannot pump hydrogen ions into the lumen against a significant concentration gradient. There are three ways in which the tubular hydrogen ion concentration is limited:

1) Bicarbonate trapping
2) Ammonia production
3) Titratable acid production

Bicarbonate Trapping

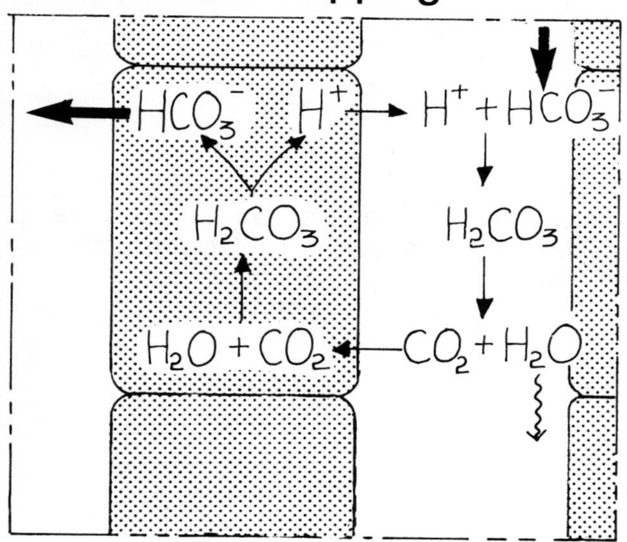

The first mechanism occurs when secreted hydrogen ion meets filtered bicarbonate and is trapped in the familiar carbonic acid buffer system. The result is the formation of water and the generation of carbon dioxide which can diffuse back into the tubular cell where it will favour the generation of more bicarbonate and more hydrogen ions.

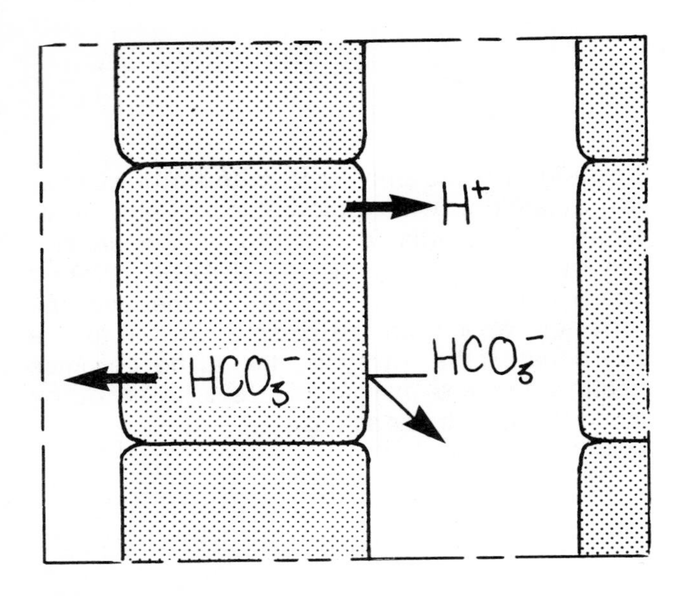

In effect, for every bicarbonate ion that is trapped by a hydrogen ion, another bicarbonate ion is generated by the tubular cell. Whilst bicarbonate ions cannot be reabsorbed *as such*, the end result is as if every secreted hydrogen ion resulted in the reabsorption of one bicarbonate ion.

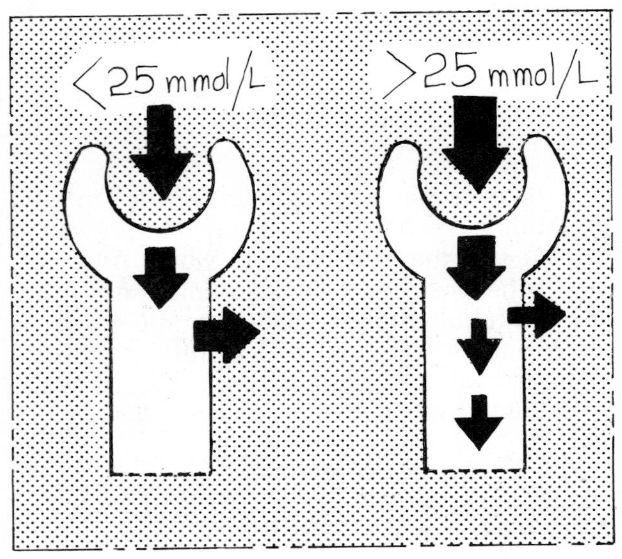

Below a plasma bicarbonate level of about 25 millimoles per litre all filtered bicarbonate is trapped in this way. Above this ''threshold'' level, bicarbonate may ''escape'' and appear in the final urine, but in these circumstances bicarbonate excretion, rather than its conservation, will serve the best interests of pH regulation for the whole organism.

Ammonia Production

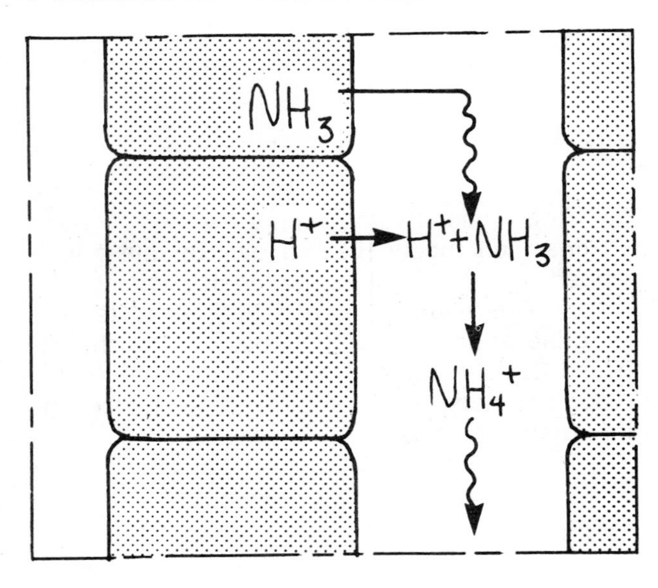

The second mechanism for handling renal tubular hydrogen ions is by the production of ammonia.

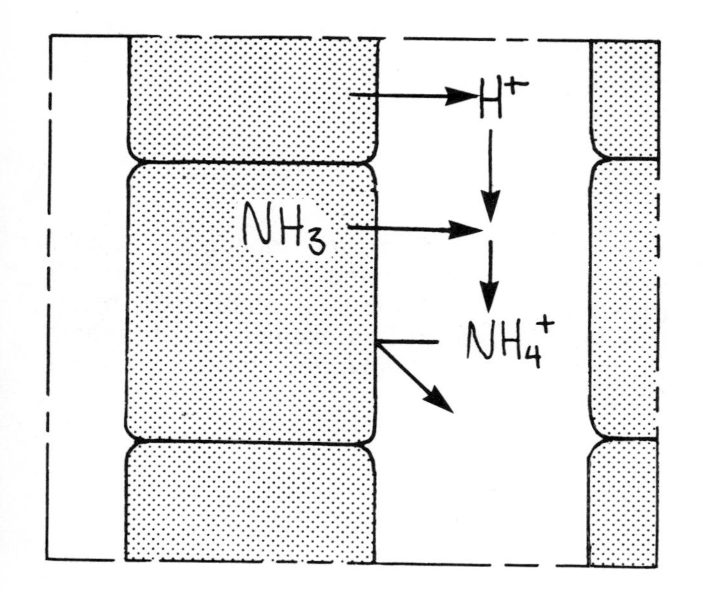

Ammonia is a highly diffusible substance. As soon as it is produced it diffuses rapidly and completely into the tubular lumen. There it meets secreted hydrogen ions and promptly traps them to form ammonium ions. Unlike non-ionic ammonia, ammonium ions cannot diffuse back into the cell, and so stay in the lumen.

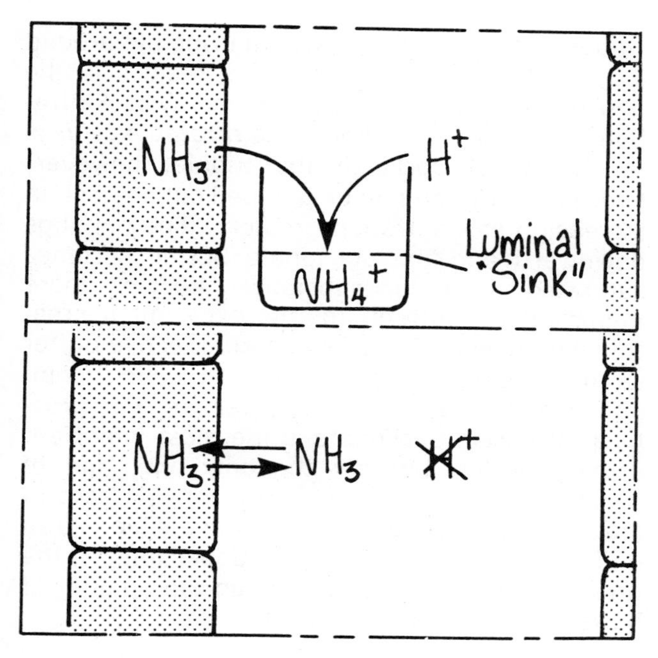

So long as ammonia is trapped by hydrogen ions there will be a gradient for diffusion of ammonia from the cell into the "sink" created by its conversion to non-diffusible ammonium ions. If free hydrogen ions are not present, then the luminal ammonia level will rise and the gradient will disappear; ammonia will therefore stay within the cells and accumulate, a process which promptly results in a shut-down of further ammonia production. Ammonia is continuously produced only when free hydrogen ions are being produced by the tubule. This means that ammonia production is closely related to the amount of hydrogen ions being produced. Thus the role of this system becomes more important as the need to excrete hydrogen ions and regenerate bicarbonate becomes more urgent.

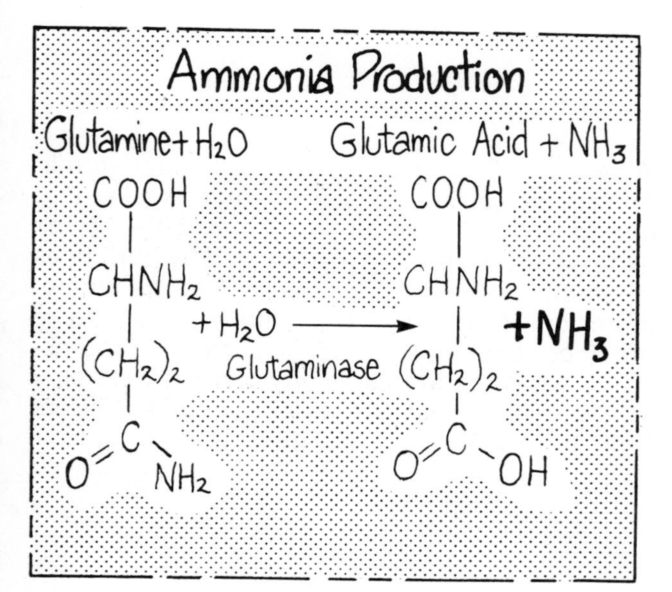

Ammonia production within the tubular cells involves a number of chemical steps which lead to the release of amino groups and to the formation of free ammonia. A final step in that process is shown here.

Titratable Acid Production

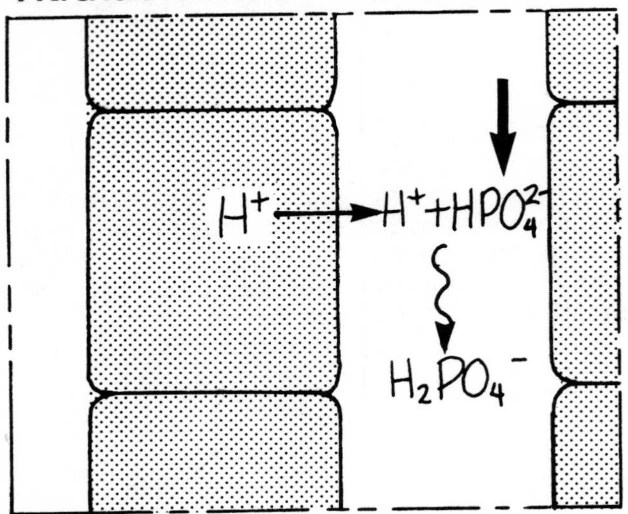

The third mechanism for managing tubular hydrogen ions is trapping them within buffers (filtered at the glomerulus) other than bicarbonate. The best known example is phosphate which is quantitatively the most important urinary buffer.

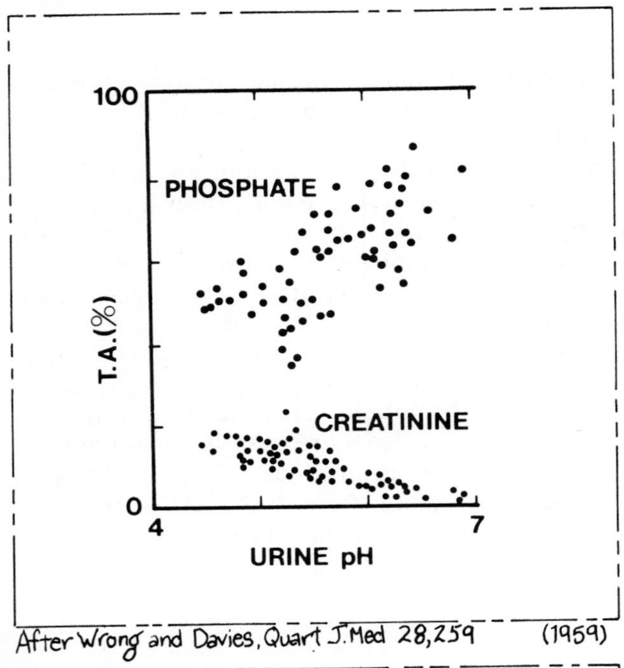

After Wrong and Davies, Quart J. Med 28,259 (1959)

Under normal conditions with a urine pH around 6.5, most of the buffering power resides in the phosphate ion and can be measured by back titration to pH 7.4. This is known as *titratable acidity.* In the distal parts of the tubule, however, hydrogen ions begin to be pumped out against an increasing concentration gradient. The pH slips downward and the phosphate buffer becomes saturated, so other substances become more important as buffers. In this experiment creatinine is shown to become a more effective buffer as urine pH falls below 5.0, at which point phosphate is becoming less effective. For practical purposes, phosphate is the most important urinary buffer, but not the only one, and of course, the titratable acid can only be accommodated by the amount of buffer available to trap it. This means that there is an upper limit to the titratable acidity depending upon the amount of buffer in the glomerular filtrate.

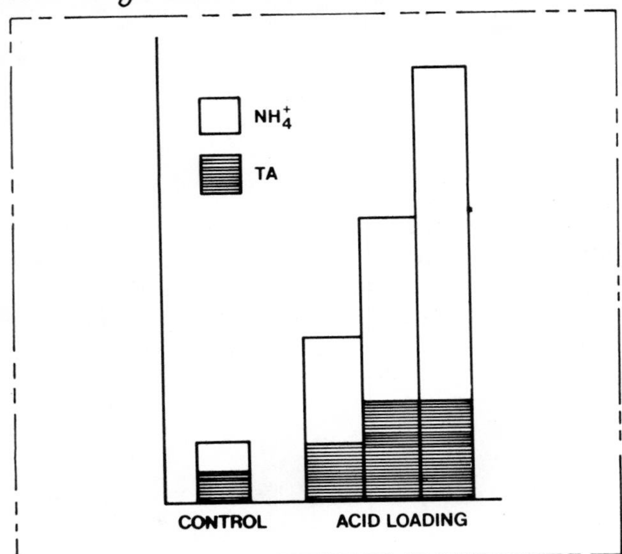

When titratable acidity and ammonium formation are compared, as shown here, it is clear that during a control period more than 50% of the hydrogen ions excreted are handled by buffers as titratable acid, whilst less is trapped as ammonium ions. *But* when a continuing acid load is provided, ammonium excretion can rise progressively whilst titratable acidity can increase only to the extent that there are suitable buffers available in the glomerular filtrate.

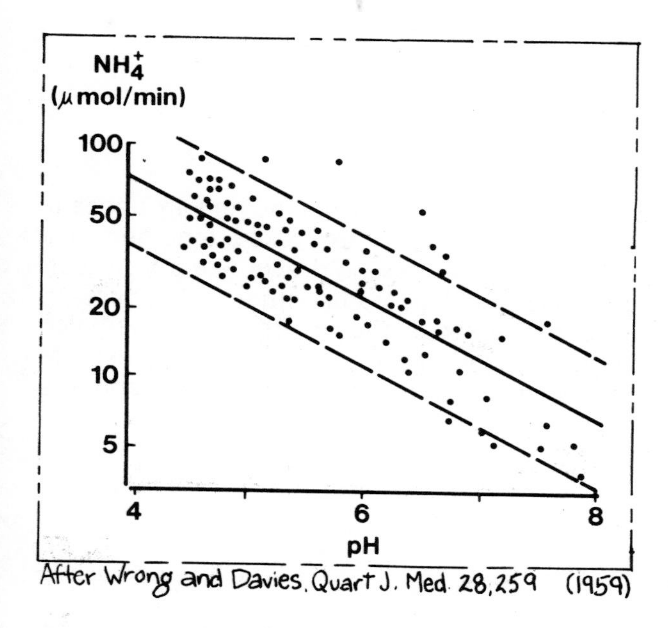

Here is an experiment that illustrates the importance of urinary ammonium; it re-emphasizes the fact that the more acid the urine (and the more saturated titratable acid becomes) the more ammonium is produced.

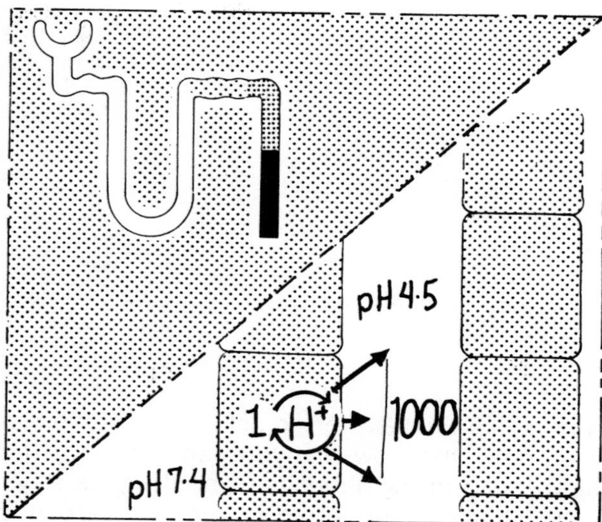

One final point about these renal mechanisms: Although hydrogen ion is secreted (and bicarbonate regenerated), all along the tubule it is only distally that a significant pH gradient develops. Proximally hydrogen ions are so rapidly and completely trapped that no gradient exists even though a large hydrogen ion load is handled. At the end of the tubule, however, hydrogen ions can be secreted against a considerable gradient as urinary buffers become successively saturated. The final urine pH in man can approach 4.5 which is three pH units removed from plasma and means approximately a thousandfold gradient for hydrogen ions. Clearly this is an effective pump which has the potential to maximize renal bicarbonate recovery.

The Link Between Lung and Kidney

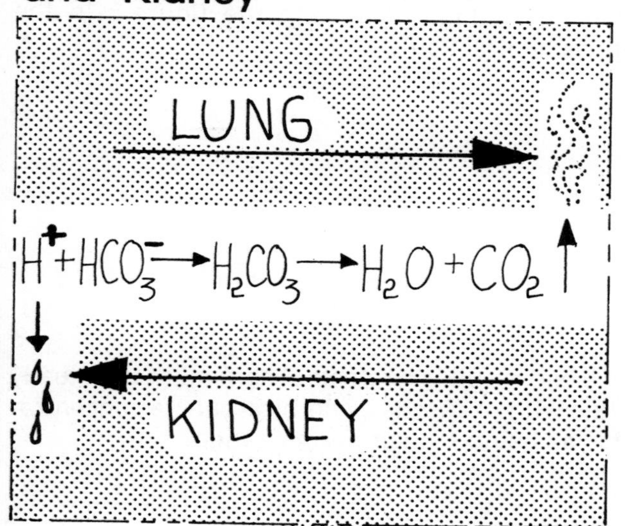

It now remains to link together the role of the kidney and the lung in the overall management of varying hydrogen ion loads.

The lung is concerned with the trapping of hydrogen ion as water and keeping things "moving from left to right". The kidney is concerned with regenerating bicarbonate and so is concerned with keeping things "moving from right to left."

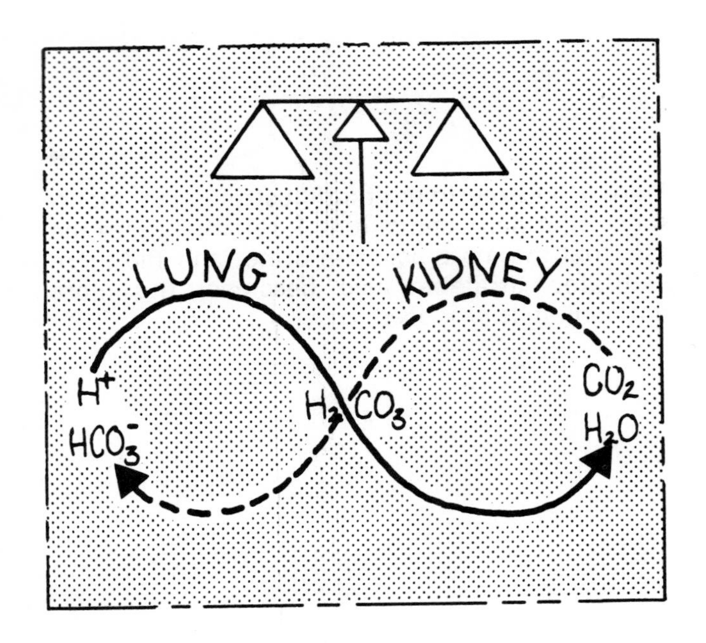

But the two organs are not pulling in opposite direction, rather they are yoked together in a continuum which may be drawn like a figure-of-eight. In terms of laboratory study, this figure-of-eight is expressed by the Henderson-Hasselbalch equation which links lung and kidney by expressing pH in terms of the lung (P_{CO_2}) and the kidney (bicarbonate concentration).

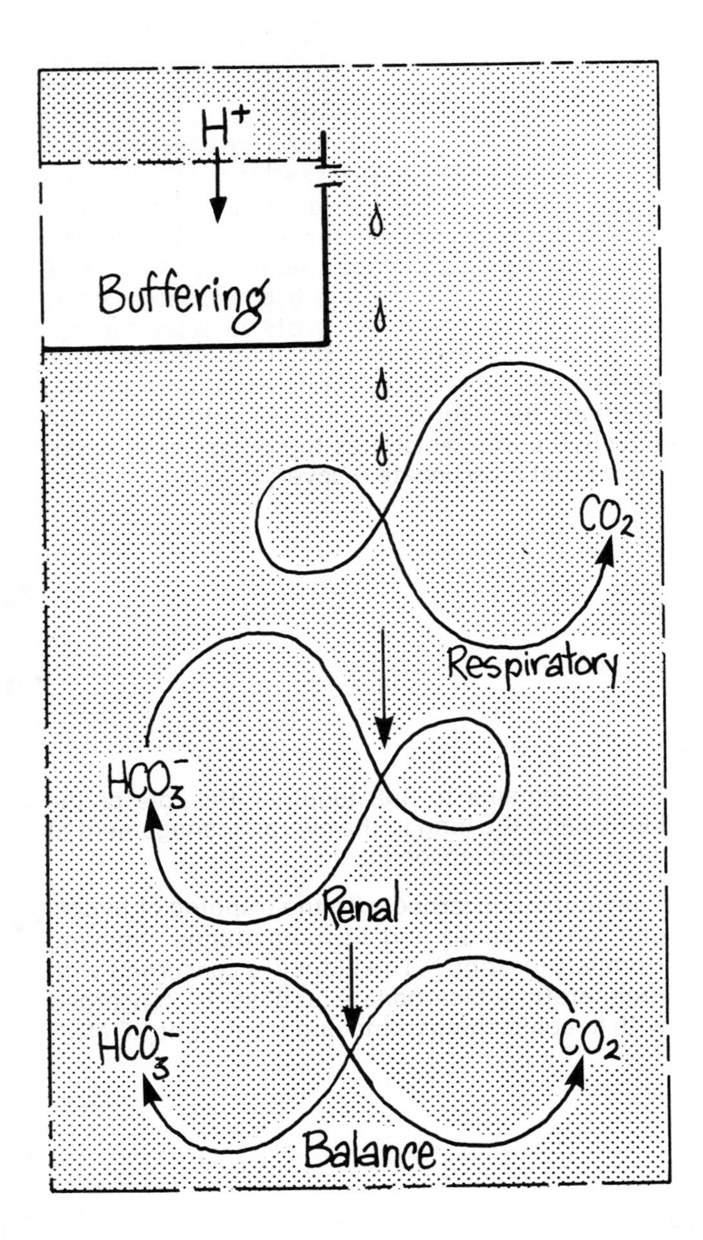

It works like this: Buffering acts as a reservoir to ''even out'' the flow of hydrogen ions into the respiratory and renal mechanisms. It is instantaneous. The relatively rapid respiratory response emphasizes the shift of the carbonic acid buffer system towards the production of carbon dioxide (distorting the figure 8). The slower renal response then emphasizes bicarbonate recovery from the kidney (distorting the figure 8 in the opposite direction).

The end result is a re-establishment of the steady state with pH, P_{CO_2}, and bicarbonate concentration all being in their normal range.

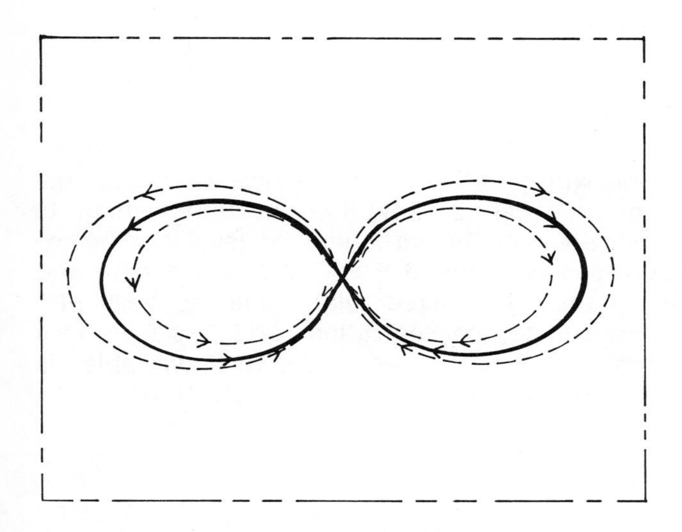

In the normal events of physiological regulation, the distortion of this system in either direction is never great and pH variation is small.

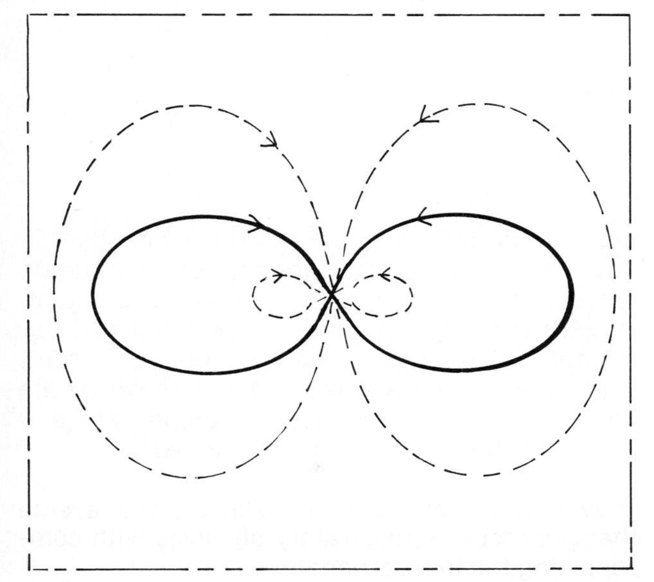

Only in the catastrophes of *abnormal* events do the distortions become so large as to produce swings of pH that go well outside acceptable limits and outside the ability of normal regulatory processes to correct. Failure of the lung or the kidney to fulfill their task in this regulatory system is often at the root of such distortions of pH regulation.

Clinical Examples of Disturbed H^+ Handling

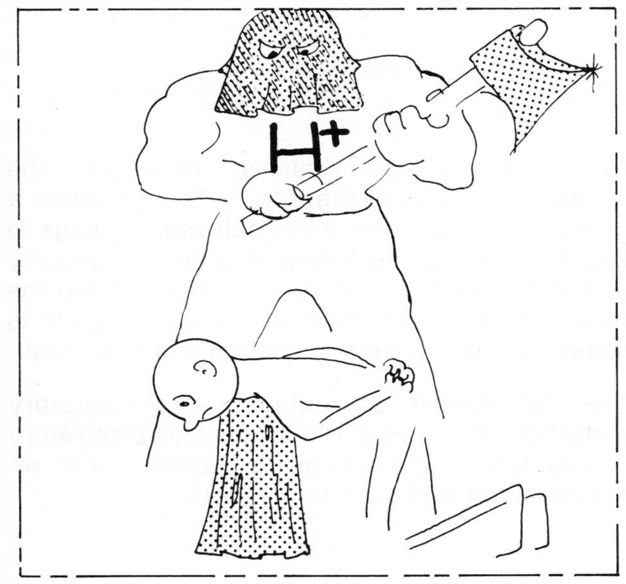

The major clinical threat is of overproduction of hydrogen ion. Acidosis is a more common problem than alkalosis.
Since the bicarbonate/carbonic acid buffer system is the most important indicator of the regulation of hydrogen ion concentration in the E.C.F., it is the most useful relationship to work with at a clinical level. It can be expressed in terms of the Henderson-Hasselbalch equation.

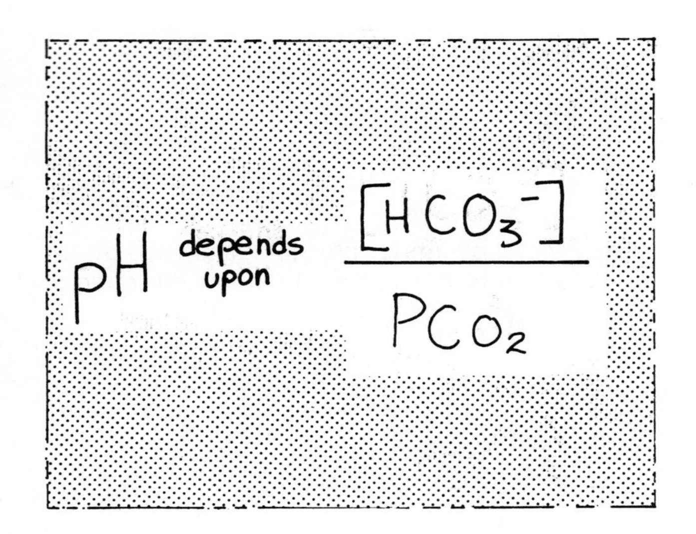

You will recall that for practical purposes, the equation can be rewritten in this fashion to indicate that the carbonic acid level can be expressed in terms of P_{CO_2} and bicarbonate and that there is a fixed relationship between pH, bicarbonate concentration and P_{CO_2}. This means that if any two of the three variables is measured then the third can be calculated.

Respiratory Disturbances

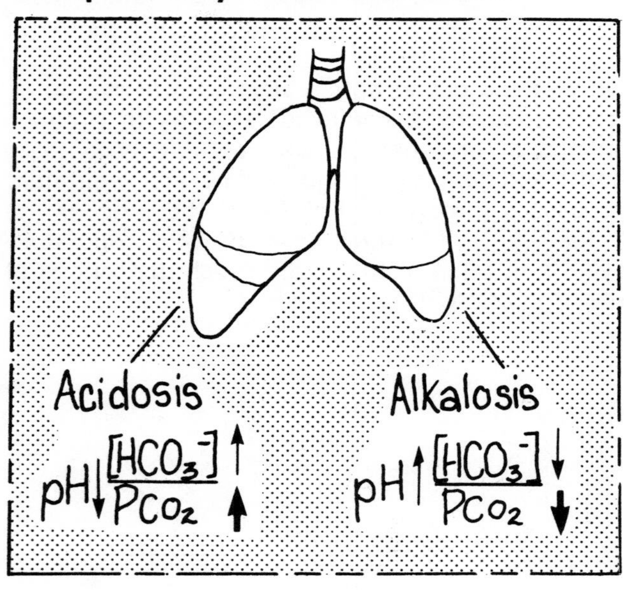

Respiratory changes will produce changes in P_{CO_2} and compensatory changes of bicarbonate concentration will occur in the same direction, thus limiting the degree of pH change. For example, in respiratory failure, the P_{CO_2} rises thus tending to lower the pH, but the bicarbonate concentration will also rise to compensate and thus correct the acidosis (at least in part).

In overventilation the P_{CO_2} falls and the reverse changes occur — respiratory alkalosis with compensating fall in bicarbonate.

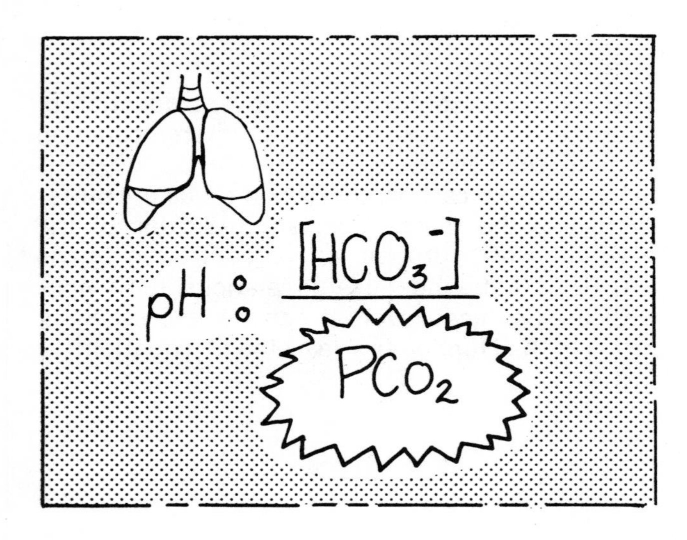

In respiratory disturbances, therefore, the primary change is of the P_{CO_2}. This produces a change in pH and then a compensatory change in bicarbonate. To the extent that the respiratory problem is not reversed and to the extent that the renal compensation (bicarbonate loss or gain) is incomplete or delayed, so the pH will rise or fall.

The commonest presentation of respiratory disturbances of pH is the acidosis of respiratory failure seen, for example, in a patient with an exacerbation of chronic bronchitis.

Metabolic Disturbances

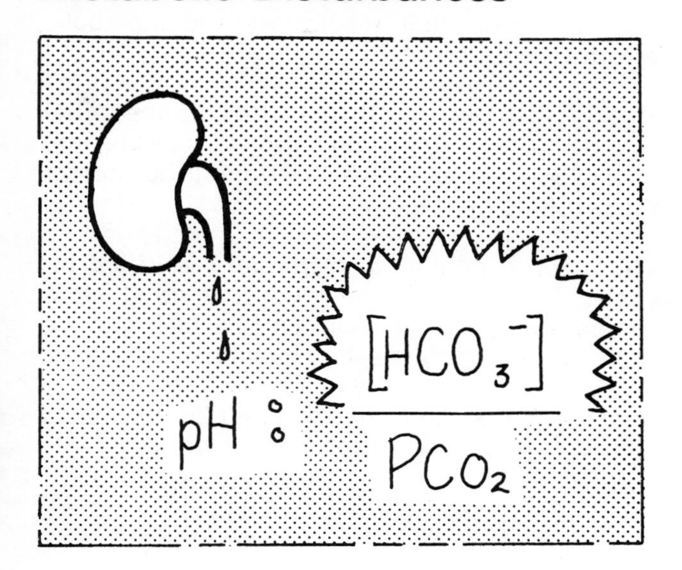

Non-respiratory, or metabolic, events involve primary changes in the levels of bicarbonate with a compensatory respiratory response. Depending on the degree to which the stress continues and the compensation is incomplete, there will be changes of pH. The limiting factor is the rate at which the kidney can manufacture or excrete bicarbonate.

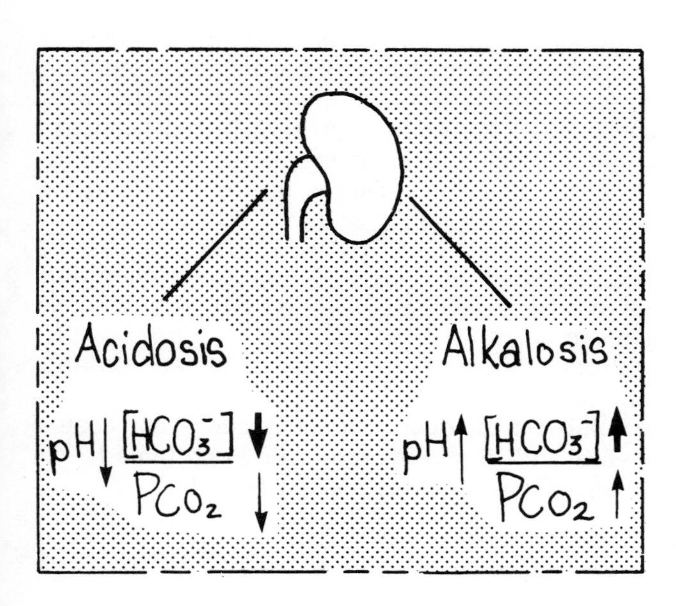

The changes of pH, bicarbonate concentration and P_{CO_2} are shown here for metabolic problems. You will recognize that for a given shift of pH the bicarbonate concentration and P_{CO_2} change in the same direction, but for metabolic disturbances that direction is the opposite of what occurs in respiratory disturbances: For a respiratory acidosis P_{CO_2} and bicarbonate concentration both rise, but for a metabolic acidosis they both fall.

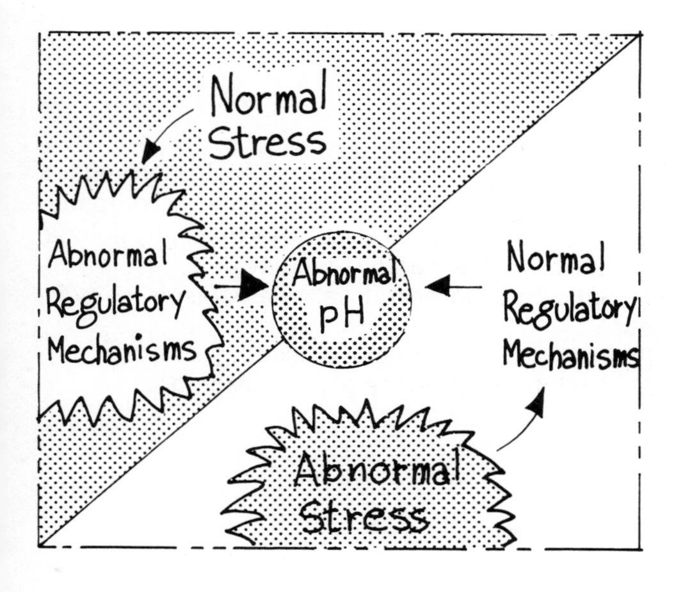

The normal pH "stresses" of everyday life are adequately handled by these physiologic mechanisms. However an abnormal degree of stress, or disordered function of lung or kidney in the presence of a normal stress, will lead to a breakdown of compensation.

	Respiratory	Metabolic
Acidosis pH ↓	$\frac{[HCO_3^-]}{P_{CO_2}}$ ↑ ↑	$\frac{[HCO_3^-]}{P_{CO_2}}$ ↓ ↓
Alkalosis pH ↑	$\frac{[HCO_3^-]}{P_{CO_2}}$ ↓ ↓	$\frac{[HCO_3^-]}{P_{CO_2}}$ ↑ ↑

In terms of the Henderson Hasselbalch relationship it is clear that respiratory and metabolic effects look different.

The thick arrows in this table indicate the initial change, whilst the thin arrows indicate the physiologic response which attempts to correct the resulting pH change.

Thus, for example, a *respiratory acidosis* involves an initial rise of P_{CO_2} followed by a compensatory rise of bicarbonate which may be considered a *compensatory metabolic alkalosis.*

Similarly a *metabolic alkalosis* will be associated with a *compensatory respiratory acidosis.*

Metabolic Acidosis

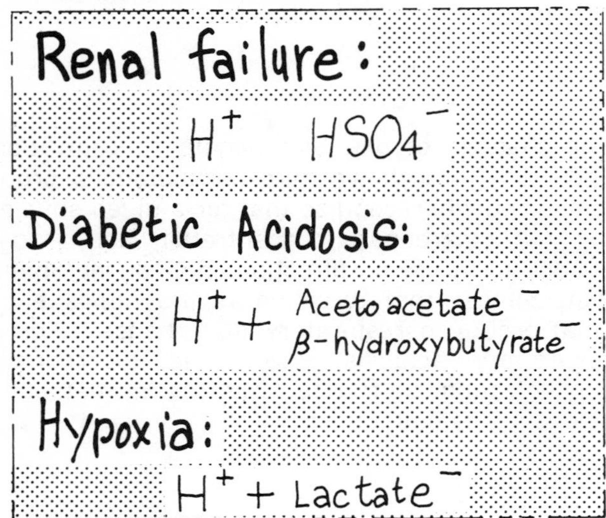

Metabolic acidosis may be due to retention of hydrogen ions or loss of bicarbonate. Retention of hydrogen ions may occur in company with some anion not usually present in high concentrations. Such situations may arise from "endogenous" disease processes or from loading by "exogenous" acids.

Accumulation of "endogenous" hydrogen ion will occur in association with a number of anions. Thus in
1) Renal failure, anions such as phosphate and sulphate are retained.
2) Diabetic acidosis, anions such as acetoacetate and **β**-hydroxybutyrate are retained.
3) Severe hypoxia, lactate is retained.

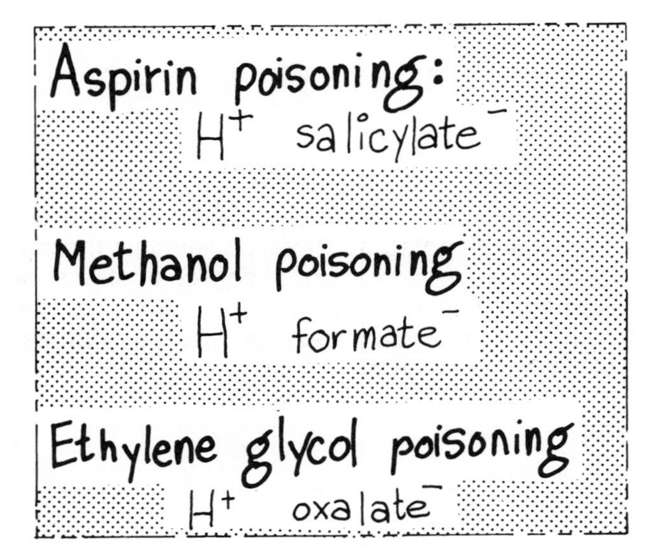

The presence of "exogenous" hydrogen ions is associated with accumulation of the accompanying anions which are not normal physiologic metabolites. Such anions include
1) Salicylate, in acetyl salicylate (aspirin) overdose
2) Formate, in methanol poisoning
3) Oxalate, in ethylene glycol poisoning.

The Anion Gap

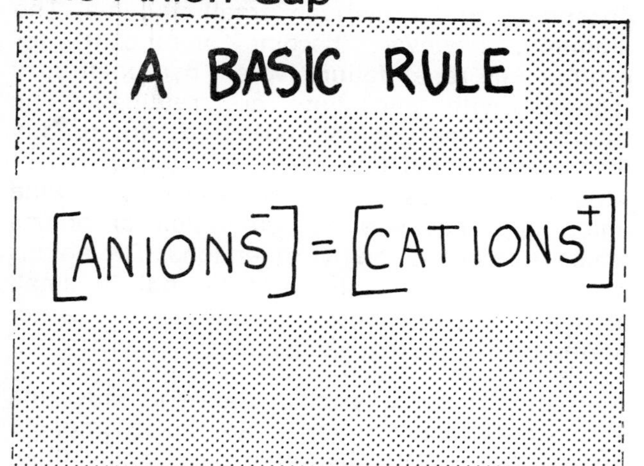

The accumulation of anions, either normal or abnormal will disturb the pattern of anions that are *usually* measured in the plasma. It is, of course, a physical and chemical reality that in the body fluids anions and cations must be present in equal concentrations.

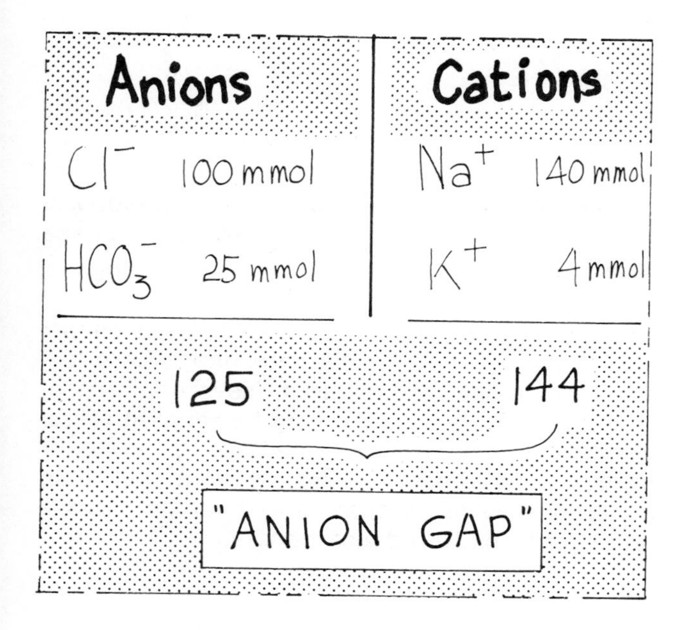

Normally the anions and cations in plasma are distributed as shown here, and you will recognize that they are apparently *not* present in equal concentrations. This is, of course, because the missing anions are comprised of many different substances which we do not normally measure, but they are all present in small amounts and together make up the *"anion gap."*

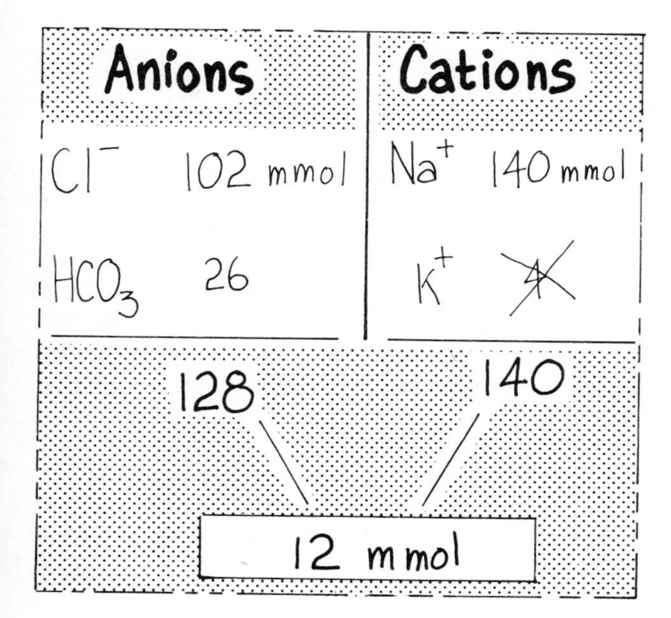

Because the potassium ion concentration is so small and will vary only a small amount it is generally excluded when the *anion gap* is calculated. Normally it is less than 15 millimoles of unmeasured anions per litre.

Metabolic Acidosis

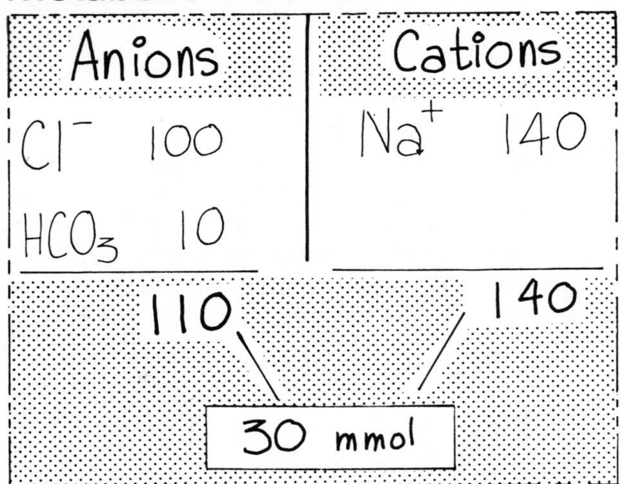

The unmeasured anions include such anions as phosphate but in relatively small amounts. The amounts, however, become significant, for example in renal failure when hydrogen ions together with phosphate are retained. The retained anions are not measured, but greatly increase the anion gap.

Metabolic acidosis with an abnormally wide anion gap means an accumulation of anions which accompany the hydrogen ions. The anions may be those normally present, but in small amounts (eg. phosphate, lactate) or those not normally present (eg. salicylate).

As the hydrogen ions accumulate and cause the bicarbonate concentration to fall, the accompanying anions "fill the gap" that bicarbonate filled before.

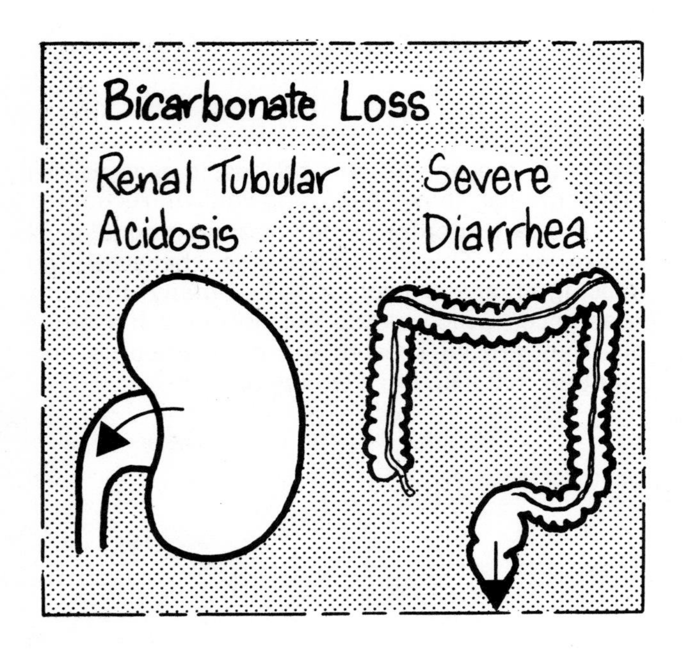

Remember that the removal of bicarbonate will cause the pH to fall just as surely as if free hydrogen ions are added.

Metabolic acidosis may be due to *bicarbonate loss.* This may be from the intestine, where losses from below the pylorus (severe diarrhea) contain bicarbonate. Or it may be renal in origin where a failure of the tubules to generate or transport hydrogen ion results in a failure to generate new bicarbonate and in some cases a failure to trap filtered bicarbonate.

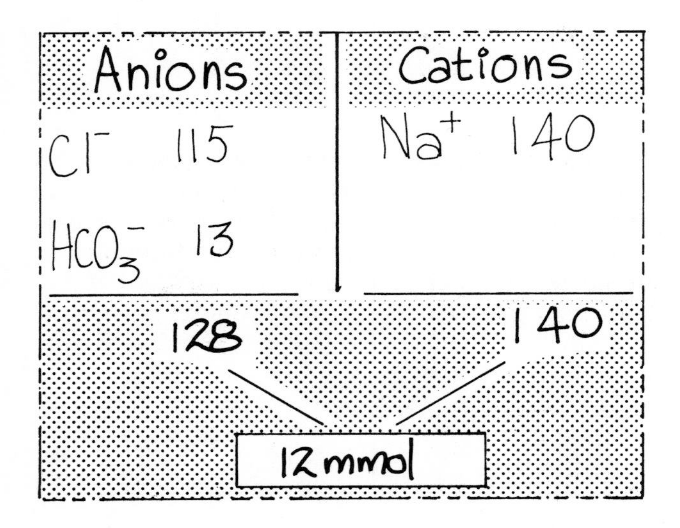

In these situations the pH falls because of the loss of bicarbonate anions. This might be expected to produce an increase in the anion gap, but since there is no accumulation of unmeasured anions the gap has to be filled by something else. The most readily available anion is, of course, *chloride* and the concentration of this anion increases to fill the gap left by loss of bicarbonate.

This can be recognized by the presence of acidosis with a normal anion gap and hyperchloremia, and is often called "hyperchloremic acidosis".

Respiratory Alkalosis

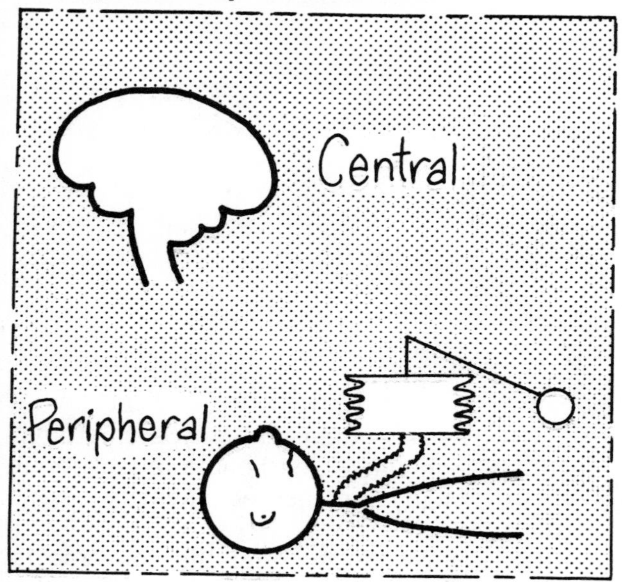

Alkalosis is generally a less common primary problem than acidosis.

Respiratory Alkalosis occurs due to over-ventilation. Many causes can be listed but can be classified as *central* and *peripheral.* Central causes include the hyperventilation of anxiety or secondary to brain stem injury by disease or drugs. Peripheral causes include uncontrolled use of artificial ventilation.

Whatever the mechanism, the resultant fall of P_{CO_2} causes the pH to rise.

Metabolic Alkalosis

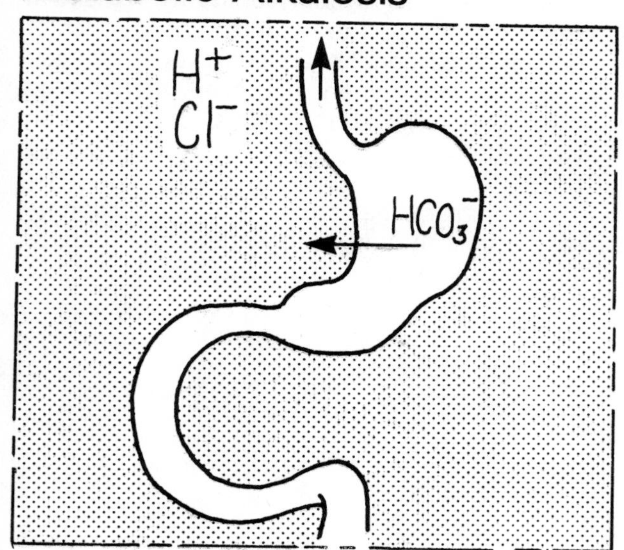

Metabolic Alkalosis may be due to ingestion of absorbable alkali (such as sodium bicarbonate) or to loss of hydrogen ion due to vomiting. Since vomiting results in loss of chloride as well as hydrogen ion the striking rise of bicarbonate will go along with a severe depression of chloride.

Whether caused by direct loss of hydrogen ions or a rising bicarbonate, the final result is a rising pH.

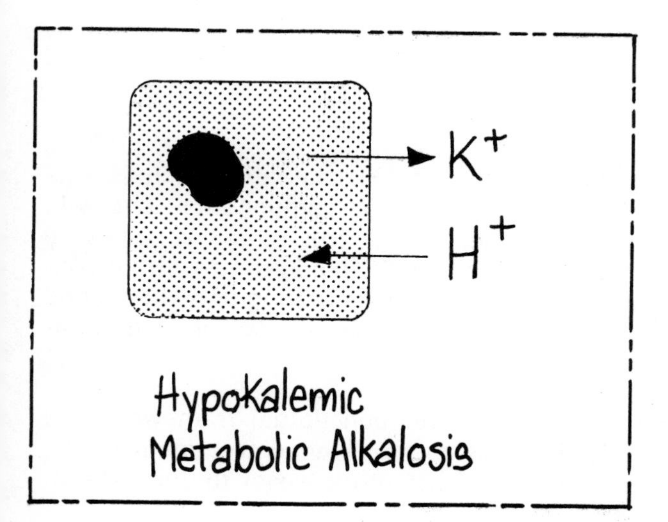

Vomiting also causes a loss of E.C.F. volume and of potassium. The former leads to aldosterone secretion which will add to the potassium loss at a renal level. Hypokalemia exaggerates the alkalosis because as potassium leaves the cells to replace E.C.F. losses, it exchanges with hydrogen ion in the E.C.F. Thus a metabolic alkalosis is exaggerated by additional *hypokalemic metabolic alkalosis* due to movement of hydrogen ion into the I.C.F.

The Importance of Chloride

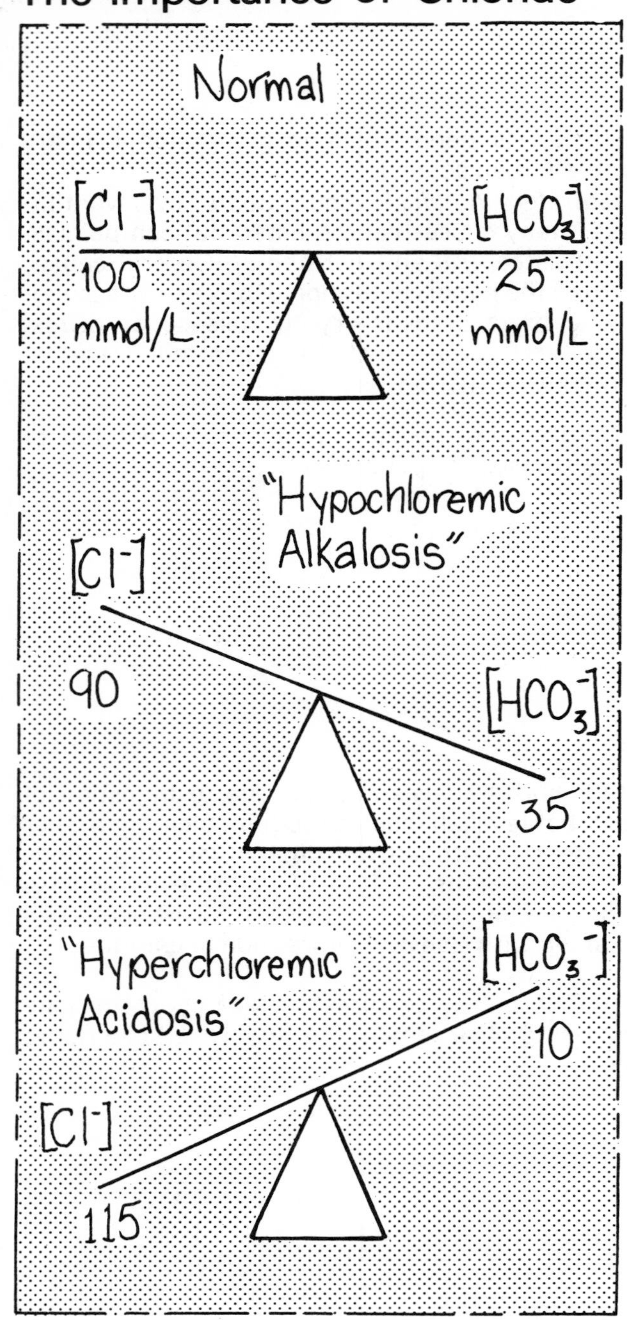

It has already been noted that ''hyperchloremic acidosis'' is seen when bicarbonate falls without the retention of any unmeasured anions.

It is often not recognized that the converse is true: If the chloride concentration rises, then the bicarbonate concentration *must* fall (provided unmeasured anions remain constant). This will result in acidosis as an inevitable consequence of the fall of bicarbonate concentration. It can be seen clincially when inappropriately large amounts of isotonic sodium chloride are given intravenously.

It will be apparent that chloride and bicarbonate will vary inversely with one another so long as the unmeasured anions do not change.

A Final Word

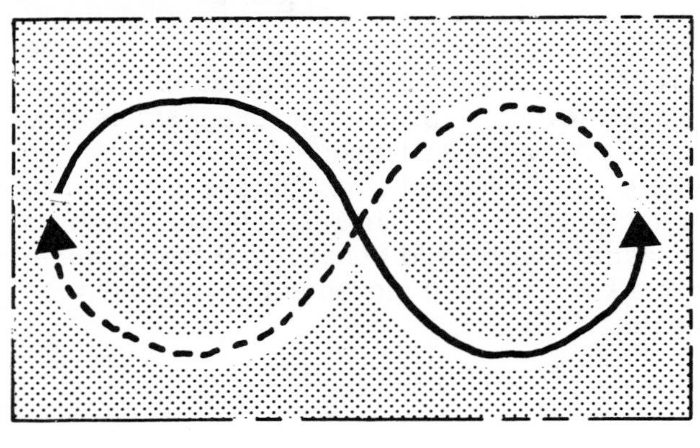

All disturbances of pH regulation include respiratory and metabolic components; one will initiate changes whilst the other will show a compensatory change. This is the essential concept linking the functions of lung and kidney and expressed by the Henderson-Hasselbalch equation.

On the basis of this conceptual framework you should find the more detailed treatment of pH regulation in the literature easier to understand.

The Regulation of Potassium

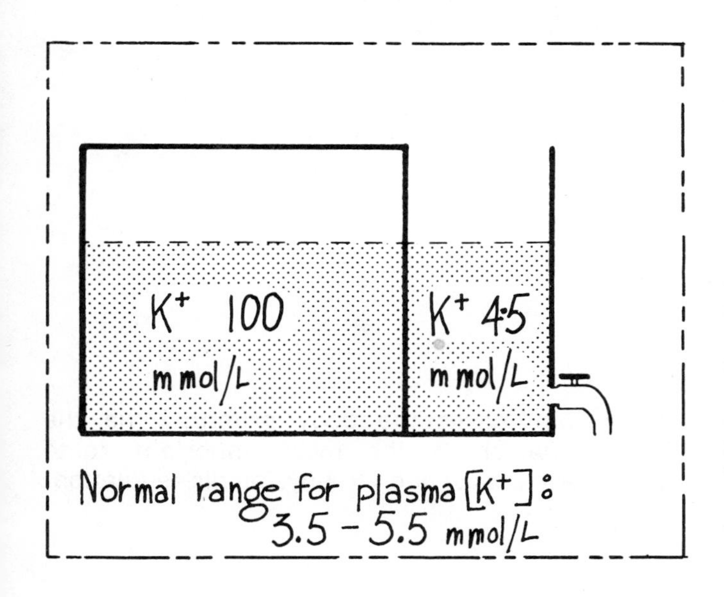

Most of the potassium in the body is within the I.C.F. Only a fraction is in the E.C.F., uniformly distributed through plasma and interstitial compartments. Because the plasma potassium concentration is low it will not vary appreciably with changes of total body water (unlike the situation with the predominantly extracellular sodium ion). Changes of plasma potassium concentration reflect

1) Changes of total body potassium
2) Shifts of potassium into or out of the large intracellular pool.

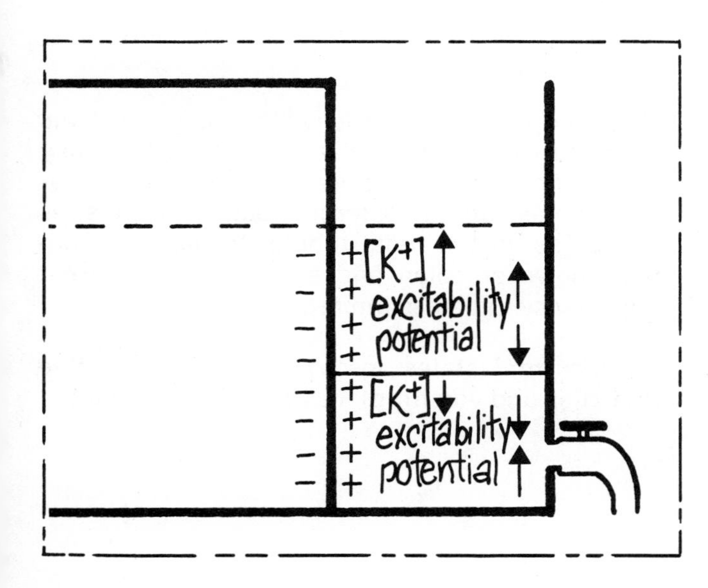

Although the concentration of potassium in the E.C.F. is small it is a critical determinant of the resting membrane potential of cells.

If potassium concentration outside the cells rises, the membrane potential falls resulting in the membrane being more readily depolarised. This means a *lower* threshold for excitation of tissues such as nerve and muscle when the E.C.F. potassium concentration rises. Conversely it means a *higher* threshold for excitation when E.C.F. potassium concentration falls.

Intake

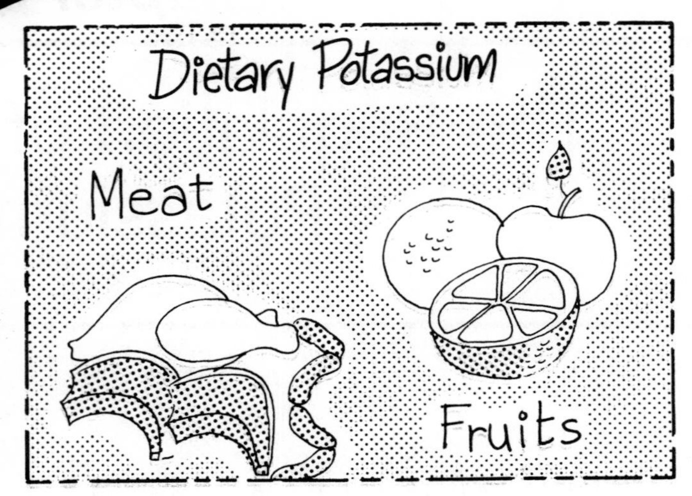

Potassium enters the body in the diet and is present in virtually all protein-containing foods, particularly meat. On a North American diet the potassium intake is directly related to protein intake and is about 60 to 80 millimoles per day. Although fruits and fruit juices are well known to be rich in potassium, they do not comprise the major source of this ion in a normal diet.

Output

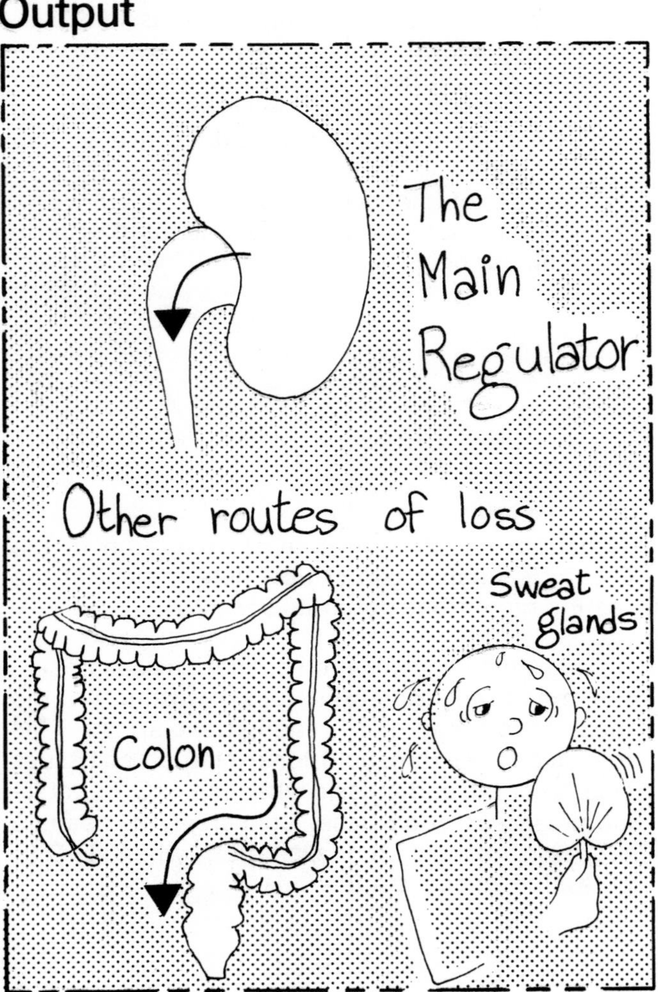

Potassium leaves the body primarily via the kidney, which is the most important route because it is most open to physiologic regulation.

However potassium can leave the body in colonic fluid and via the sweat glands; routes which may be significant at times since the hormonal regulator of potassium (aldosterone) acts at these sites as well as on the renal tubule. In *normal* situations however colon and skin are not of great importance in potassium regulation.

In *abnormal* states losses from the intestine (diarrhea and vomiting) may be very important.

Potassium and the Kidney

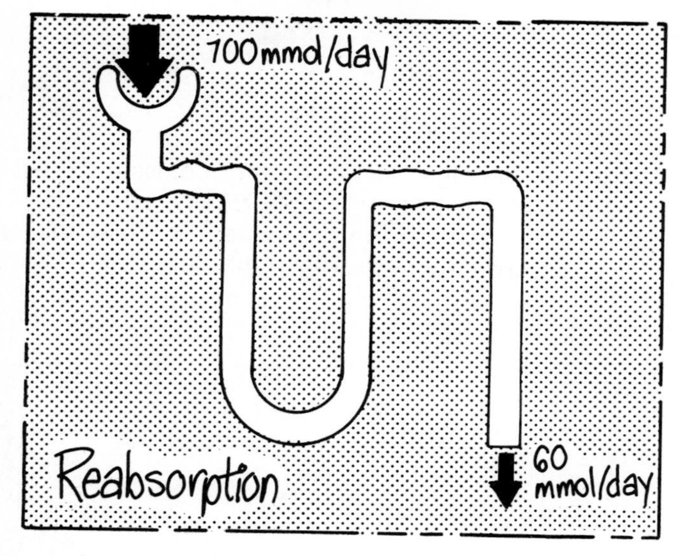

Because the plasma concentration of potassium is low, as compared with sodium, the amount filtered at the glomerulus is low, about 700 millimoles per day, far less than the 25,000 millimoles of sodium that is filtered. Normal subjects on a normal Canadian diet take in and excrete about 60 millimoles per day. Thus, with 700 millimoles filtered and only 60 millimoles excreted there must usually be net tubular *reabsorption*.

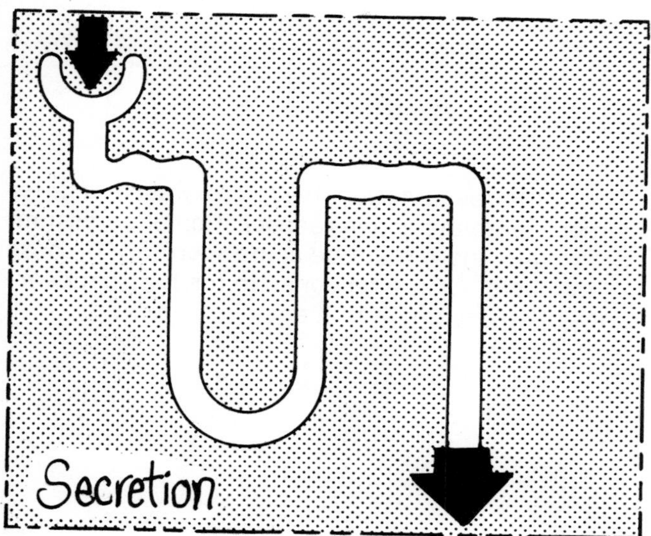

Whilst this is the normal pattern, conditions can occur in which potassium excretion is higher than the filtered load of potassium. This must, of course, be an indication that the tubule is capable of potassium *secretion* as well as reabsorption.

The Proximal Tubule

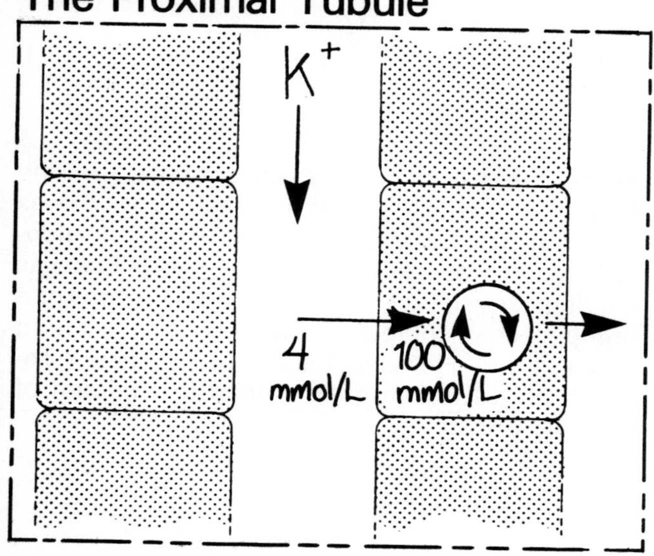

Micropuncture studies have confirmed these assumptions. By the end of the loop of Henle at least 90% of filtered potassium in the tubular lumen has been reabsorbed. Since the potassium in the lumen is about 4 millimoles per litre and in the tubular cell about 100 millimoles per litre, this reabsorption must be active and require some kind of pump.

It appears to be relatively fixed and does not vary to any degree with changing physiologic events.

The Distal Tubule

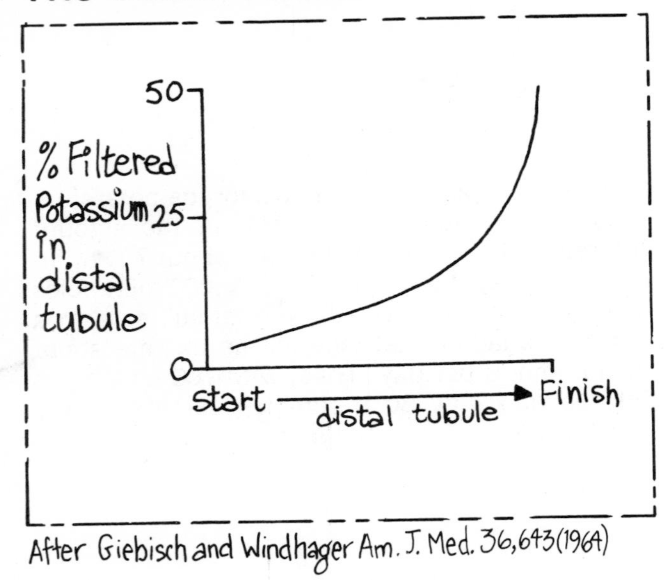

After Giebisch and Windhager Am. J. Med. 36, 643 (1964)

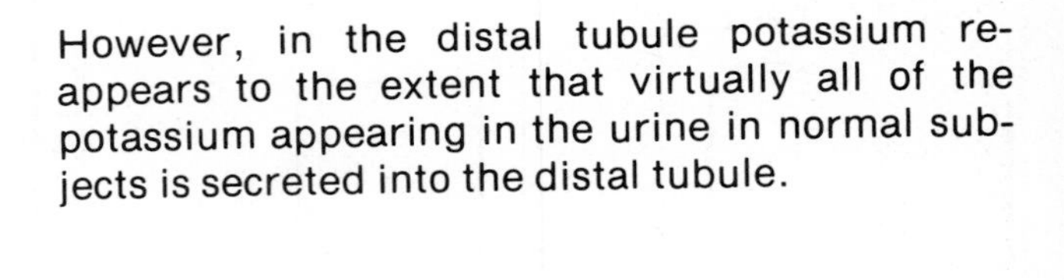

However, in the distal tubule potassium reappears to the extent that virtually all of the potassium appearing in the urine in normal subjects is secreted into the distal tubule.

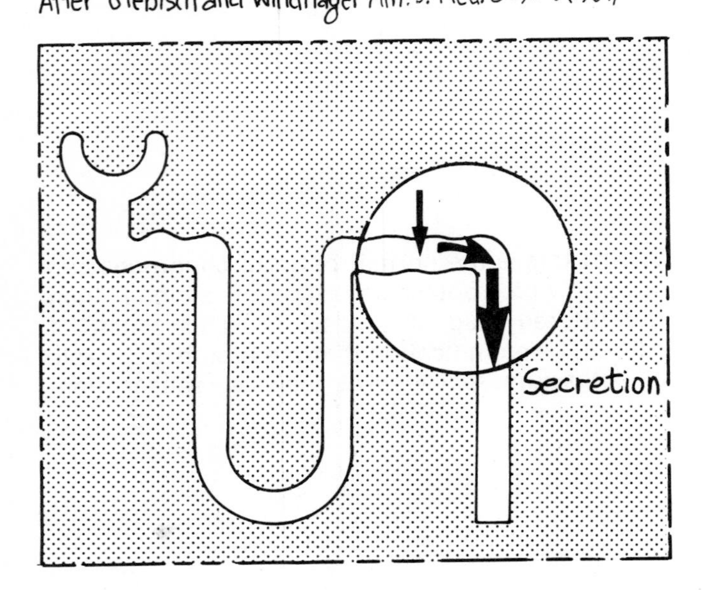

The initial assumption therefore was that after almost complete active reabsorption of potassium *proximally* there must be some secretory process in the distal tubule responsible for the reappearance of potassium at that site.

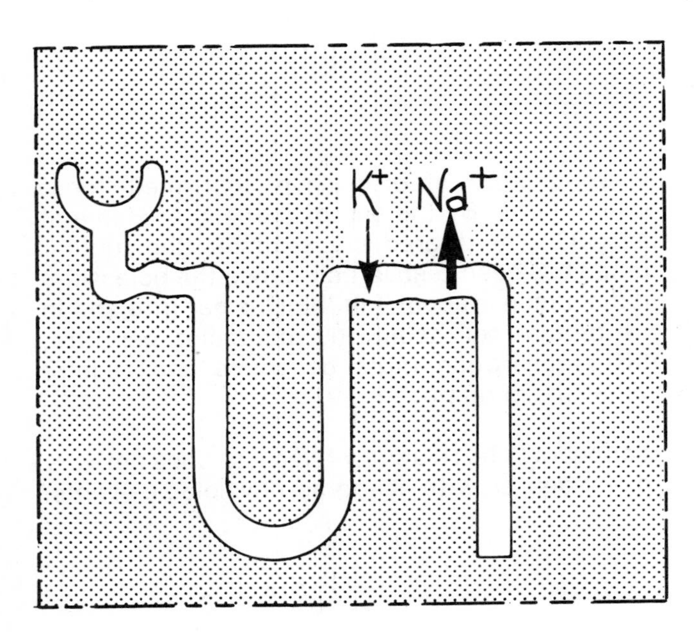

Since this secretion occurred at a part of the tubule where active sodium reabsorption was known to occur under the influence of aldosterone, it was considered that potassium was moving *actively* from cell to lumen by means of a pump mechanism that might be linked to sodium reabsorption. However the amount of sodium reabsorbed was much greater than the amount of potassium secreted.

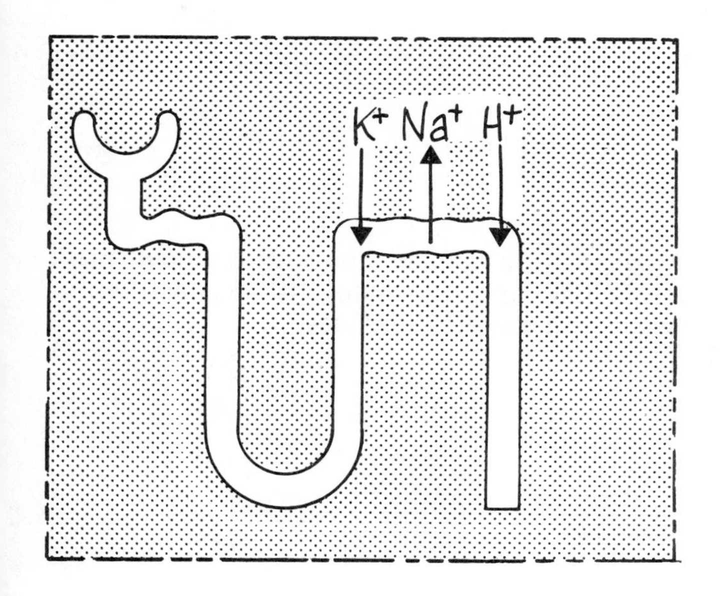

At this site much of the sodium is reabsorbed in exchange for actively secreted hydrogen ions and it was thought that potassium and hydrogen were competing in some way for the same transport system which in turn was linked to sodium transport in the reverse direction.

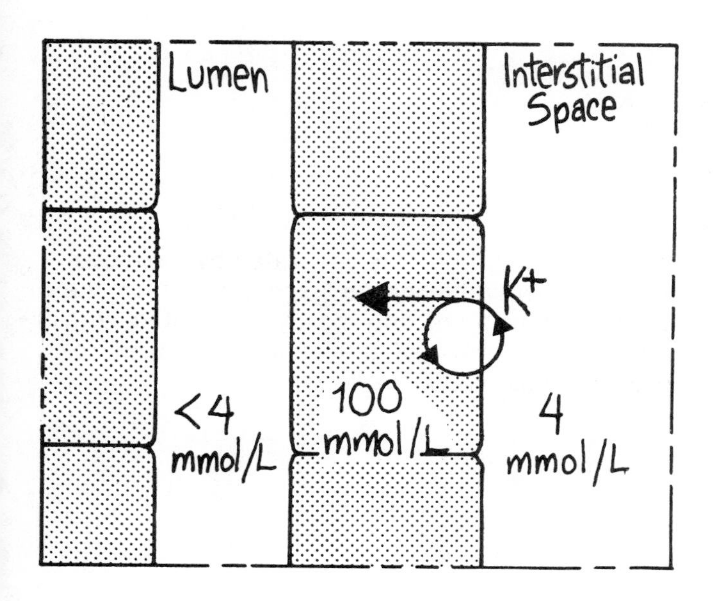

It is now recognized that no active *luminal* secretory process for potassium is present. The movement of potassium into the tubular lumen works like this:
The *active* step in potassium transport is from the interstitial space, where its concentration is low, into the tubular cell, where its concentration is high. This involves a pump.

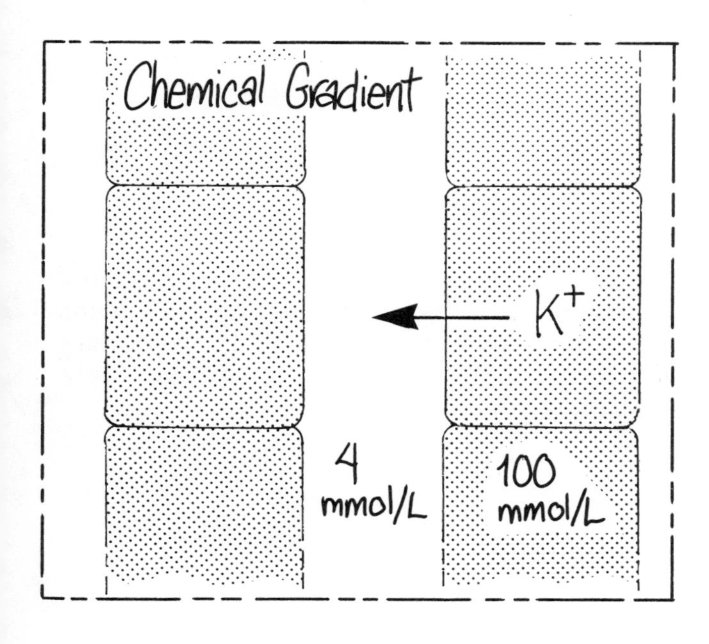

A chemical gradient is thus set up between the tubular cell and the fluid in the lumen, where the potassium concentration is as low as it is in the interstitial space. Potassium can thus move into the lumen by diffusing down its concentration gradient without the need to postulate any active pump mechanism on the luminal side of the cell.

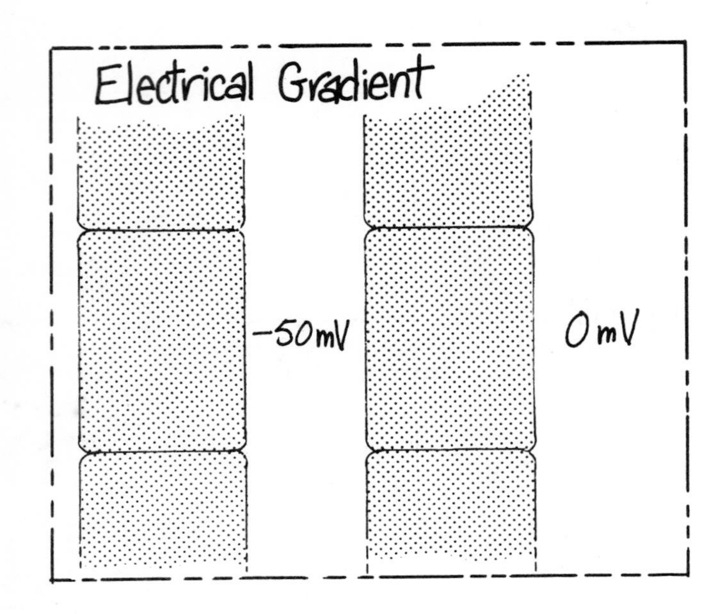

In addition the tubular lumen at this point has been found to be 50 millivolts negative, with respect to the interstitial space.

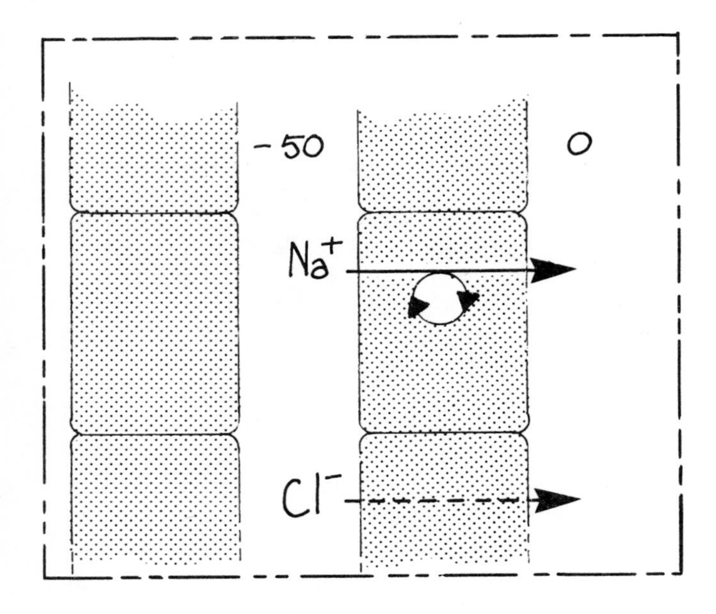

This electrical gradient is generated by the movement of positively charged sodium out of the lumen, a movement only partly associated with chloride since most of that anion has been reabsorbed proximally.

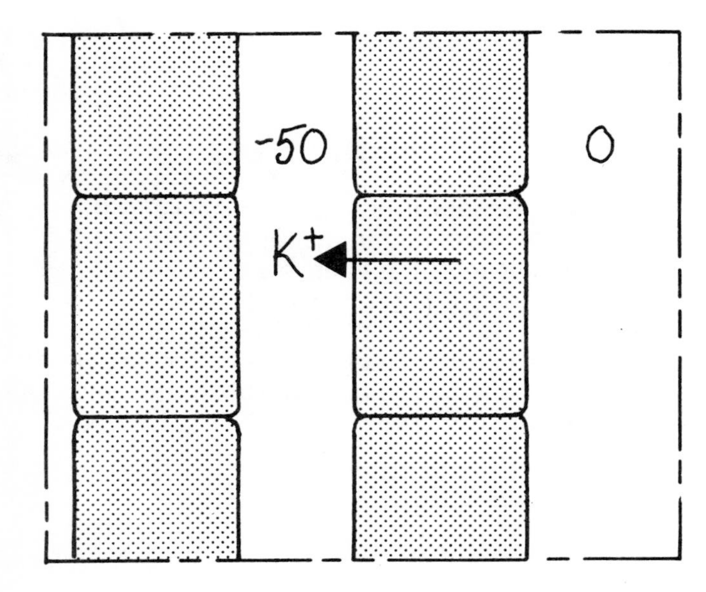

Clearly there is now an electrical gradient favoring the movement of potassium into the lumen. Conditions are thus favorable both chemically *and* electrically for potassium to move passively into the tubule.

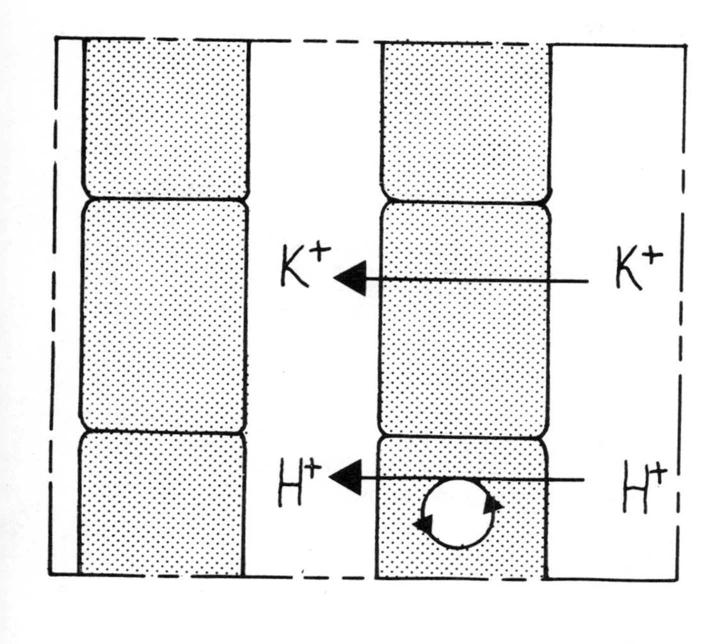

A complication to this process is that in the distal tubule hydrogen ion is actively secreted, and in fact a significant concentration gradient can be generated by this pump as the urine becomes acidified.

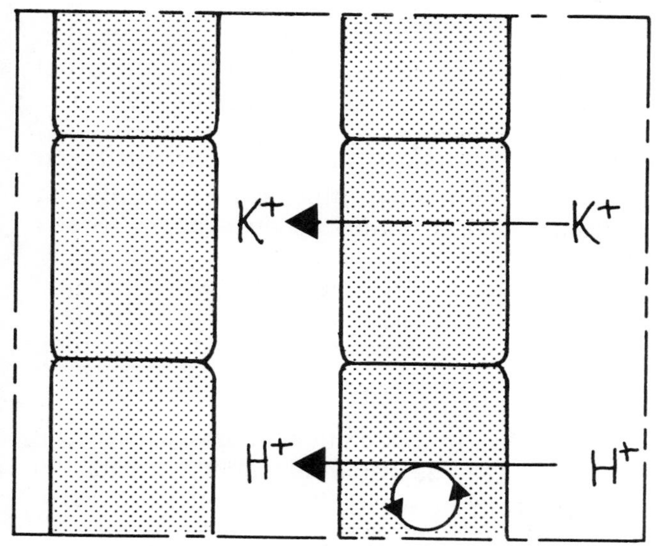

The more hydrogen ion that is pumped into the lumen in exchange for sodium being absorbed, the more the electrical gradient will tend to be reduced. The ''downhill'' gradient for potassium diffusion will thus be limited.

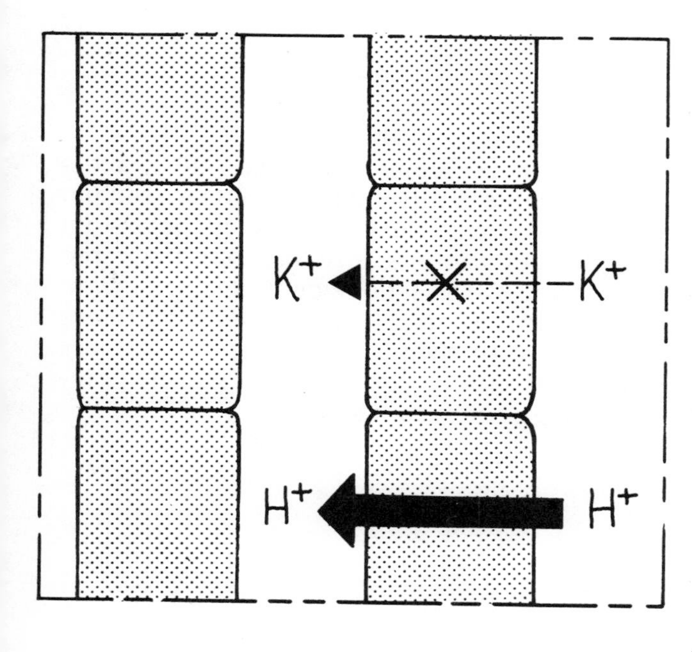

If, therefore, there is a surge of hydrogen ion secretion into the distal system, as would be the case in an acidotic patient, less potassium might be excreted in the urine. This, of course, contributes to the hyperkalemia of acidosis. Conversely, if in the face of alkalosis, less hydrogen ion is being secreted, there would be a tendency for more potassium to enter the lumen, thus contributing to the hypokalemia of alkalosis.

Distal Tubule - Summary

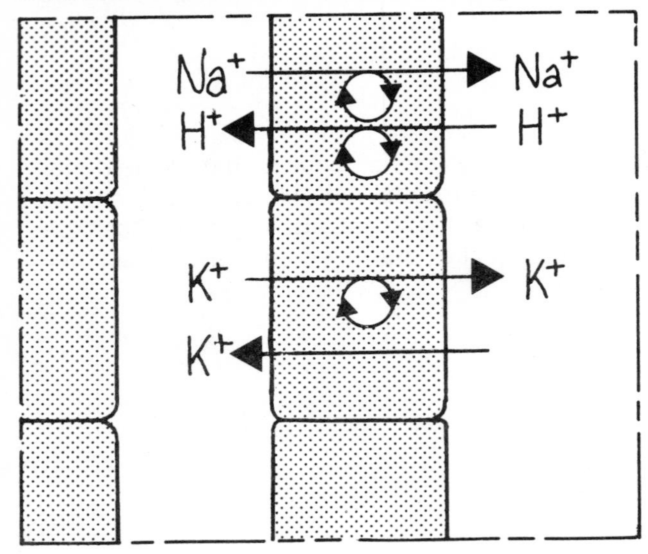

In summary, therefore, potassium is handled in two ways in the distal tubule
1) The active reabsorption, which occurs in the proximal tubule, also continues to operate distally
2) Alongside it, potassium *secretion* can occur and depends upon passive chemical and electrical gradients between cell and lumen

Factors which favor *secretion* include
1) Increased sodium reabsorption
2) Increased intracellular potassium
3) Decreased hydrogen ion secretion

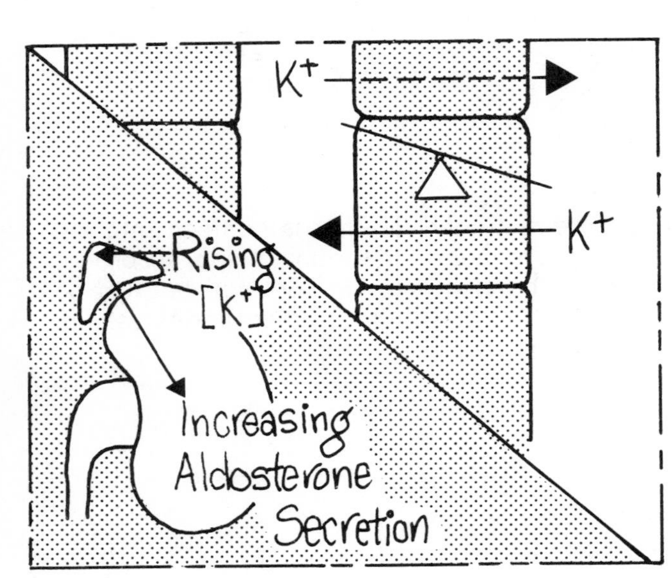

The balance between reabsorption and secretion of potassium determines whether there is a net loss or a net gain of potassium at distal sites.

The rate of sodium reabsorption in the distal tubule can be directly influenced by the plasma potassium concentration because an increased plasma potassium concentration can directly stimulate aldosterone release from the adrenal gland. This is a *direct* reaction which is independent of the renin-angiotensin-aldosterone system.

In states of sodium depletion aldosterone release will, of course, be renin-dependent and is the major pathway by which distal tubular reabsorption of sodium is increased.

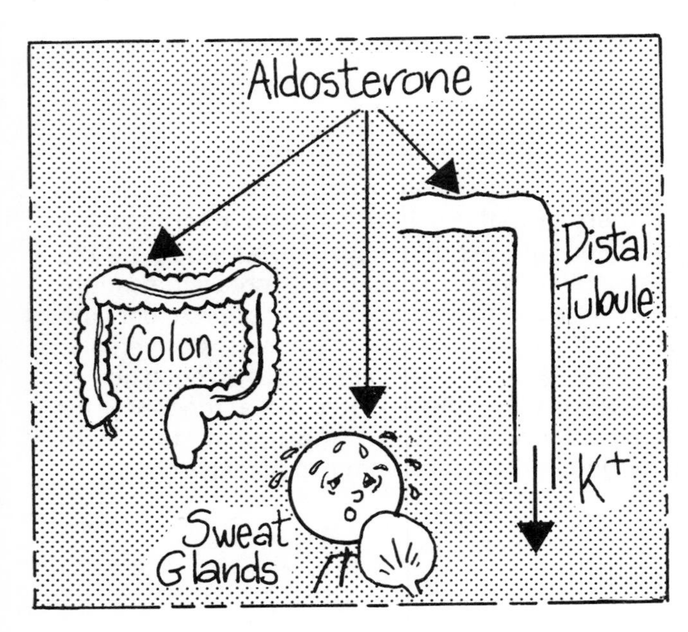

Whatever the stimulus to its secretion, aldosterone results in increasing tubular sodium reabsorption and thus increasing potassium secretion. Aldosterone also acts on the large intestine and the sweat glands, reducing sodium loss but increasing potassium loss by these routes. In patients with severe renal failure, loss via the colon may be an important factor in limiting the degree of hyperkalemia.

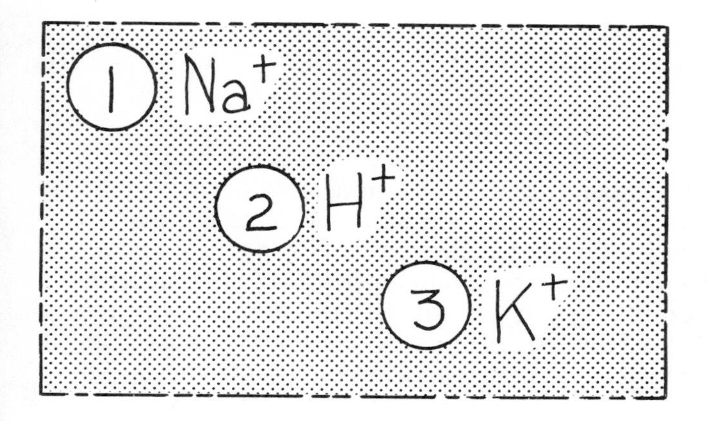

It is clear that the kidney handles sodium, hydrogen and potassium ions in a closely linked manner. As a general rule, for the whole organism, renal handling of sodium seems to take priority over regulation of the other cations. Regulation of sodium balance will often occur at the expense of hydrogen or potassium ions when external stresses demand it.

Similarly, provided sodium balance is satisfied, pH regulation often will be at the expense of E.C.F. potassium regulation.

Presumably the reason for potassium regulation tending to be lower on the list of priorities as far as the E.C.F. is concerned is that potassium is predominantly the property of the I.C.F. It is the cells that have an efficient pump to keep potassium inside and sodium outside and therefore it is the cells as a whole that contain and regulate the bulk of body potassium.

Clinical Examples of Disturbed Potassium Metabolism

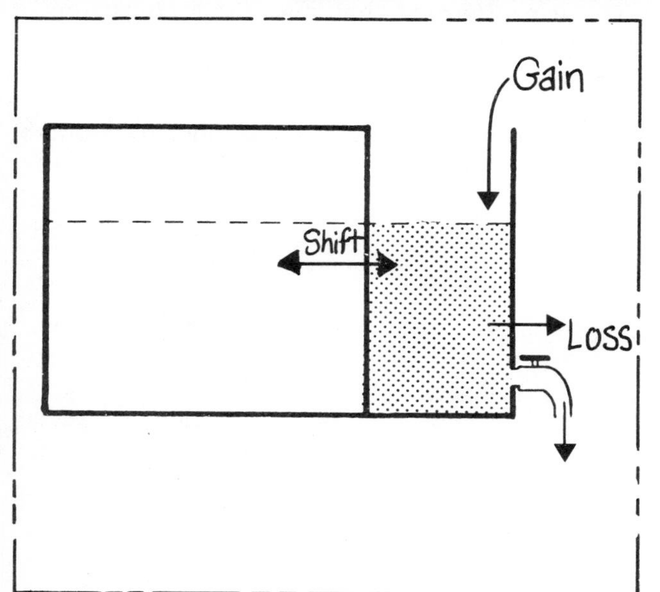

Changes in the plasma potassium may reflect
1) Loss or gain of total body potassium.
2) Shifts of potassium into or out of the cells, often dependent upon reciprocal movements of hydrogen ion.

"False" Hyperkalemia

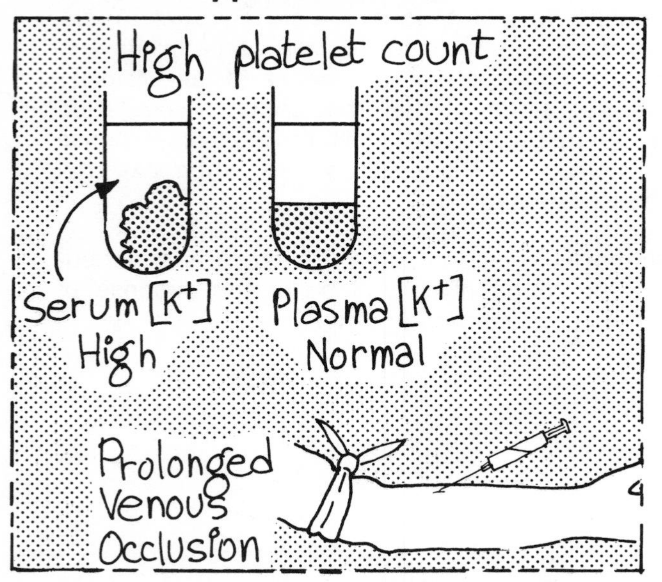

A "false" elevation of potassium concentration in the *serum* may occur if the platelet count is very high since potassium is extruded from platelets trapped in blood clot. In such situations the *plasma* potassium will be normal. Similarly the prolonged occlusion of veins with a tourniquet can cause a false elevation of the plasma potassium concentration due to purely local factors.

Hemolysis of blood after venepuncture, or delayed separation of cells and plasma may also give false plasma levels.

Hyperkalemia

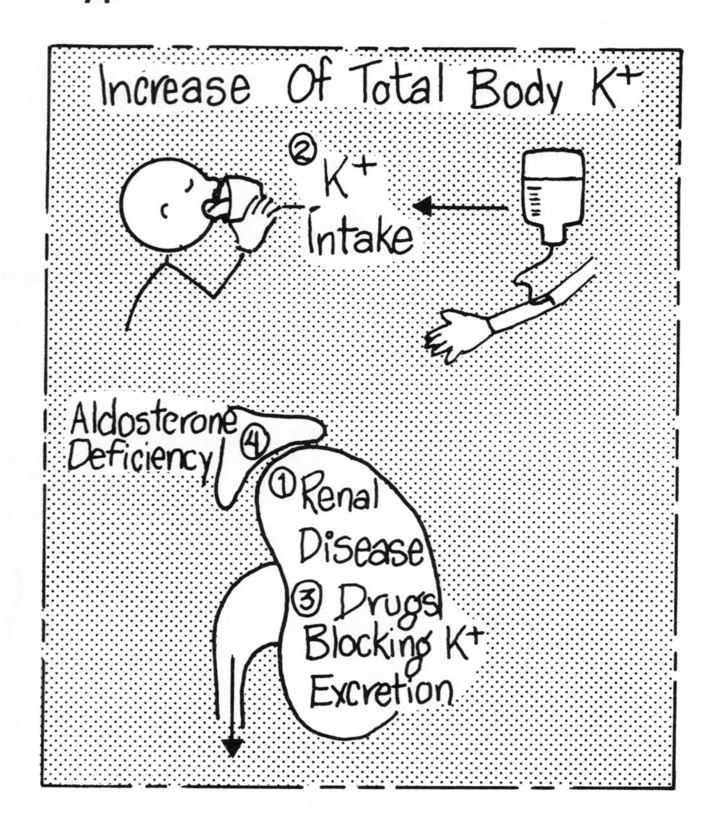

Hyperkalemia is a dangerous and life threatening metabolic emergency.

It may be due to a gain in total body potassium. Factors involved may be

1) Intrinsic renal disease.
2) Excessive potassium intake by mouth or other route.
3) The use of drugs that may limit potassium secretion in the distal tubule.
4) Rarely primary aldosterone deficiency.

Decreased renal excretion is the most frequent cause.

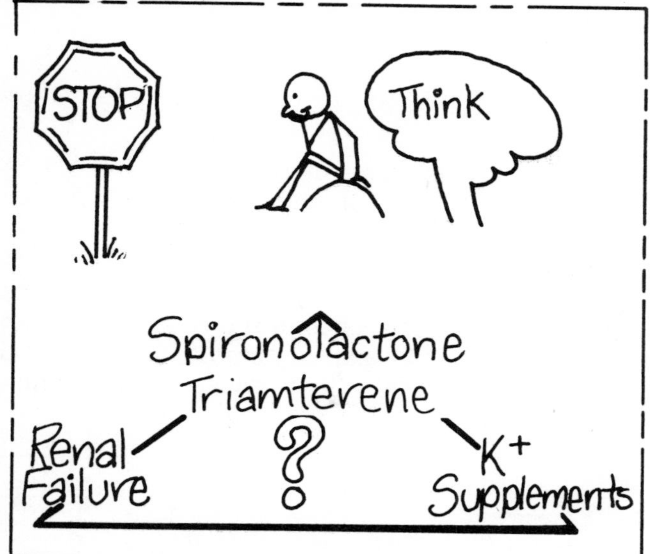

Remember that *spironolactone* and *triamterene* block both sodium reabsorption and potassium secretion in the distal tubule.

Never use these diuretics in renal failure without careful thought.

Never use them together with potassium supplements without careful thought.

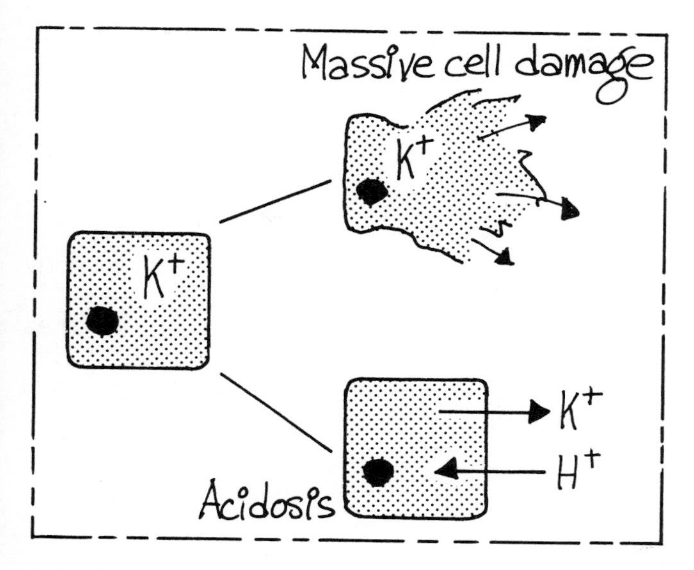

Hyperkalemia may also be due to a rapid shift of potassium out of the cells into the E.C.F. such as may occur:

1) In exchange for hydrogen ion in severe acidosis.

2) Secondary to massive cell damage, such as a crush injury or massive hemolysis.

A Dangerous Duo

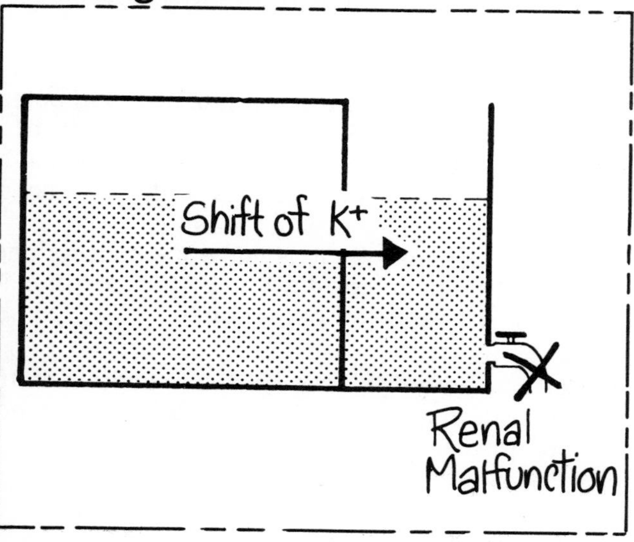

Both of these mechanisms are often associated with impaired renal function which limits the rate of excretion of potassium into the urine and therefore increases the rate of rise of the plasma potassium concentration. Thus:

Acidosis may be secondary to primary renal disease.

Massive tissue injury is often associated with transient acute renal failure.

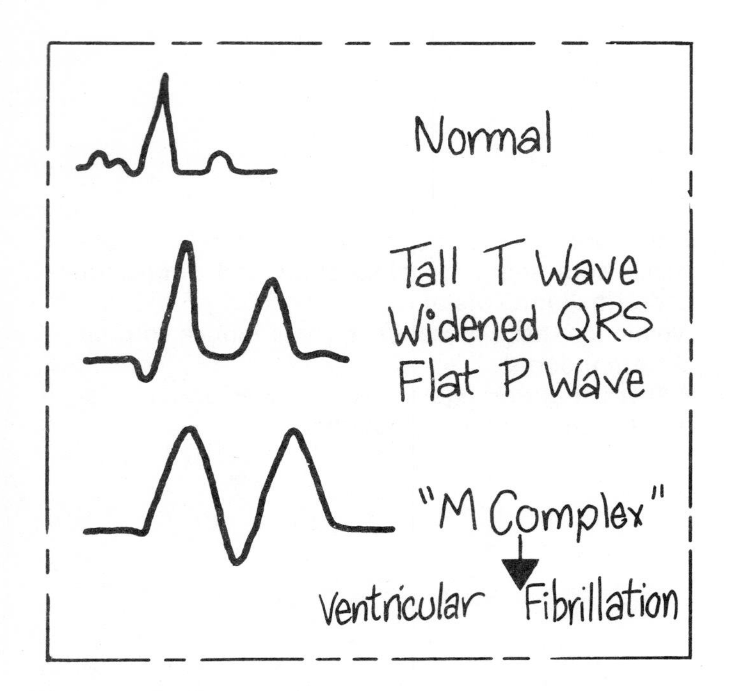

Clinically, hyperkalemia produces:
Weakness and irritable "twitchy" muscles
Fearfulness and parasthesiae
Cardiac Arrhythmias
The EKG may show a sequence of changes as the plasma potassium concentration rises; they can result in effective cardiac standstill at plasma potassium concentrations above 7.5-8.0 millimoles per litre.

Hypokalemia

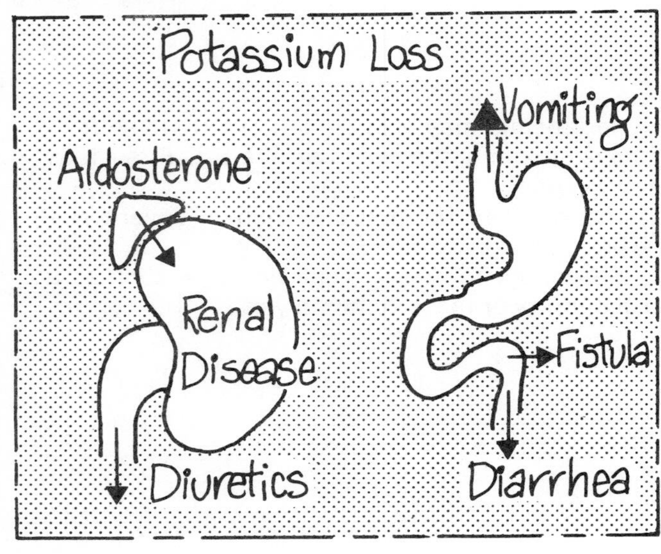

Hypokalemia may be due to loss of total body potassium
1) From the intestine due to vomiting, diarrhea or a surgical fistula.
2) From the kidney due to renal disease, diuretic administration or increased aldosterone production.

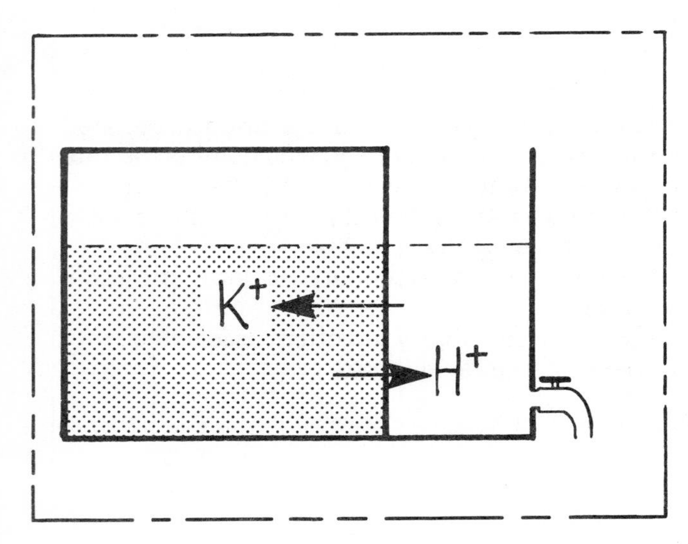

Hypokalemia may be the result of a shift of potassium into the I.C.F. This is usually due to a primary alkalotic state or due to the correction of high blood sugar in diabetics with the use of insulin.
In alkalosis hydrogen ions leave the cells as part of the attempt to correct the E.C.F. pH and in doing so exchange for potassium.

"Periodic paralysis" can be due to sudden shifts of potassium into the cells. Its mechanism is not understood and it is very rare.

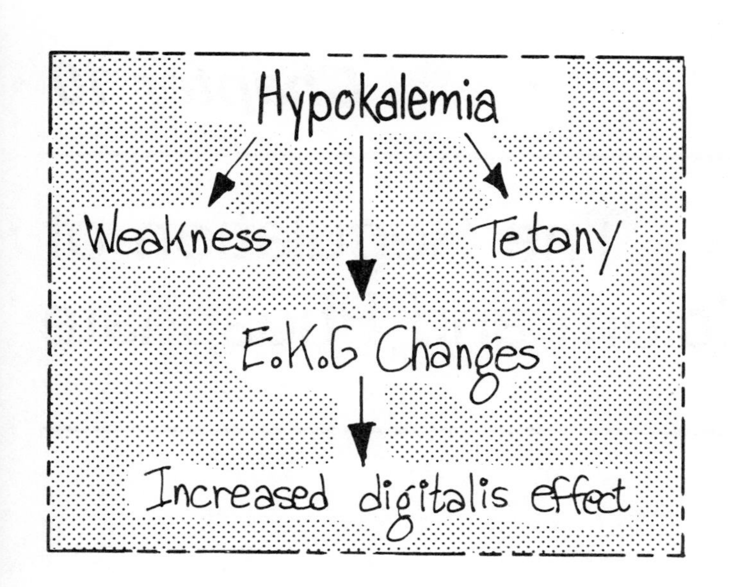

Clinically hypokalemia produces:
Severe weakness
Tetany
Cardiac arrhythmias.
EKG changes are not as dramatic as in hyperkalemia and include S T depression, flat T waves and marked U waves. The effects of digitalis become more marked in hypokalemia. Prolonged hypokalemia can cause secondary renal tubular damage which may be permanent.

Chapter 10

The Regulation of Calcium and Phosphate

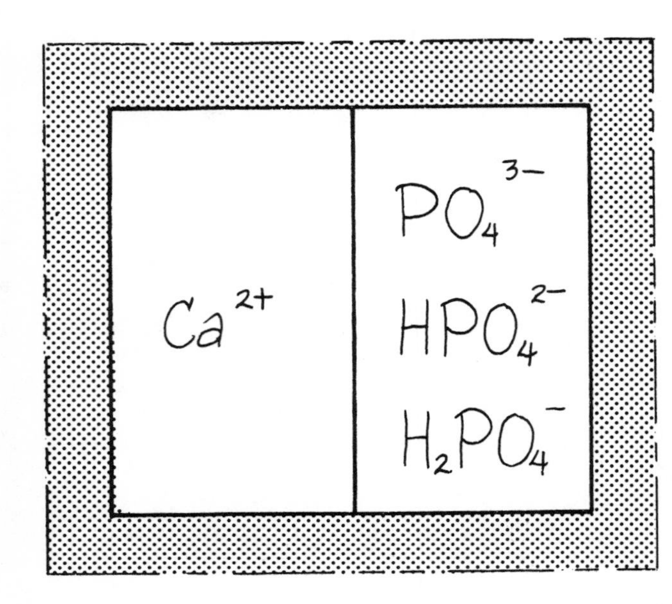

There is more calcium than any other cation in the human body, amounting to about 25,000 millimoles (1 kilogram) for an average man. Almost all of it is within the structure of bone, in company with phosphate.

Both calcium and phosphorus also have important intracellular roles, with phosphorus having particular importance because of its function in storing energy in organic phosphate compounds such as adenosine triphosphate (ATP).

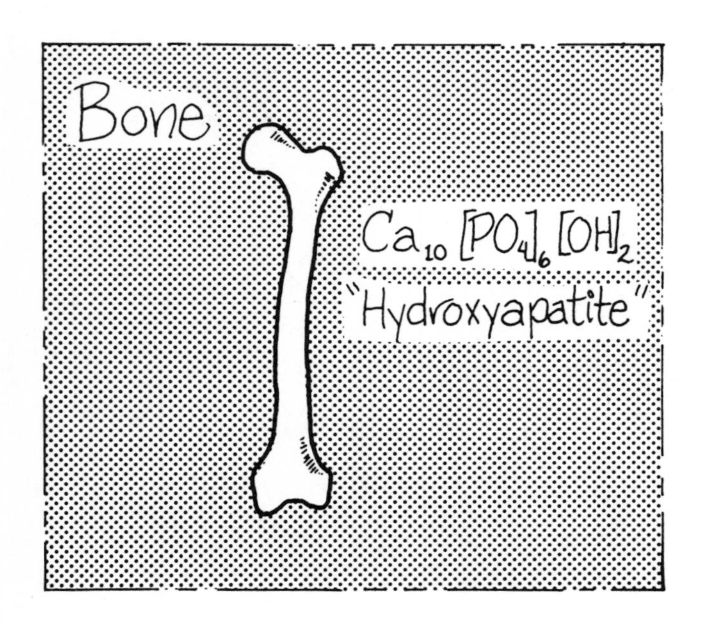

In their inorganic forms calcium and phosphate are locked together in the crystalline structure of bone; for this reason it is impossible to consider calcium as an extracellular cation without considering phosphate anions at the same time.

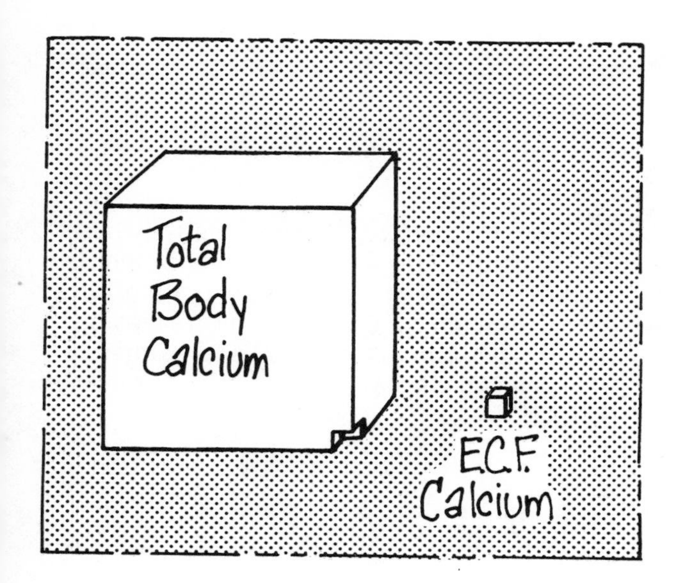

Precise regulation of calcium concentration in the E.C.F. is demanded because of the effects of extracellular calcium on nerve and muscle function, (both skeletal and myocardial). This poses problems because less than 1% of the total body calcium is in the E.C.F., the rest being within the bone. In spite of this wide difference in distribution, the E.C.F. calcium concentration remains precisely regulated.

E.C.F. Calcium

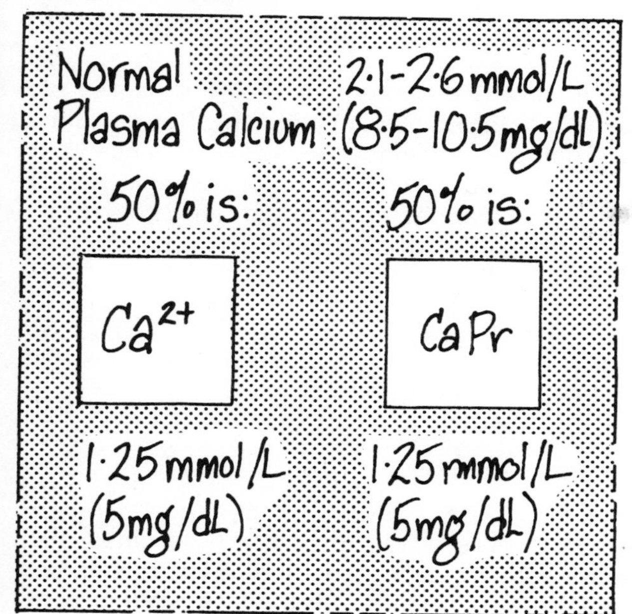

Within the E.C.F., calcium exists in two forms:
1) Free, ionized calcium
2) Bound calcium, predominantly attached to proteins, particularly albumin. A small amount is bound in a complex with organic acids.
In the *plasma* compartment only half the calcium exists as free ionized calcium. The other half is bound to albumin.

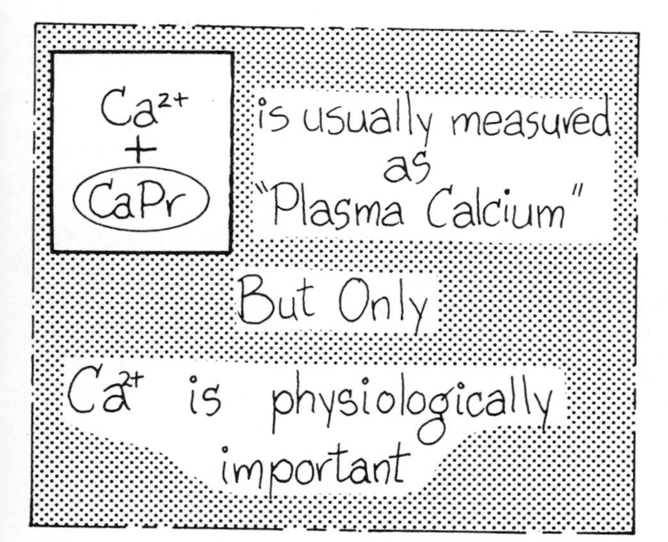

It is the ionized calcium that is physiologically important, although in most clinical settings it is *total* plasma calcium that is measured.

Interstitial Calcium	Plasma Calcium
Ca^{2+} 1·25 mmol/L	Ca^{2+} 1·25 mmol/L
CaPr Zero	CaPr 1·25 mmol/L
1·25 mmol/L	2·5 mmol/L

In the *interstitial fluid* compartment, where albumin concentrations are small, the calcium is predominantly in the ionized form, in equilibrium with the calcium ion concentration in the plasma. Thus the total *plasma* calcium is approximately twice the concentration of *interstitial* calcium.

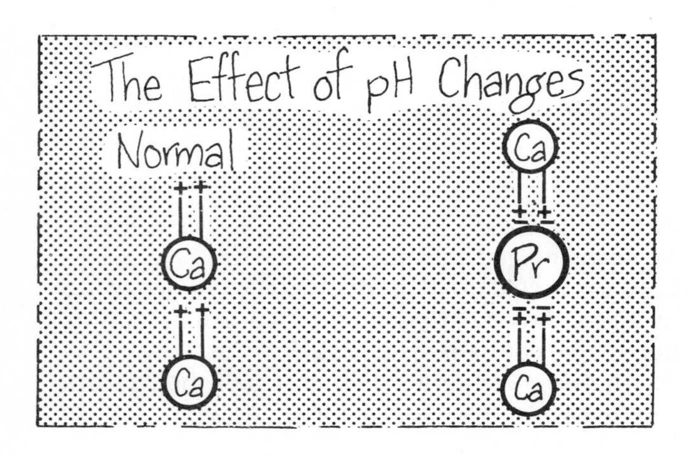

It is important to remember that the degree of protein binding of calcium will vary. In particular the binding is dependent upon pH, since the number of negative charges on protein molecules varies with changes of hydrogen ion concentration. Proteins, as we have seen, act as one of the important buffers in the body.

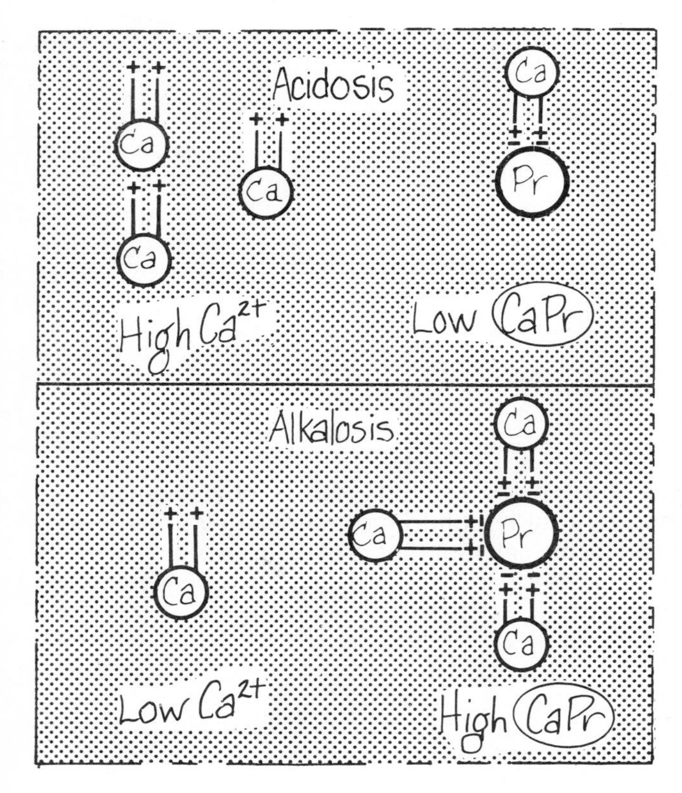

Thus, the more acidic the plasma becomes the fewer negatively charged binding sites will be available to fix calcium. So the free calcium will tend to rise.

Conversely the less acidic the plasma the more calcium will be protein bound. In states of alkalosis the amount of free calcium will fall.
This is important to the clinicians who may see signs of hypocalcemia occurring in alkalotic patients whose total plasma calcium is normal, but whose free ionized calcium has fallen.

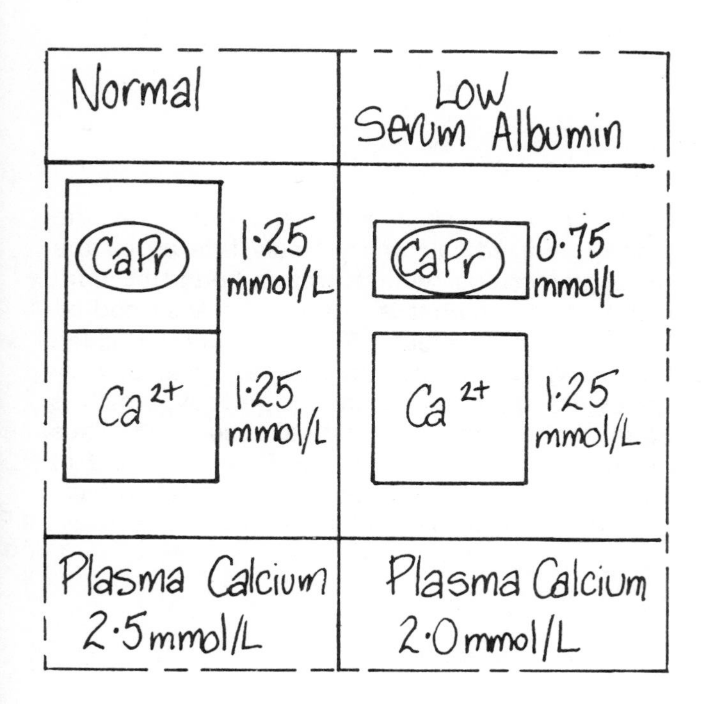

In patients with a low serum albumin the total plasma calcium will be low but the free ionized calcium will be normal. The patient will have no signs or symptoms of hypocalcemia, because these depend upon the ionized calcium concentration and not the total calcium.

Calcium Absorption

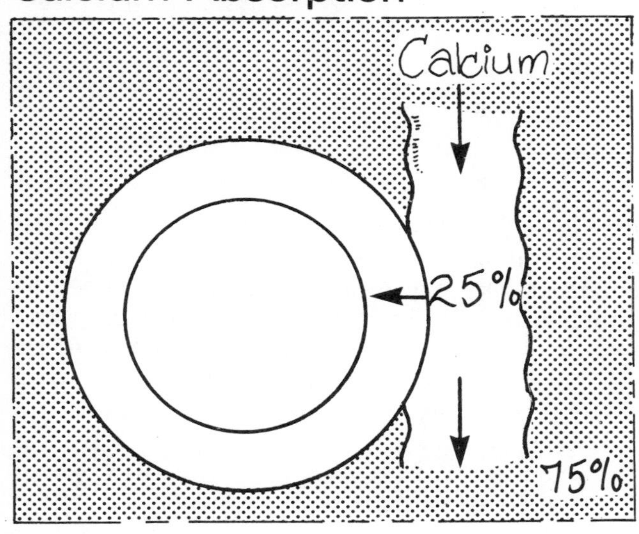

Calcium enters the body via the intestine. About 25 millimoles is ingested each day, but only about 25% of this is absorbed, the rest appearing in the feces.

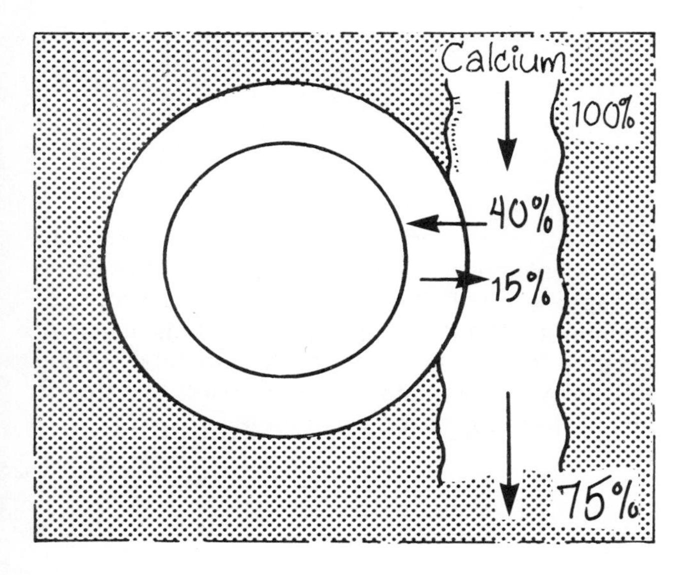

A 25% figure for absorption in the intestine is not quite accurate. In fact about 40% of the intake is absorbed, but an amount, equal to about 15% of the intake, is secreted into the intestine as part of the digestive juices. This results in a balance that is equivalent to only 25% absorption.

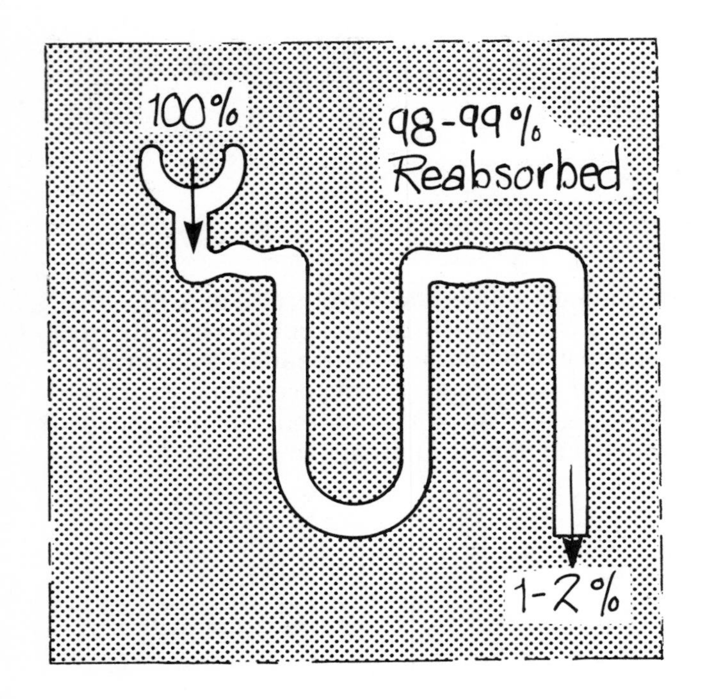

Some calcium can leave the E.C.F. via the kidney. Ionized calcium is freely filtered by the renal glomerulus (whilst protein-bound calcium is not). Most of what is filtered is reabsorbed by the tubule, only about 1-2% of the filtered calcium being lost in the final urine. Total urinary losses are about 4-6 millimoles per day, which is equal to the daily net intestinal absorption. Thus normal adults are in balance, with an intake of about 25 millimoles per day and a combined fecal and urinary loss of about 25 millimoles per day.

E.C.F. Phosphate

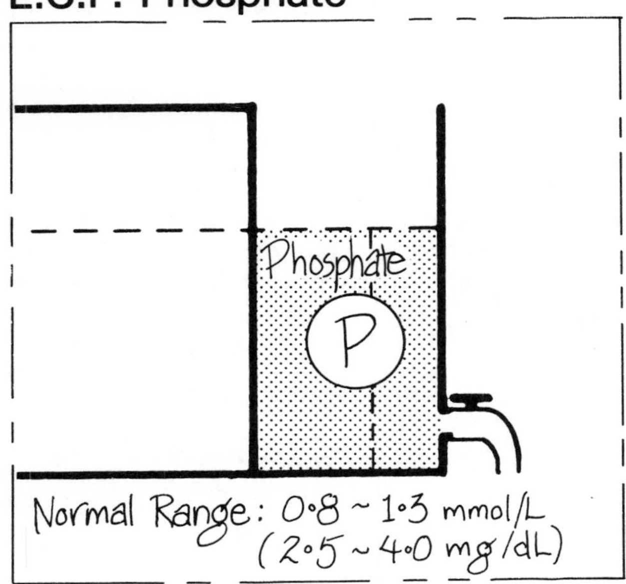

Inorganic phosphate exists in the E.C.F. in two major forms, but the amount of phosphate present is usually expressed in terms of elemental phosphorus (P). In the cells, of course, phosphorus also exists in organic forms such as ATP.

The two major forms of phosphate are monohydrogen phosphate (HPO_4^{2-}) and dihydrogen phosphate ($H_2PO_4^-$). Although a third form of phosphate (PO_4^{3-}) occurs, at physiologic levels of pH this form does not exist. In a clinical setting the term ''phosphate'' is often used to refer to any of these forms indiscriminately.

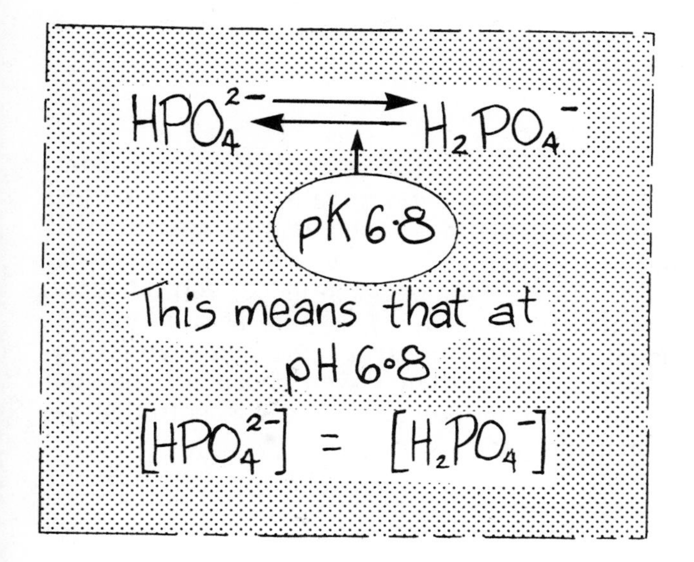

The relationship between the two major anionic forms of phosphate is determined by the pH of the E.C.F. The pH at which the two forms are in equilibrium is 6.8; that is, the reaction has a pK of 6.8. Thus at a pH of 6.8 the concentration of monohydrogen phosphate equals the concentration of dihydrogen phosphate.

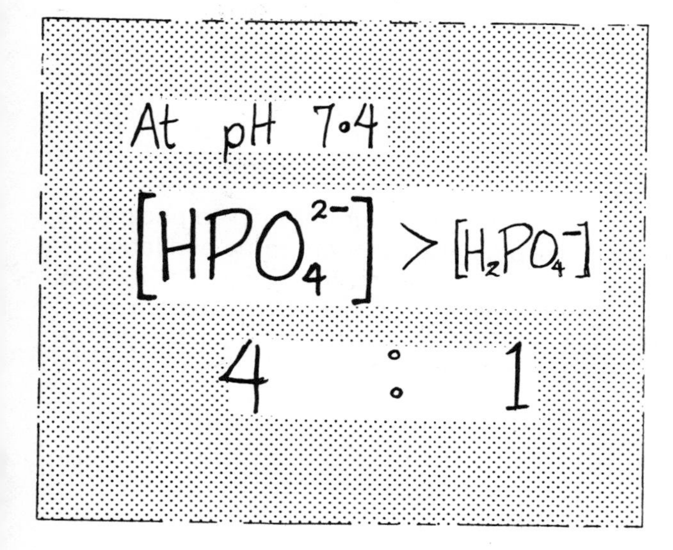

At the pH of extracellular fluid (7.4) about 80% of phosphate will exist as monohydrogen phosphate and 20% as dihydrogen phosphate.

Urinary Phosphate

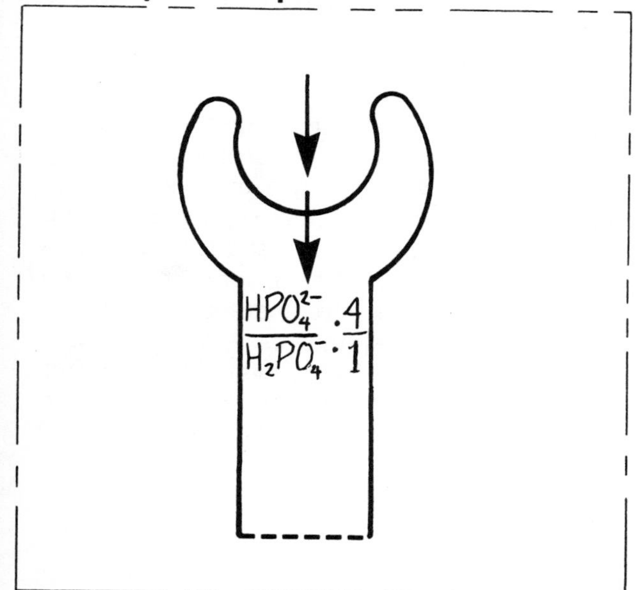

Phosphate is filtered at the glomerulus and in the filtrate the ratio of monohydrogen phosphate to dihydrogen phosphate is the same as in the plasma.

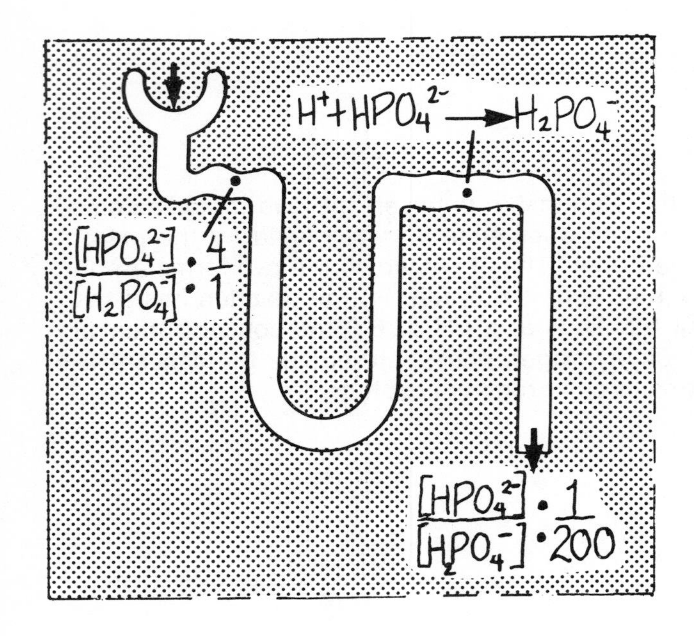

But as the filtrate moves down into the distal tubule the filtered phosphate acts as a buffer which traps hydrogen ions as they are pumped out of the tubular cells into the lumen. As this continues the ratio between mono- and dihydrogen phosphate changes. In very acid urine (pH about 4.5) with almost all phosphate buffering power saturated, dihydrogen phosphate may be up to two hundred times the concentration of monohydrogen phosphate.

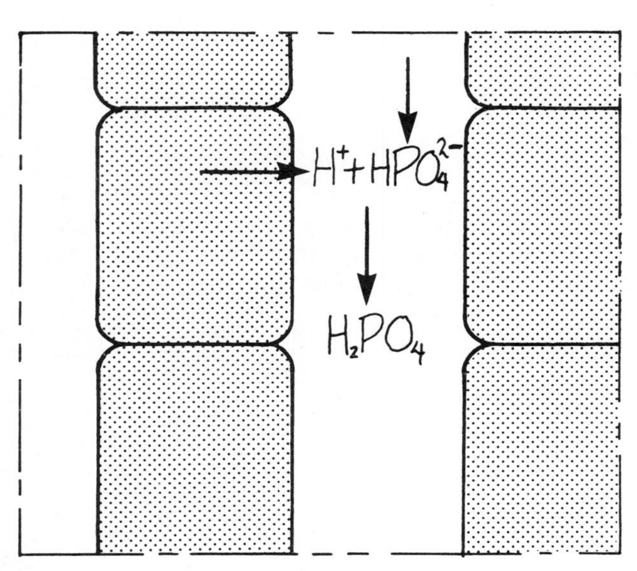

The relatively large amount of filtered phosphate allows a surplus of phosphate to be available for excretion through the renal tubule. On its way out in the urine this becomes the single most important buffer for ''titratable acid'' in the urine. (See Chapter 8)

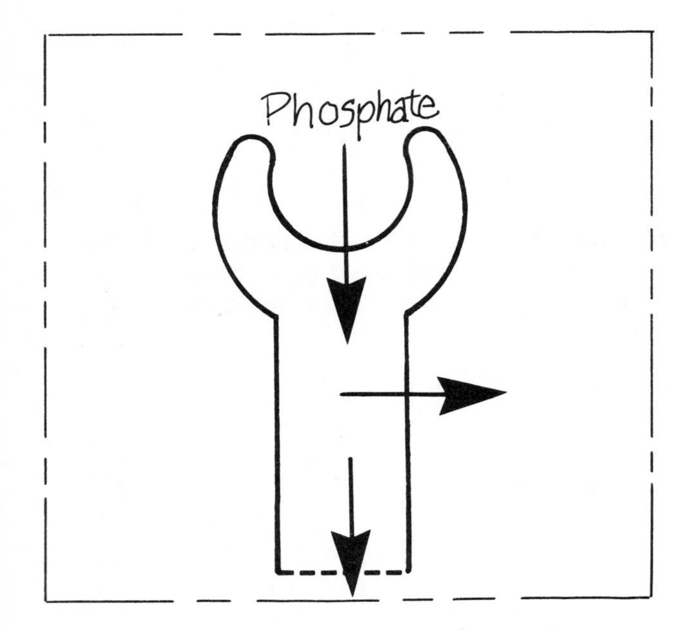

Reabsorption of filtered phosphate by the tubule occurs, but the reabsorptive capacity is normally at a low level so that changes in plasma phosphate levels (and thus in filtered phosphate) can change excretion markedly. Changes in glomerular function can also alter plasma phosphate markedly. In normal circumstances there is plenty of filtered phosphate escaping reabsorption in the tubule and being available to buffer secreted hydrogen ions. Thus the kidney plays an important role in determining plasma phosphate levels.

Calcium and Phosphate

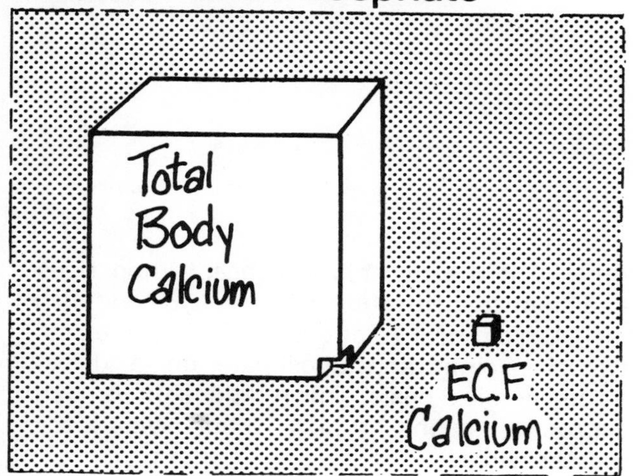

Because the E.C.F. concentration of calcium is small in relation to total body calcium, and because it has such marked physiological effects it is important to understand the intricate mechanisms which have evolved to ensure that this narrow range is not violated. A few millimoles lost or gained from the vast body store of calcium is neither here nor there, but a few millimoles lost or gained from the E.C.F. pool of calcium may be critical to survival.

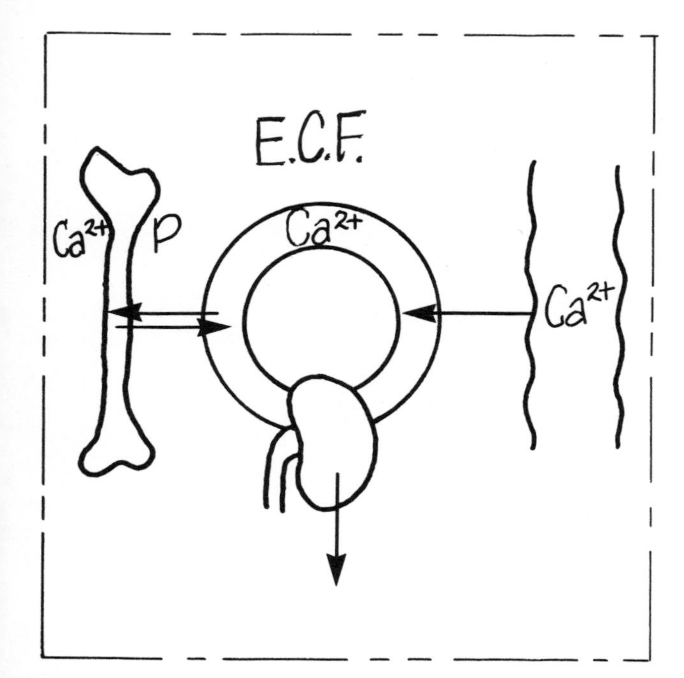

Clearly calcium can enter the E.C.F. in two ways.
1) From the intestine
2) From the store in bone
And it can leave the E.C.F.
1) Into the bone
2) Via the urine
These facts underlie the mechanisms for regulation of E.C.F. calcium concentration.

Clinically it can be useful to express the intimate relationship between calcium and phosphate in terms of the product of the concentrations of the two in the plasma (the ''solubility product'').

Normally this product remains roughly constant. If it rises then calcium and phosphate will tend to precipitate out of solution, probably mainly into bone but sometimes (with abnormally high products) into soft tissues.

Ordinarily if the phosphate concentration rises the calcium concentration will fall . . . . and vice versa. A reciprocal relationship therefore exists between calcium and phosphate on a direct physico-chemical basis. This relationship forms a background against which other regulatory processes will act.

This reciprocal relationship is seen most obviously with elevations of E.C.F. phosphate concentration and it is important to note that many more complex factors come into play to determine the precise relationship between calcium and phosphate in any given situation.

Regulators

1. Parathyroid Hormone (PTH)
2. Vitamin D (Vit D)

Two major factors which regulate calcium and phosphate concentrations are both humoral. They are

1) Parathyroid hormone
2) Vitamin D

Parathyroid Hormone

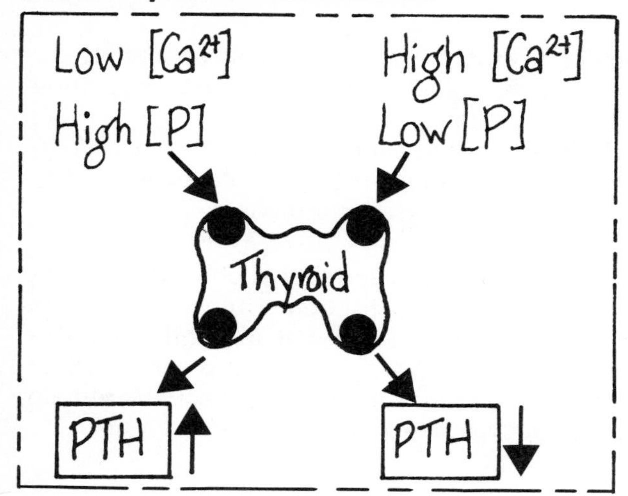

Parathyroid hormone (PTH) is a peptide hormone secreted by the four parathyroid glands in the neck. Secretion is increased by a fall of plasma ionized calcium concentration and decreased by a rise of plasma ionized calcium concentration. Primary changes of plasma phosphate concentration will also modulate parathyroid hormone secretion presumably through their effects upon plasma calcium concentration.

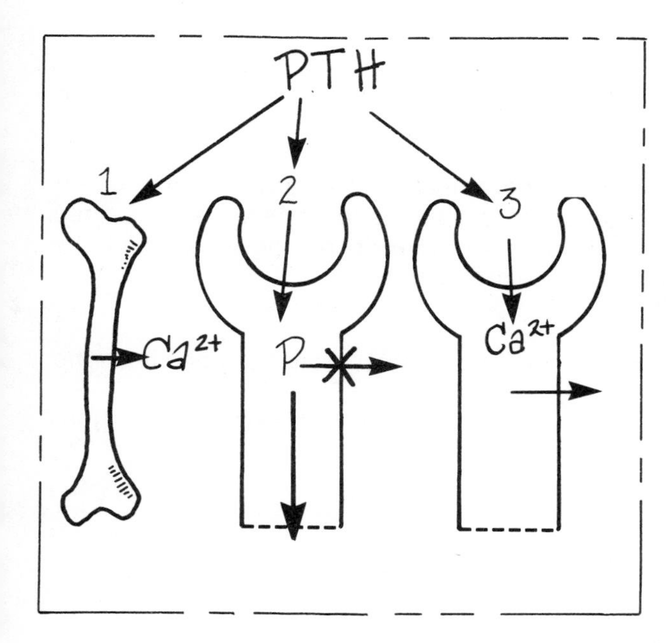

Parathyroid hormone probably has three main actions:

1) Release of calcium from bone (In the presence of Vitamin D).
2) Increase in renal tubular loss of phosphate
3) Increase in renal tubular reabsorption of calcium.

Vitamin D

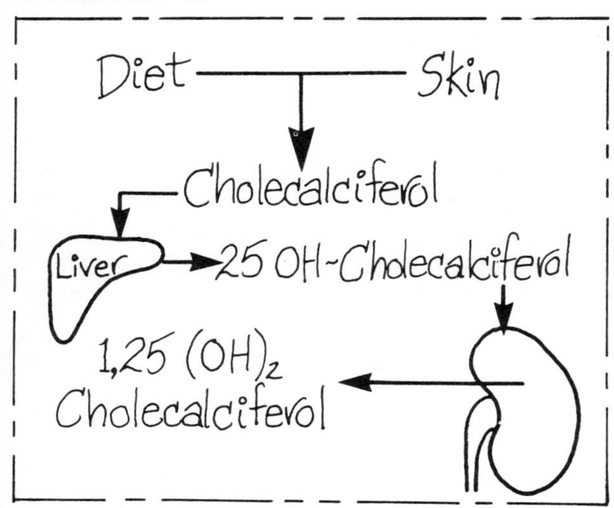

Vitamin D is a steroid hormone, cholecalciferol, derived from precursors either ingested in the diet or produced by ultraviolet light acting upon the skin. Metabolic steps in the liver and the kidney produce the active hormone 1,25 dihydroxycholecalciferol.

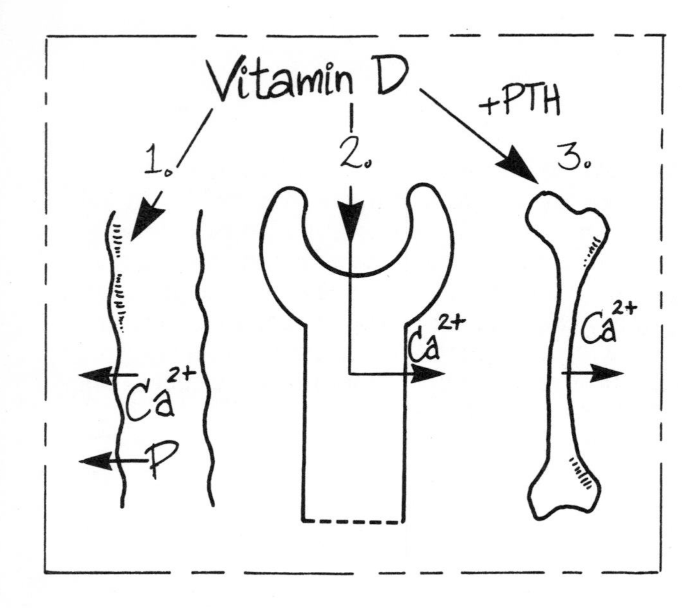

Vitamin D probably has three main actions. It
1) Enhances calcium and phosphate absorption from the intestine
2) Enhances renal tubular calcium reabsorption
3) Enhances calcium release from bone (in the presence of parathyroid hormone).

Other actions may include renal phosphate retention, a direct role in bone mineralisation in certain situations (somewhat in conflict with 3) above) and possibly a direct suppressive effect upon PTH release. All these other actions remain incompletely worked out.

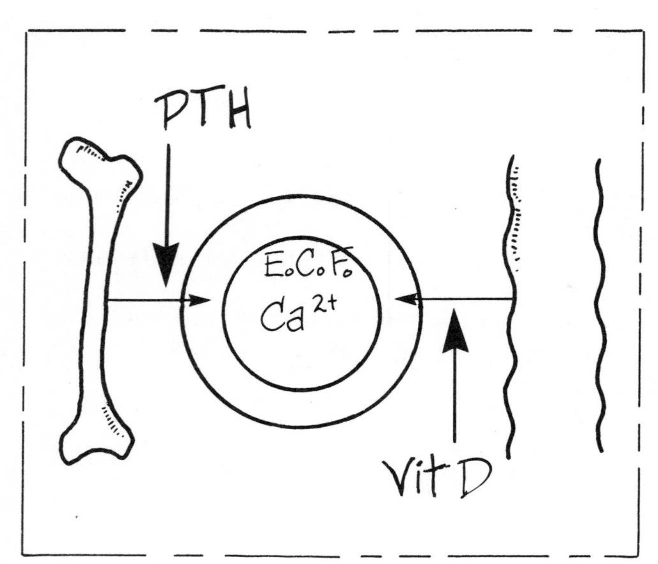

These two hormones are complementary in the regulation of calcium and phosphate and often the full effects of one do not occur without the co-existence of the other.
But ...

The *major* effect of parathyroid hormone is on the bone.
The *major* effect of vitamin D is on the intestine.

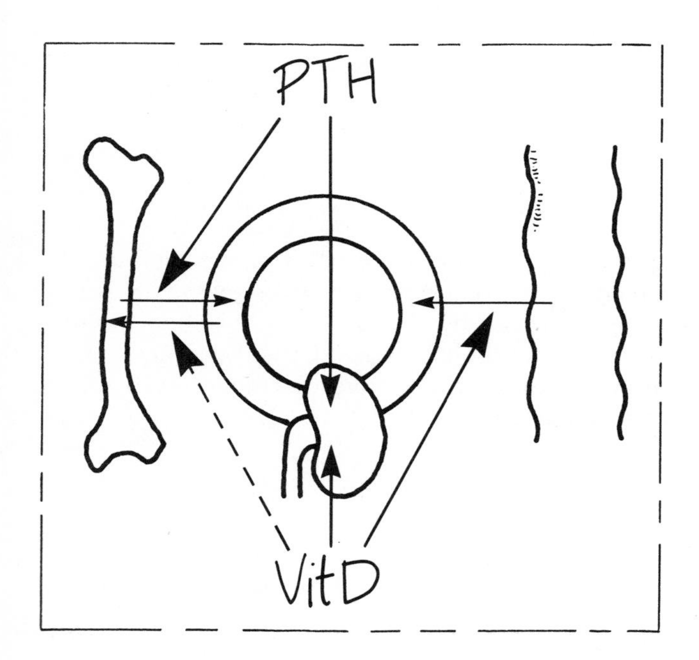

The precise nature and degree of collaboration between parathyroid hormone and vitamin D is not yet fully unravelled, but it is much more complex than can be shown here and is a fascinating story of modern biochemical research.

Calcitonin

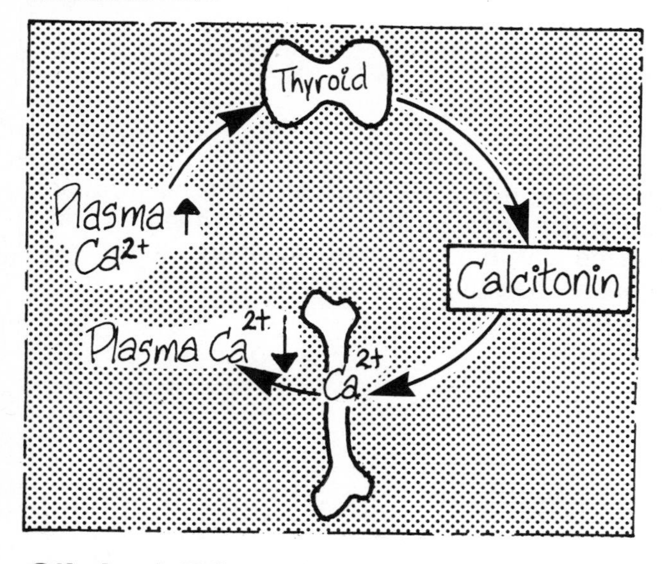

A word about calcitonin, which is a peptide hormone secreted by the parafollicular cells of the *thyroid* gland. Calcitonin lowers E.C.F. calcium by reducing the release of calcium from bone. Its secretion is probably regulated directly by serum calcium concentrations. Its role in calcium regulation appears to be minor in man.

Clinical Disorders of Calcium and Phosphate - Hypocalcemia

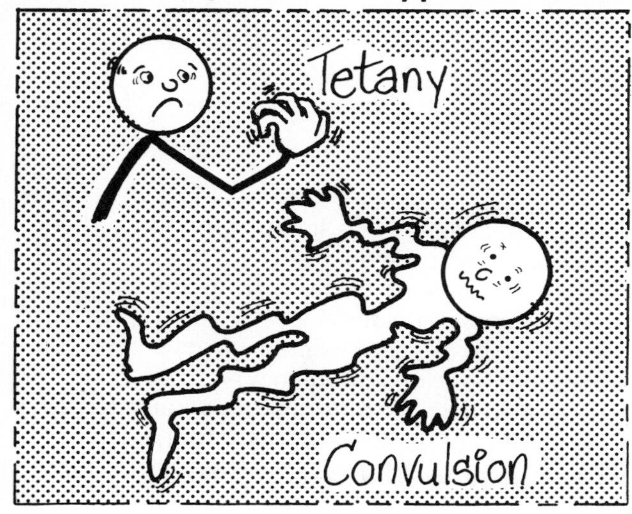

A low plasma ionized calcium produces neuromuscular signs and symptoms. The best known sign is *tetany* characterized by muscle cramps involving the hands and feet. Sometimes convulsions may occur.

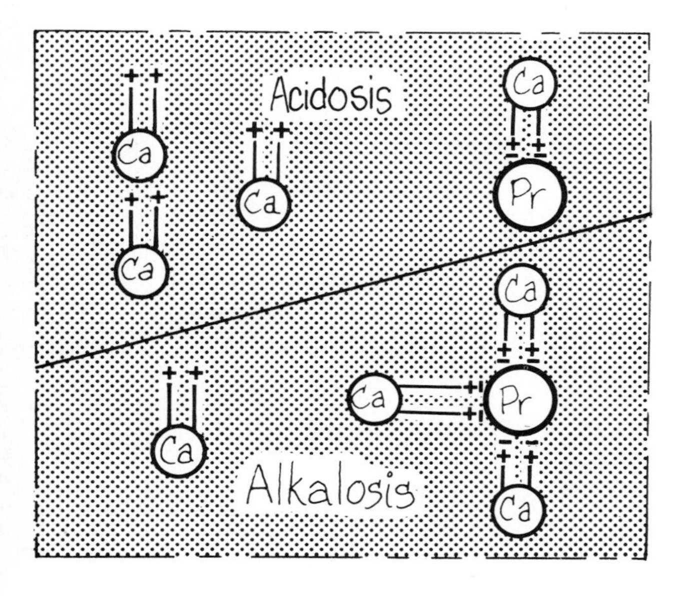

Tetany can occur without a measured fall in *total* serum calcium if a sudden rise in pH occurs, thus suddenly increasing the degree to which calcium is bound to proteins. This may occur if metabolic acidosis is corrected too rapidly by the intravenous infusion of bicarbonate.

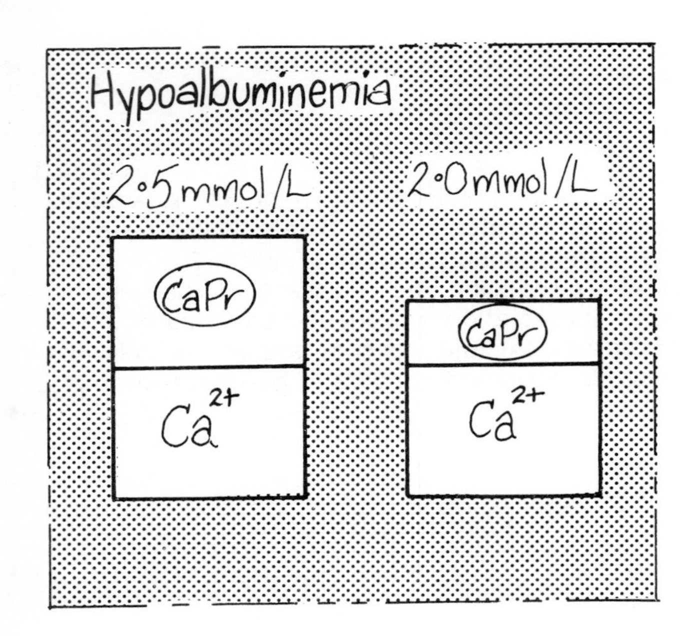

Remember that a low *total* calcium concentration can occur in hypoalbuminemia *without* a low ionized calcium. In this situation, therefore, there are no signs or symptoms.

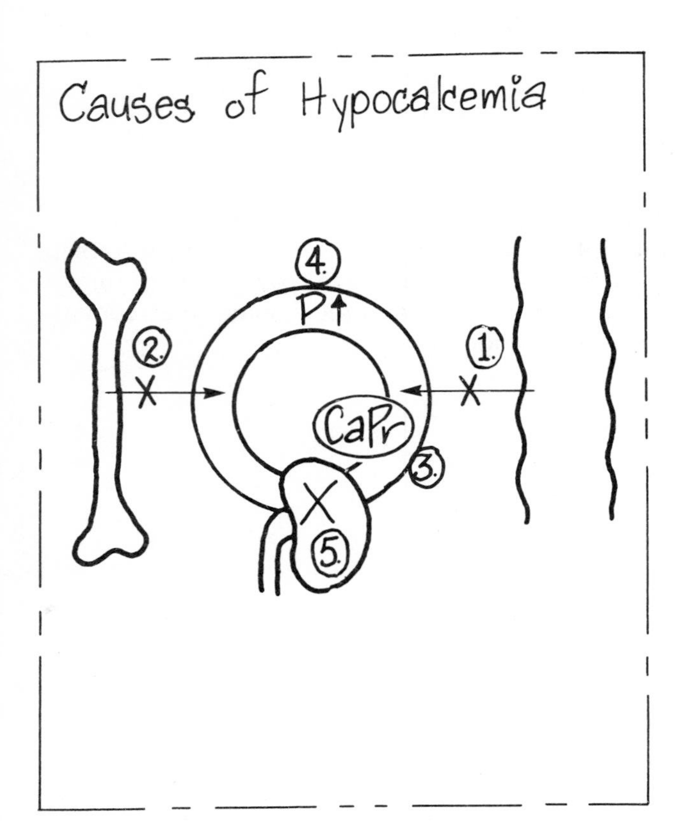

Amongst the causes of a low plasma ionized calcium concentration are

1) Failure to absorb calcium from the intestine (eg. Vitamin D deficiency)
2) Failure to release calcium from bone (eg. parathyroid hormone deficiency)
3) Increased calcium binding (eg. alkalosis)
4) Hyperphosphatemia (eg. renal failure)
5) Renal failure (failure to hydroxylate Vitamin D)

Hypercalcemia

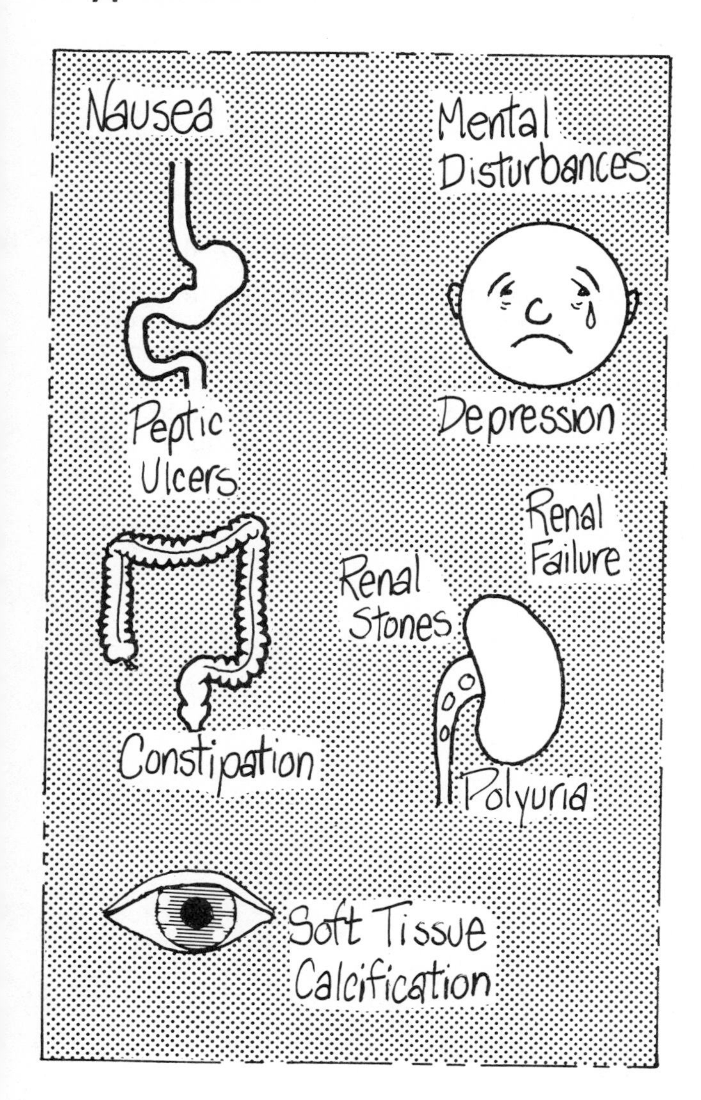

Hypercalcemia causes disordered intestinal motility, disturbances of higher cerebral function and renal damage (due to a direct toxic effect of calcium on the renal tubule). Prolonged hypercalcemia can lead to deposition of calcium salts in soft tissues; this may be seen as a ring at the edge of the cornea. Prolonged hypercalcemia can also lead to the formation of kidney stones.

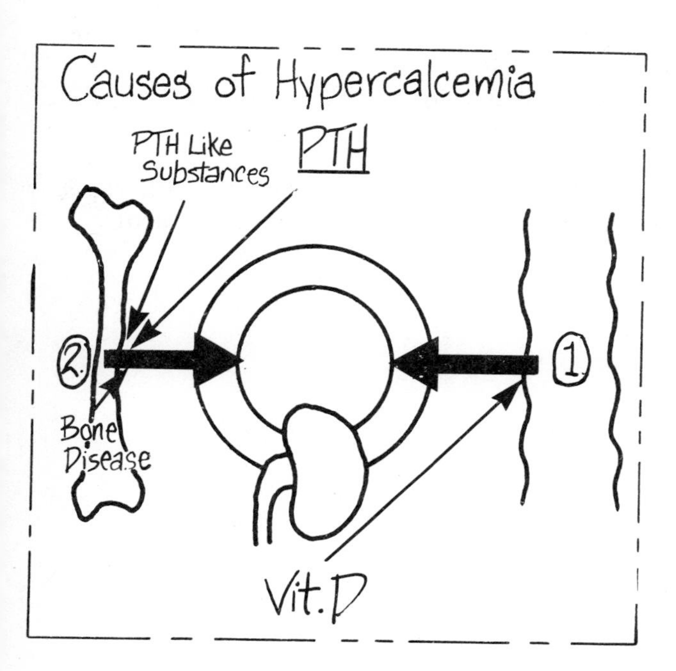

Some causes of hypercalcemia are:

1) Increased intestinal absorption of calcium (eg. an excess of Vitamin D)
2) Increased release of calcium from bone:
   a) Due to the action of parathyroid hormone (eg. parathyroid tumors)
   b) Due to disease of the bone (eg. metastatic cancer)
   c) Due to the action of PTH-like substances released from tumours (eg. carcinoma of the bronchus)

Hyperphosphatemia

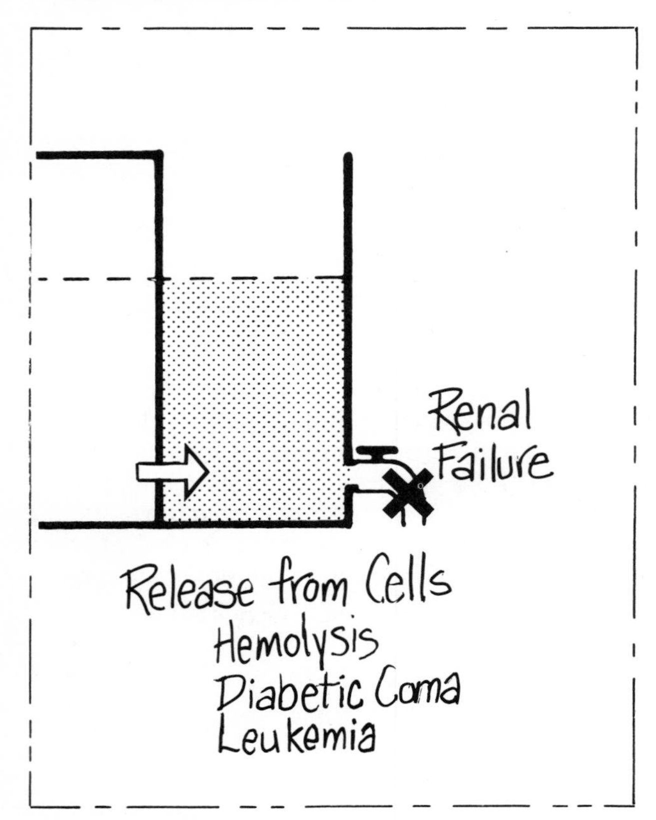

A high phosphate level in the E.C.F. is usually due to renal failure which limits phosphate excretion. Occasionally it will occur in association with massive cell lysis (releasing phosphate from the cells). The major effect of persistent hyperphosphatemia is the deposition of ''calcium phosphate'' in soft tissues. It will also produce ''reciprocal'' hypocalcemia.

Hypophosphatemia

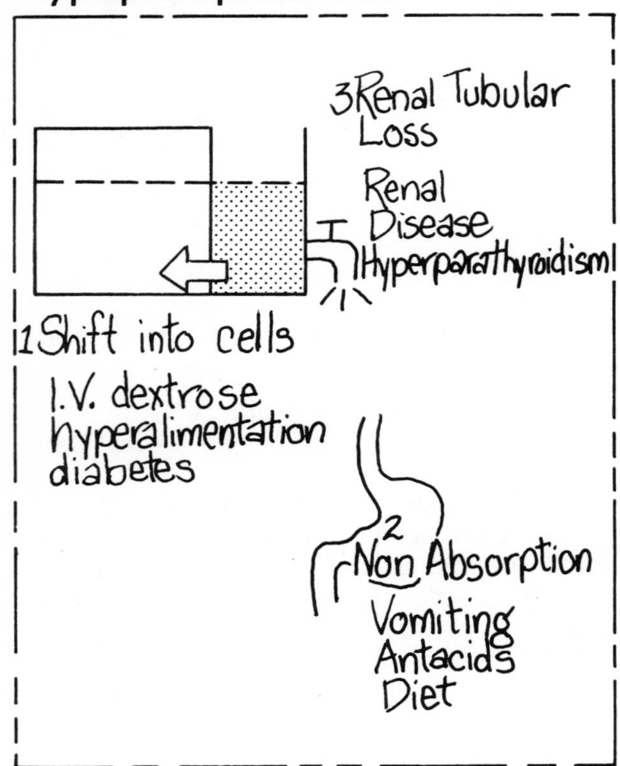

Causes of hypophosphatemia include:

1) Shift of phosphate into cells along with dextrose during the rapid correction of diabetic ketoacidosis or in association with intravenous hyperalimentation

2) Failure of intestinal absorption due to poor diet, alcoholism, vomiting or the use of phosphate-binding antacids

3) Renal tubular loss in association with renal disease or secondary to hyperparathyroidism

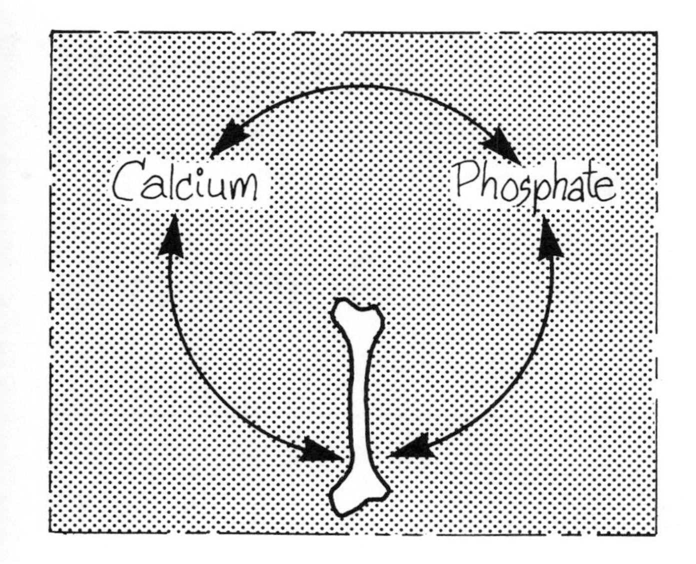

It is evident that calcium and phosphate metabolism cannot be considered without recognising its close linkage to the behaviour of bone. For this reason long-standing disorders of regulation of calcium and phosphate may become clinically apparent as disease of bone.

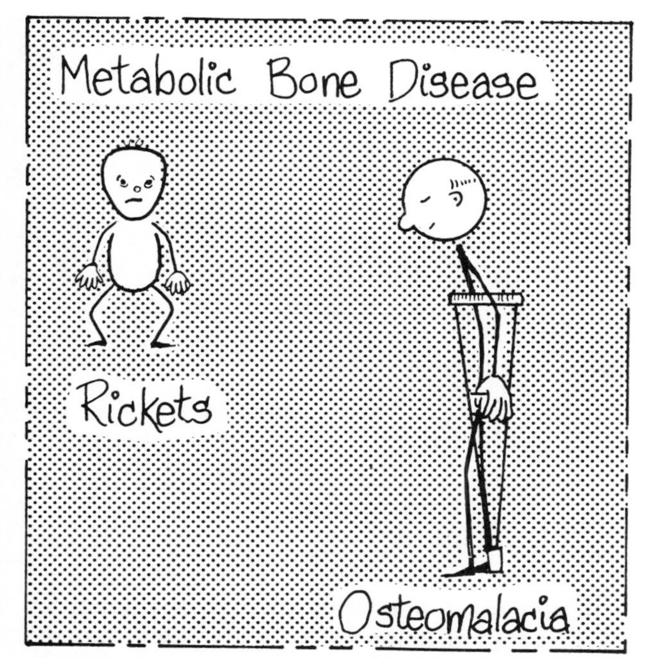

The failure to mineralize bone results in striking disability because of an inadequate skeleton. This is most commonly due to a deficiency of Vitamin D, or a failure in its activation. Occasionally it may be due to calcium or phosphate deficiency.

In the growing child the failure to mineralize bone results in the picture of *rickets* with characteristic deformities of growing bones. In the adult it presents as *osteomalacia* with pathological ''pseudofractures'' of weight-bearing bones with pain and a characteristic ''waddling'' walk.

The underlying defect of calcium metabolism is reflected by a low plasma calcium which often leads to stimulation of the parathyroid glands and resultant ''secondary'' hyperparathyroidism.

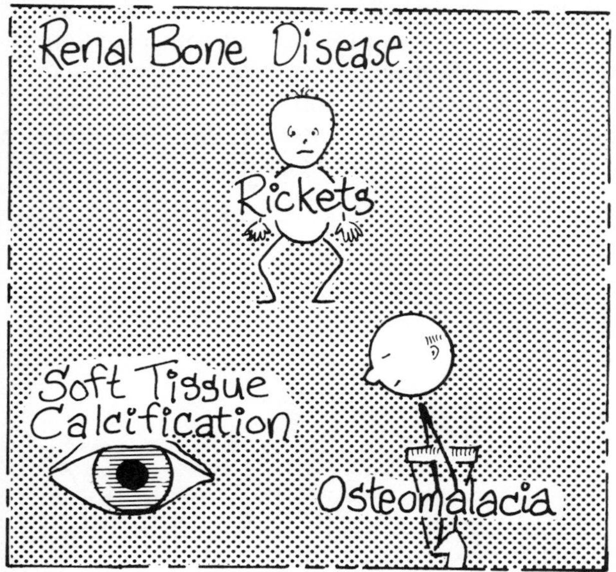

The bone disease of long-standing renal failure is an example of the complexities of calcium and phosphate regulation. It presents as the clinical picture of *rickets* or *osteomalacia* often with associated soft tissue calcification. Factors which are involved include

1) Failure to hydroxylate 25-hydroxycholecalciferol
2) Persistent hyperphosphatemia
3) Metabolic acidosis with consequent loss of calcium from bone in exchange for hydrogen ions
4) Secondary hyperparathyroidism

Chapter 11

Some Therapeutic Guidelines

Treating disorders of fluids and electrolytes should be an exercise in applied physiology rather than the use of a "recipe book" approach. Some specific notes on therapy have been given in previous chapters, and here the discussion will be confined to some general principles which should encourage confidence that fluid therapy is not a great mystery but can be planned safely and logically using the basic concepts developed in this book. More advanced texts provide greater detail and more comprehensive guidance.

Getting Organized

In planning fluid and electrolyte management avoid the common temptation to start with laboratory data. Begin with a careful review of the history preceding the patient's current problem. Then examine him as carefully as you would in any other clinical situation. Only when these steps have been taken should the laboratory reports be brought into the picture.

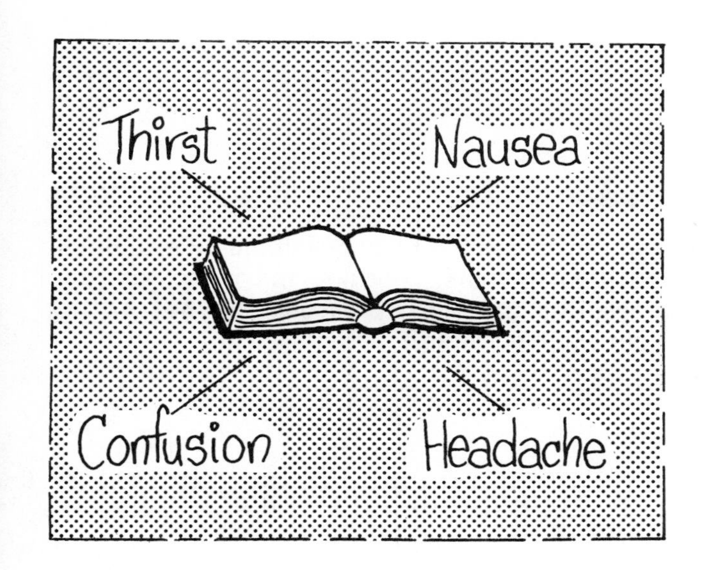

History taking will involve a search for volume loss, such as vomiting or diarrhea. In a conscious patient symptoms may include *thirst* (which may be caused by water depletion or E.C.F. volume depletion), *nausea* (occurring in E.C.F. volume depletion) or *headache* and *confusion* (indicating I.C.F. volume changes.)

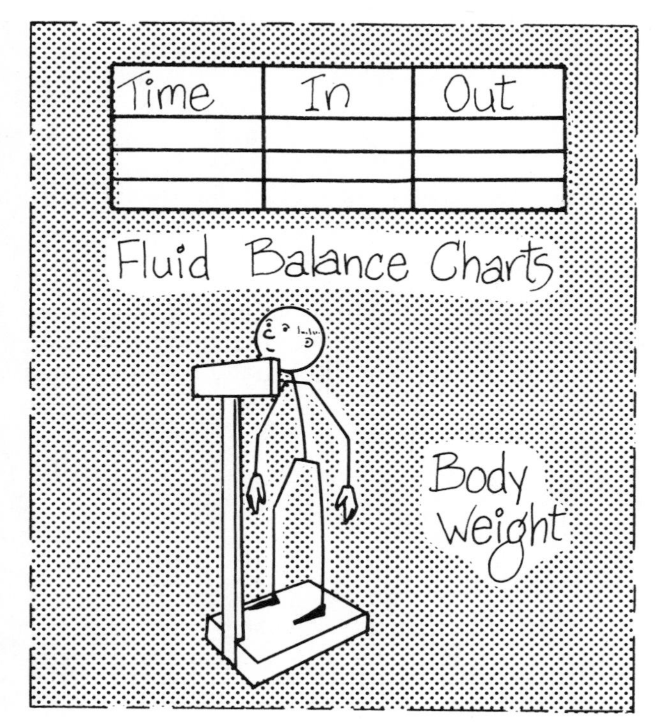

The patient's record may be helpful, and fluid balance charts are valuable. However, remember that fluid charts can be misleading if they are not kept accurately. Daily records of body weight are extremely useful as an index of changing body water, and you should search for them.

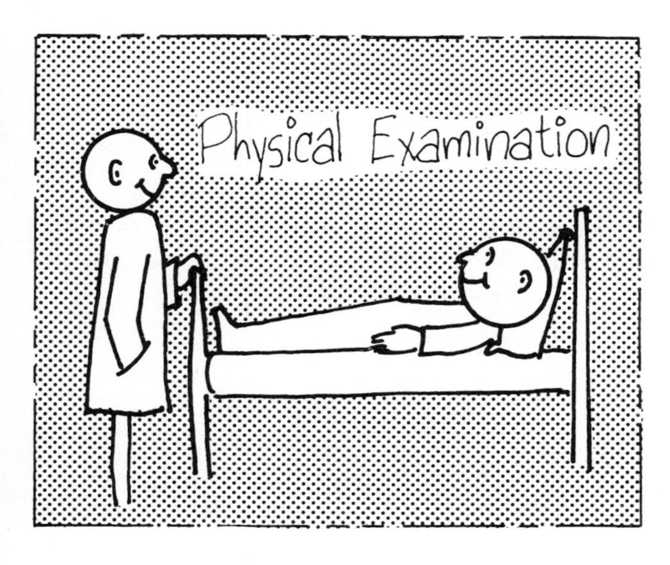

Physical examination should be meticulous. Most of the physical signs you will find will relate to the E.C.F. compartment. Remember that the I.C.F. is not readily accessible to clinical examination.

Assessment of Volume

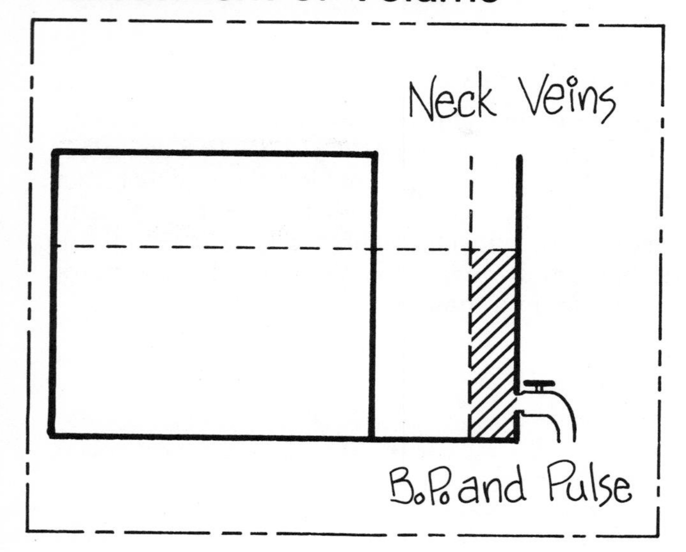

You can assess the plasma volume by examining the jugular veins, checking the pulse rate and measuring the blood pressure (supine *and* upright when possible.)

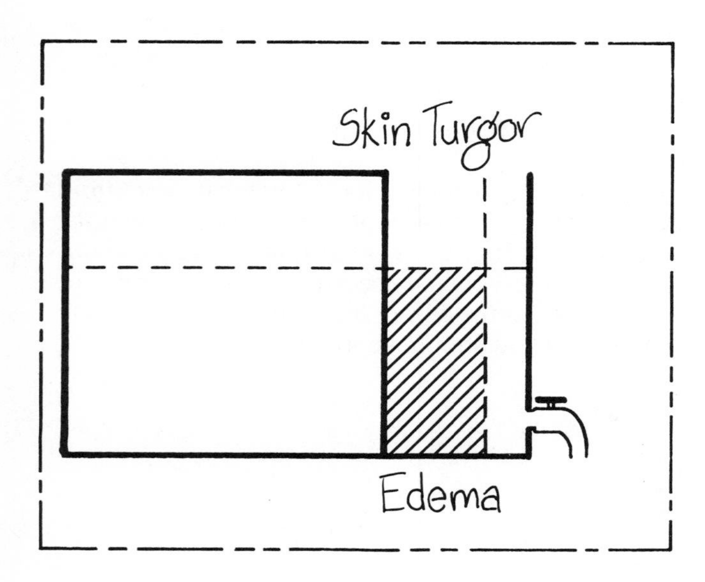

You can assess interstitial volume by the presence or absence of edema, the degree of skin turgor and the state of the usually moist mucous membranes.

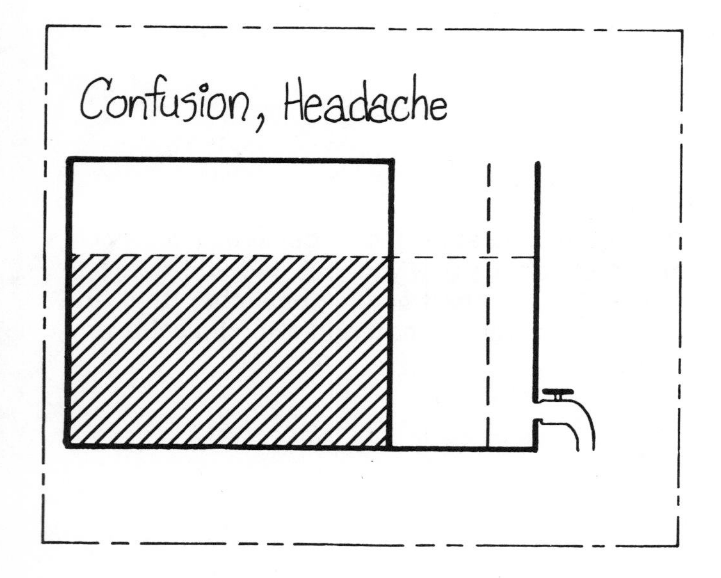

Changes of I.C.F. volume are less easy to detect and are usually reflected by disordered cerebral function. (Review Chapter 5.)

Losses of Body Water

Remember:
Losses of total body water are often underestimated. Major routes of increased loss include:

(1) The skin: in fever with increased sweating, and with high environmental temperatures.

(2) Respiratory, with increased respiratory rate and loss of dead space in patients with endotracheal tubes.

(3) Evaporative losses from operation sites. In prolonged surgical procedures these may be very significant.

Losses of E.C.F.

Remember:
Losses of E.C.F. volume may be obvious as with external losses due to vomiting or diarrhea.

But they may be hidden from view in a number of ways, for example:

(1) Massive pooling of fluid may occur sometimes after surgery, in a dilated and paralyzed intestine (''paralytic ileus''), or in the peritoneal cavity in peritonitis or if the bowel has an impaired blood supply.

(2) Sudden increases of capillary permeability may cause a shift of a large volume of fluid out of the plasma compartment into the interstitial space. This can occur in septicemic states.

The "Third Space"

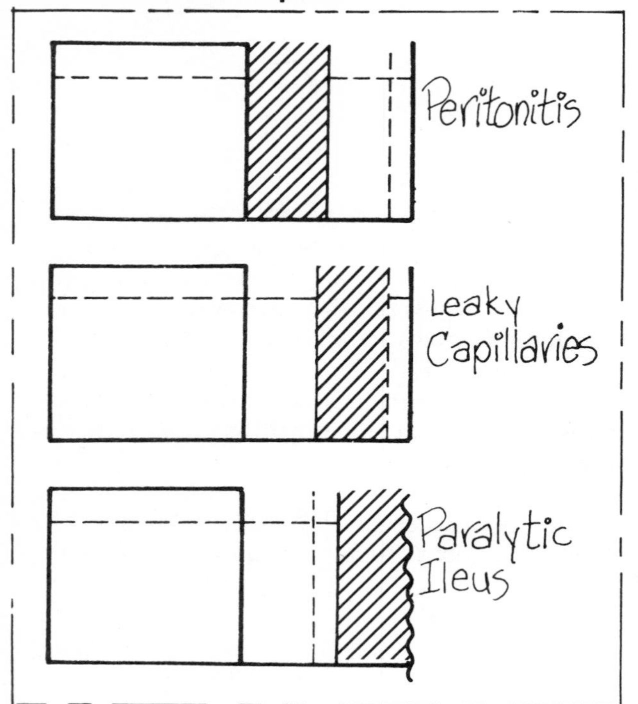

In such situations total volume is not diminished, but shifts into some newly appearing "space" which (at least temporarily) is in communication with the E.C.F. It is as if the boundaries of the E.C.F. have suddenly expanded and as a result the normal volume does not fill all the space available. This is sometimes called the *"third space" phenomenon*.

Filling of such a "third space" can be a cause of all the signs of E.C.F. volume depletion, without obvious external loss. When such a space disappears (eg. when paralytic ileus resolves) the volume may return rapidly into the E.C.F. and cause volume overload, particularly if renal function is impaired.

Laboratory Tests

Only after a full assessment of the clinical background should the laboratory results be considered in detail.

Here are a few general rules about plasma electrolyte concentrations that may be helpful.

These are some of the more obvious changes seen in plasma electrolytes, and you should look for them. The list is not all inclusive, but will be a guide in handling the figures.

Plasma Sodium

The plasma sodium is an indicator of E.C.F. *water* content rather than E.C.F. *sodium* content. Profound E.C.F. volume depletion can occur with a normal plasma sodium because fluids lost by many routes are isotonic. Before diagnosing a low plasma sodium as due to sodium depletion, make sure the patient has clinical evidence of E.C.F. volume depletion.

Plasma Potassium

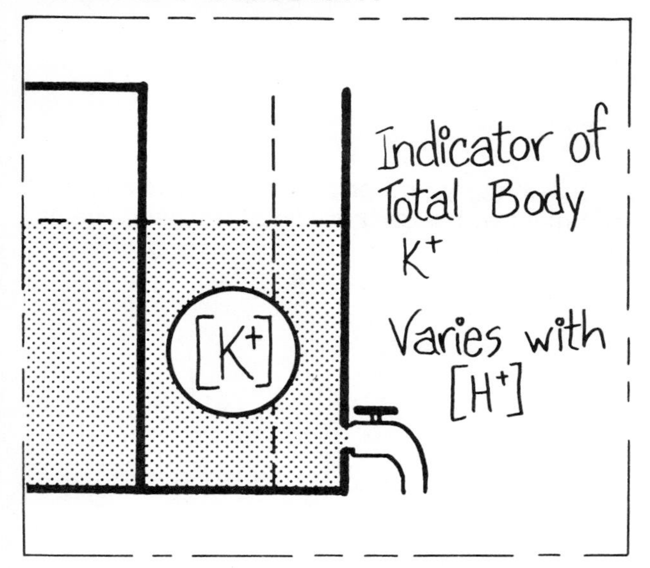

The plasma potassium will change either as a reflection of changes of total body potassium or of shifts of potassium into, or out of, the cells.

Changes of plasma potassium often reflect changes of pH status, or of blood sugar, rather than gains or losses of body potassium.

Anion Gap

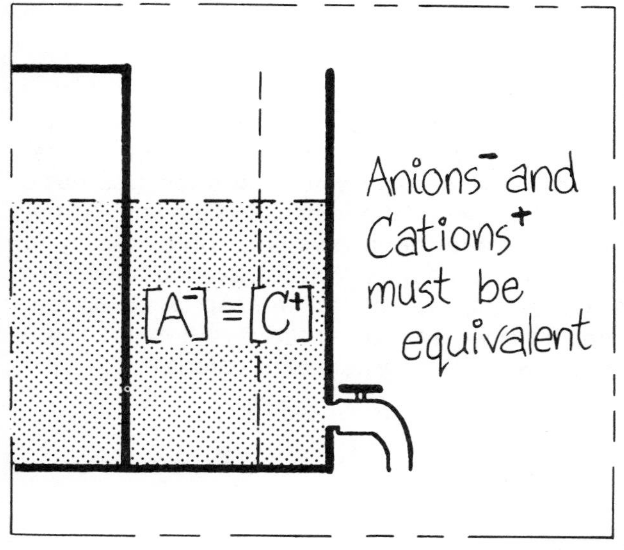

Plasma anions and cations will *always* be present in equal concentrations, but since not all anions are measured there will appear to be a gap between cations and anions. (Refer to Chapter 8)

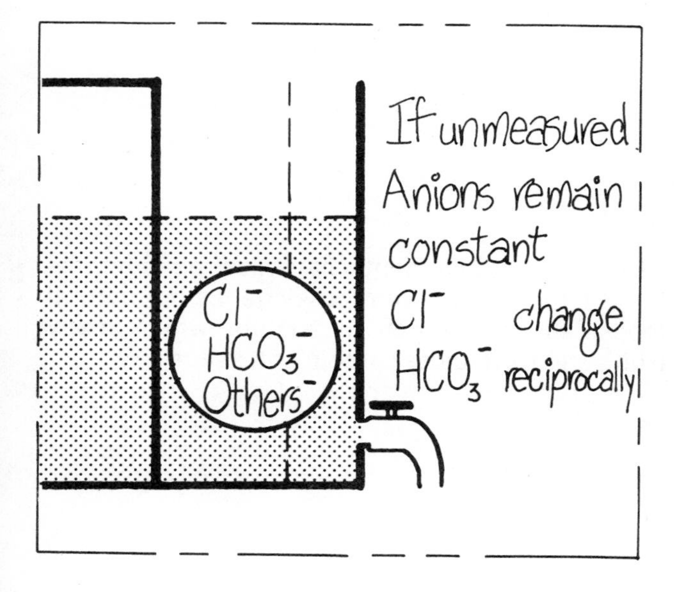

If the Anion gap remains "constant" then changes in chloride and bicarbonate will go hand in hand. A low bicarbonate will cause a high chloride (hyperchloremic acidosis) but equally a high chloride will cause a low bicarbonate. Thus whether chloride or bicarbonate changes first, the other will show a reciprocal change.

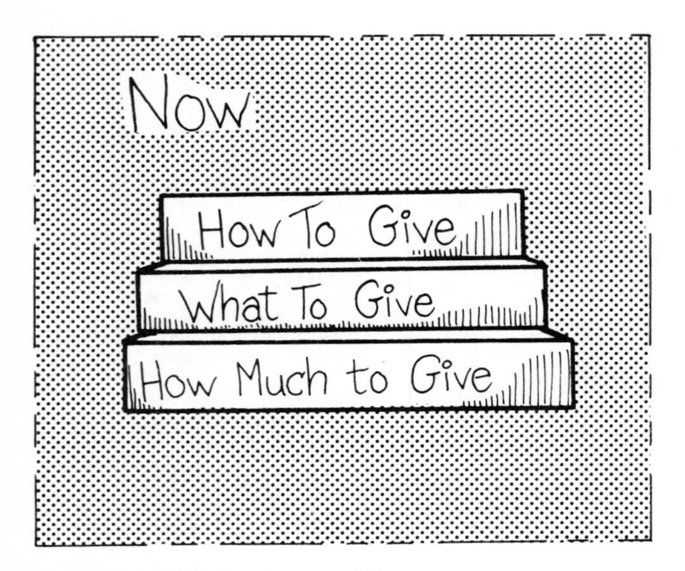

Now that history and physical examination have been linked with laboratory data you can move on to the next step:

What Route to Use

Whenever possible, fluid replacement should be administered by the normal physiologic route, i.e., orally. It is surprising how often this basic rule is forgotten. This is because modern equipment has made the intravenous route so easy to use; but do remember that intravenous lines are an increasing source of in-hospital infection. Other rarely used parenteral routes are subcutaneous or intraperitoneal infusions.

Electrolyte supplements can also be administered orally. This can be done using natural foods; soups are rich in sodium whilst many fruits and fruit juices contain potassium.

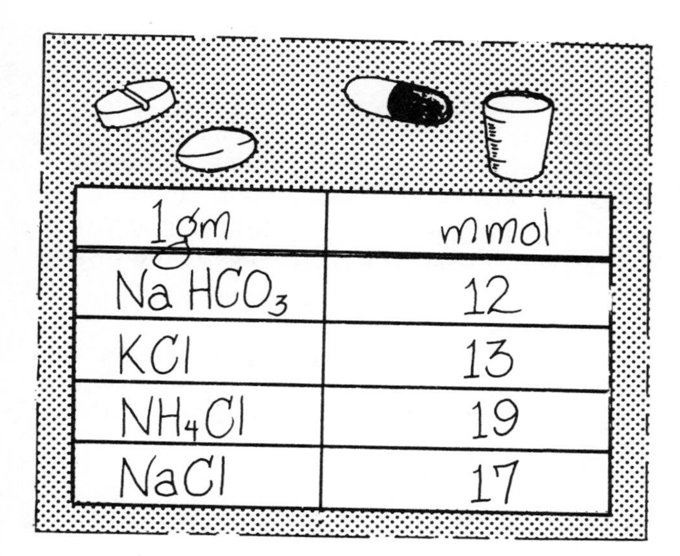

1gm	mmol
$NaHCO_3$	12
KCl	13
NH_4Cl	19
NaCl	17

When supplements are added in artificial forms it is more realistic to think in molar terms than in terms of weight. Thus 1 gram of sodium chloride contains 17 millimoles of sodium chloride whilst 1 gram of potassium chloride contains 13 millimoles of potassium chloride.

What to Give

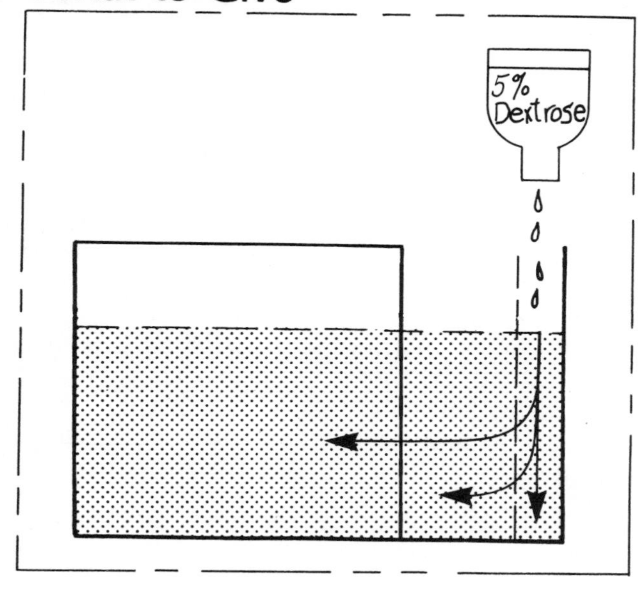

When intravenous (I.V.) fluid therapy has been decided upon, certain general principles determine the kind of fluid to be used.

(1) *water* has to be given as "5% dextrose in water" since pure water would hemolyze the red cells as it enters the vein. Adding dextrose renders the water isotonic, but the dextrose is rapidly metabolized, leaving water.

It will be distributed evenly through all body compartments and will contribute to both E.C.F. and I.C.F. If you recall the relative size of these compartments you will recognize that two thirds of any water load will enter the I.C.F. and only one third will remain in the E.C.F. Thus three litres of 5% dextrose and water would theoretically be needed to expand the E.C.F. by one litre. This fluid therefore is designed to replace deficits of total body water and not E.C.F. volume.

(2) Isotonic sodium chloride will be distributed throughout the E.C.F. and will not enter the I.C.F. Such a solution is designed to replace deficits of E.C.F. volume. Thus 1 litre of isotonic sodium chloride will expand the E.C.F. by one litre; it will contribute to both plasma and interstitial compartments but because of their relative size, one litre of isotonic saline will theoretically only expand the plasma volume by one quarter of a litre.

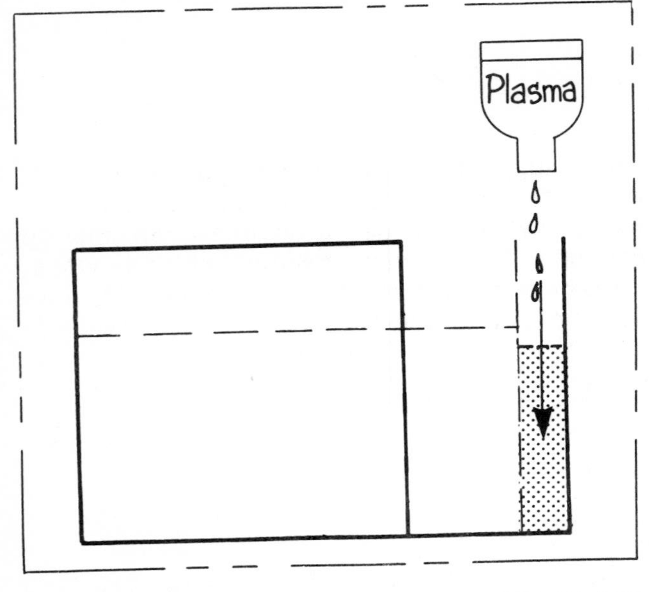

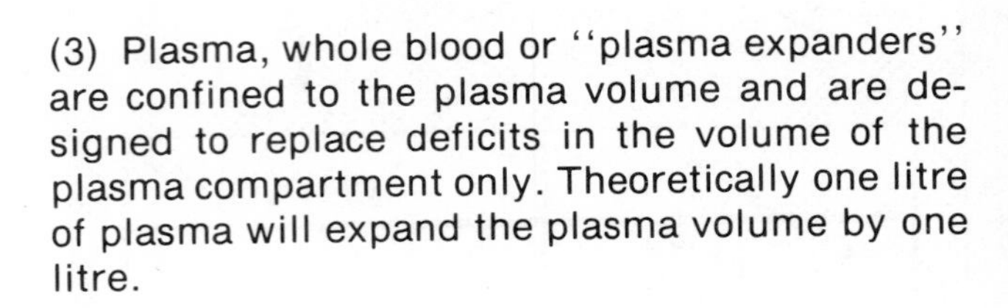

(3) Plasma, whole blood or ''plasma expanders'' are confined to the plasma volume and are designed to replace deficits in the volume of the plasma compartment only. Theoretically one litre of plasma will expand the plasma volume by one litre.

The risk of hepatitis from pooled plasma has unfortunately limited its usefulness and plasma substitutes, rather than plasma itself, are therefore often preferred.

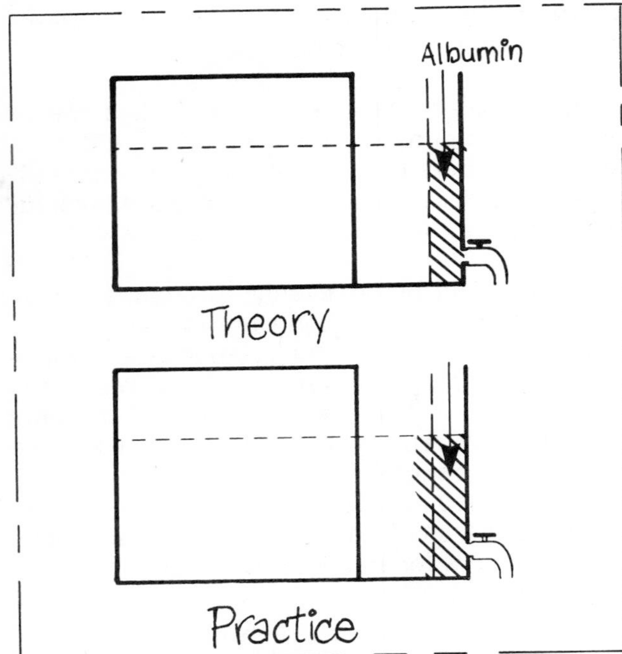

The precise distribution of fluids may not accord with theoretical ideals. Albumin, for example, is often considered a ''plasma expander'' when given intravenously, but in the long term probably becomes distributed throughout a much wider volume than that of plasma alone. In the short term, however, theory and practice coincide well enough for albumin to function as a substitute for plasma.

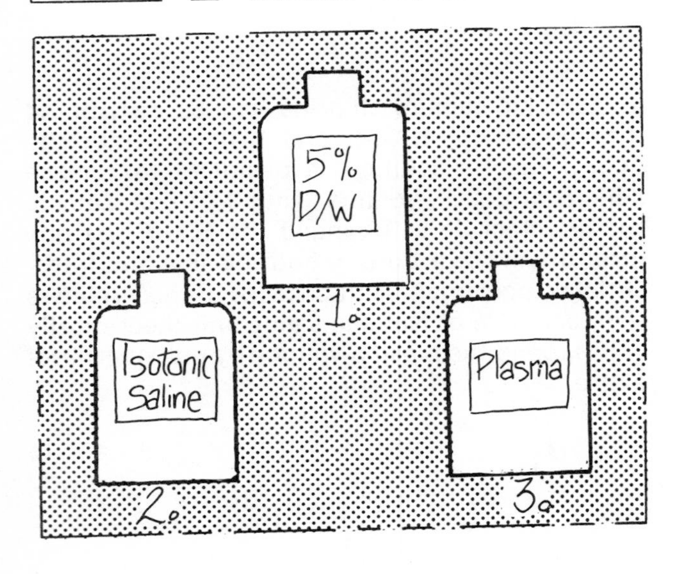

In Summary:
For practical purposes three intravenous fluids can form the basis of any volume replacement regimen. These fluids are

(1) 5% dextrose in water

(2) Isotonic sodium chloride

(3) Plasma (or plasma substitute)

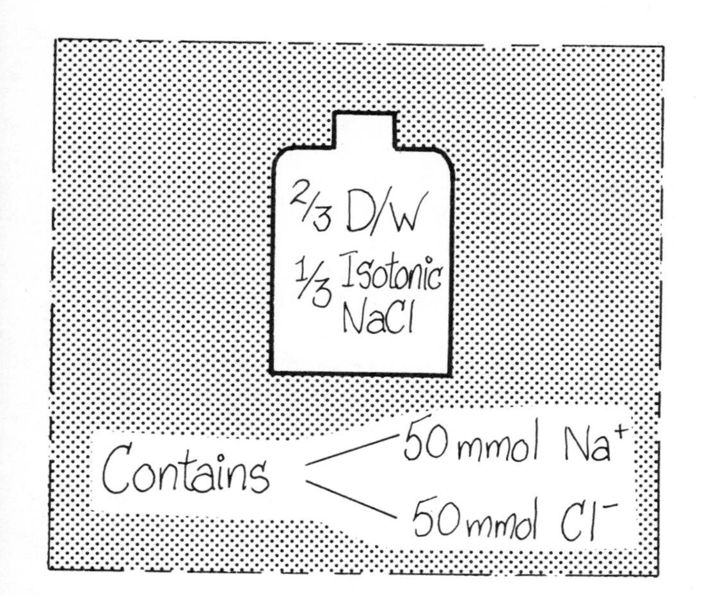

Many other intravenous solutions are available in containers suitable for immediate intravenous infusion. Many of them are widely used; in particular, mixtures of dextrose and saline such as "two thirds dextrose, one third isotonic saline". A fluid such as this is useful in "maintenance" fluid replacement (see below) and is a convenience rather than a necessity. It is hypotonic, as far as the E.C.F. is concerned, and its uncritical use can be a cause of dilutional hyponatremia.

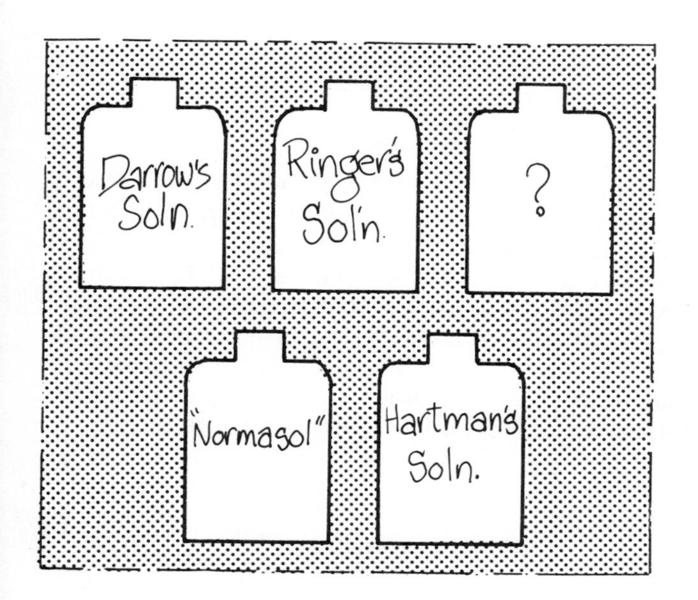

A number of more complex solutions is available and often sanctified by the term "physiological", which can be misleading.

They should be used after careful thought and appropriate experience. Some of them have trade names which may give little clue to their exact composition.

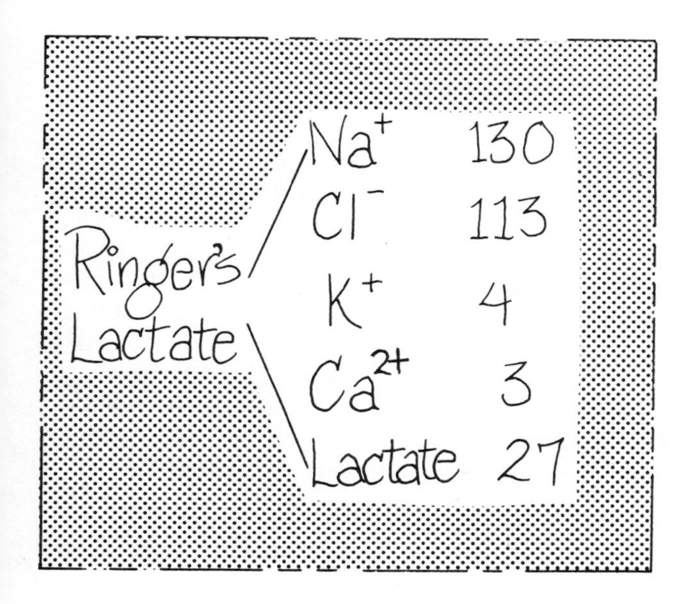

They include "balanced salt" solutions such as Ringer's lactate solution. Remember this contains potassium and calcium as well as lactate which is converted to bicarbonate in the body.

Such a fluid administered to a patient with lactic acidosis can, of course, be dangerous since failure to convert lactate to bicarbonate will make matters worse. In addition, the presence of potassium and calcium in the fluid is not indicated by its name and may be forgotten.

So with these predesigned ''solutions'' make sure that you know (1) what they contain and (2) what their special risks may be before you use them.

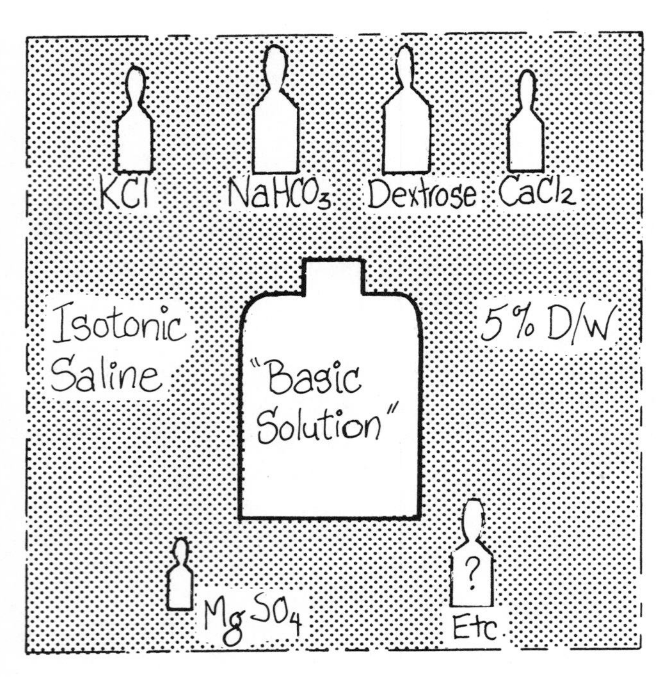

An alternative to such solutions is to use a ''basic'' fluid (5% dextrose or isotonic saline) and add concentrated solutions to it from prepacked ampoules. The advantage of this is that each addition is made consciously (rather than automatically by a distant manufacturer) and that bicarbonate (which is easy to sterilize commercially in small containers) is available. Some of the more useful additives are shown here.

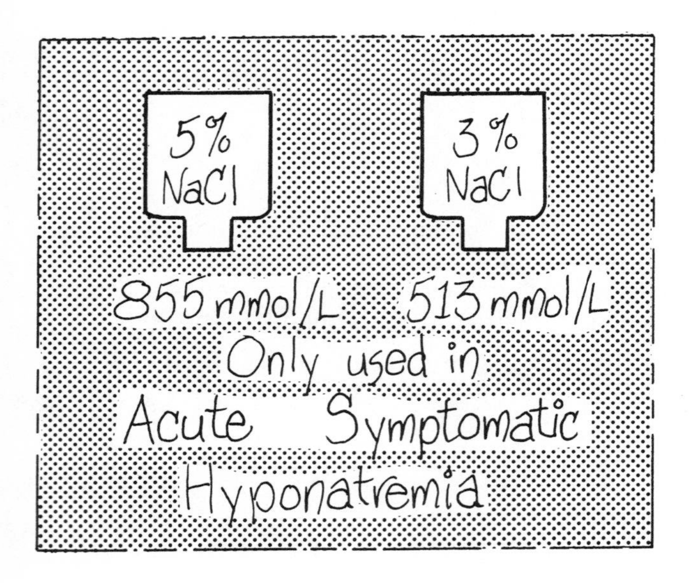

A word about ''hypertonic saline'': It is only rarely indicated for severe and rapidly developing symptomatic hyponatremia. Remember that it contains a great deal of sodium and a proportionately larger amount of chloride, and should be used with great care. Since most hyponatremic patients are edematous, (and therefore have already got an excess of body sodium) further sodium loading will increase the edema, and water restriction is the correct approach to most asymptomatic hyponatremic patients.

How Much to Give

In any fluid therapy there are two main objectives.

(1) To make up the losses that have already occurred.

(2) To keep up with the losses that are occurring whilst fluids are being given.

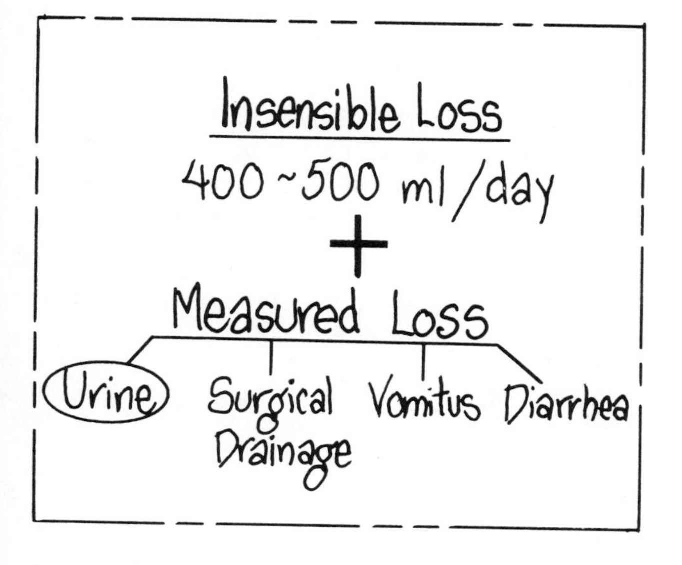

Some general rules may be of help.

(1) Basic unmeasurable and ''insensible'' losses that will occur in all patients amount to 400-500 millilitres in 24 hours for a normal adult. This is made up of losses from skin and lung and a small amount of water in normal feces.

As mentioned earlier, both of these routes of loss may increase greatly in abnormal conditions or environmental stress.

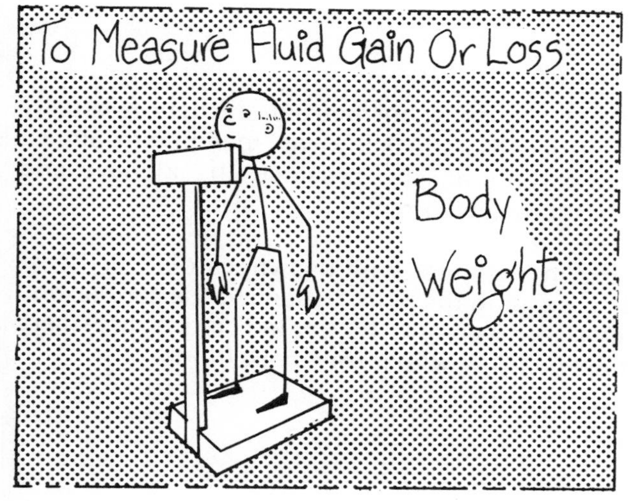

(2) Serial measurements of body weight are extremely useful indicators of loss or gain of fluid. Knowledge of the patient's weight before sustaining a volume loss may be available and will be a good guide to the volume of fluid needed to replace the loss.

(3) The loss of symptoms such as thirst and nausea will be a guide towards the degree of reversal of volume loss that has been achieved.

Similarly careful clinical assessment will indicate such things as the return of blood pressure and pulse to normal. The appearance of moist sounds in previously dry lung bases indicates that volume requirements are over corrected.

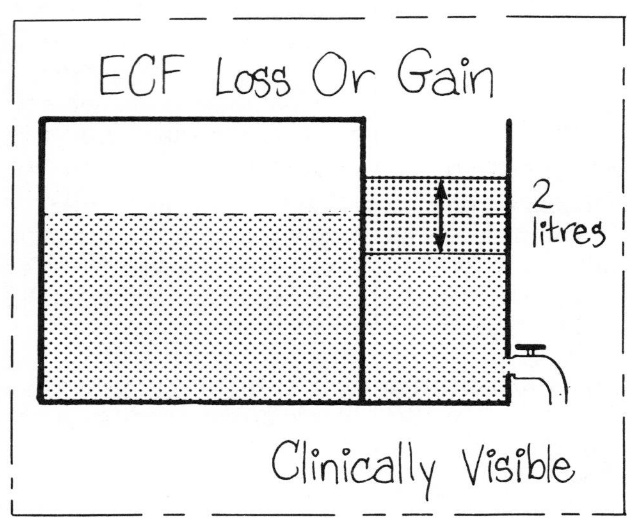

(4) In order to show clear signs of volume depletion or overload an adult is usually at least two litres up or down as far as the E.C.F. is concerned. Often, of course, the size of the deficit, or excess, will be much greater than this. Smaller losses will produce more subtle clinical signs which may easily be missed.

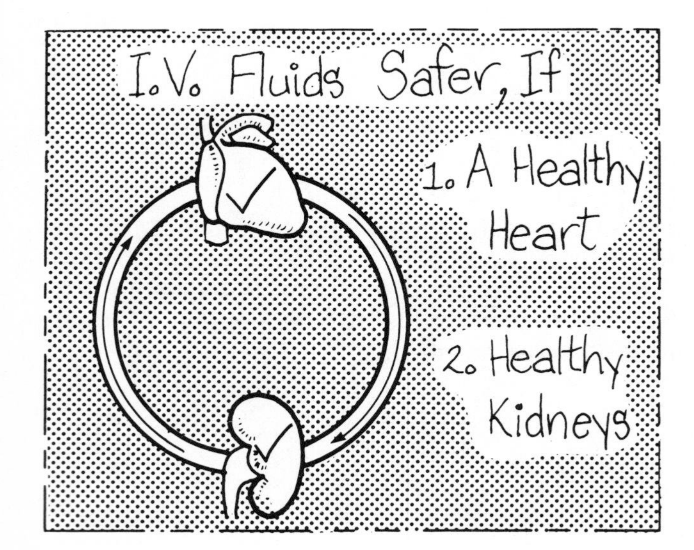

(5) As clinicians, our margin of safety is usually wide because the body can manipulate almost any parenteral load provided the kidneys and heart function normally.

But, in patients with known cardiac or renal disease, or in the older patient whose cardiac and renal reserves may be limited, much more care must be taken with intravenous fluids.

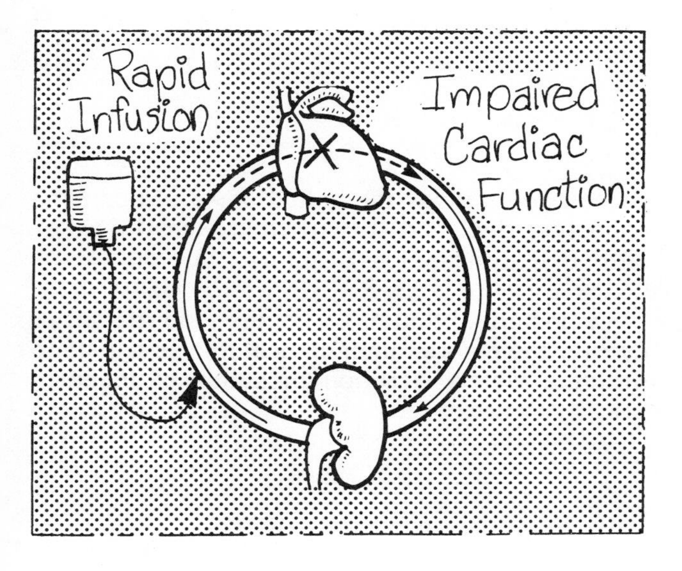

The rate of infusion is as important as the total amount that is given. Thus the rapid infusion of volume into the venous side of the circulation in a patient with heart disease may temporarily exceed the ability of the heart to distribute the added volume, and pulmonary edema may develop before a deficit of E.C.F. volume has been fully corrected.

(6) Variations of body size and weight must be considered. Fluid replacement in infants and children requires special care and experience. Remember from Chapter 3 that special factors occur in small infants, although the same general principles apply at any age and sex. The management of major fluid and electrolyte problems in infants requires specialized skill and experience.

A Note About Diuretics

1. Osmotic Diuretics
2. ADH Antagonists
3. Transport Inhibitors

Diuretics are important agents which act upon the renal tubule to increase urine flow. They find their major therapeutic role in states of sodium and water retention associated with edema.

They can be divided into the three types shown here; transport inhibitors are by far the most important in a clinical setting.

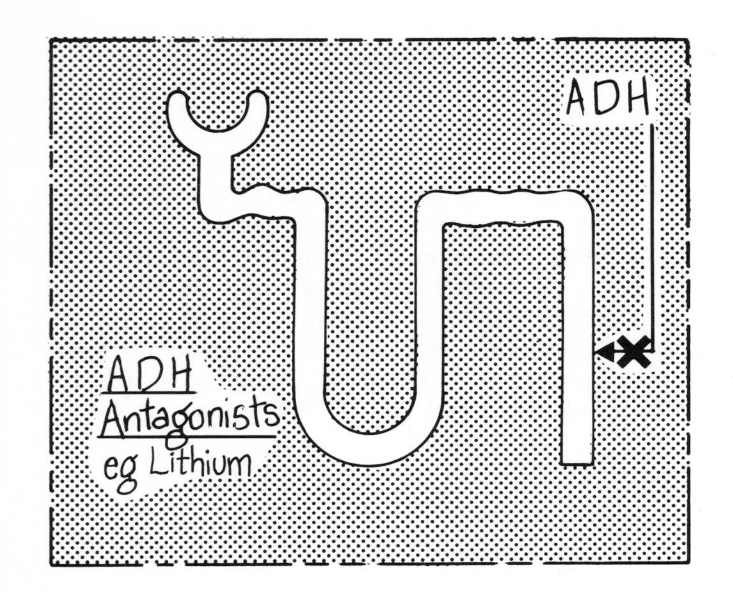

Some agents block the action of ADH upon the collecting duct and thus produce loss of water without significant loss of solute.

They include lithium carbonate and demeclocycline and have a minor clinical role as agents to promote water excretion in states of inappropriate ADH release.

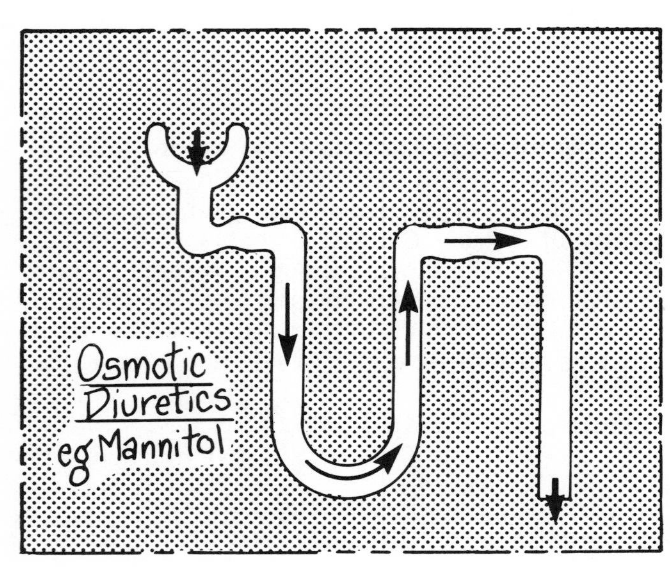

Some agents act as diuretics by being non-absorbable solutes which, once filtered, move through the tubule without any absorption and thus generate osmotic effects which keep water and solutes within the tubule.

In diabetes mellitus the high filtered concentrations of glucose can act in this way.

Mannitol is sometimes used clinically to promote rapid urine flow, but it is of limited usefulness.

Thiazides
Aldosterone Antagonists
Spironolactone
Triamterene
Carbonic Anhydrase Inhibitors
Acetazolamide
Loop Diuretics
Furosemide
Ethacrynic Acid
Mercurials

The major diuretics are all inhibitors of ionic transport processes at one or more sites in the tubule. Their precise mode of action is not always well understood, but some general comments can be made.

i) Carbonic anhydrase inhibitors block the generation of bicarbonate and hydrogen ions in the tubule, and limit sodium reabsorption by limiting the availability of hydrogen ion for exchange.

ii) Diuretics acting in the loop of Henle interfere with chloride and sodium transport and secondarily interfere with urine concentration.

iii) Diuretics acting on the distal tubule include antagonists of aldosterone and block sodium reabsorption whilst encouraging retention of potassium.

iv) Some diuretics (like the thiazides) seem to act at more than one site and their action is incompletely understood.

Risks Of Diuretics

1 Volume Depletion

2 Potassium Depletion

3 Magnesium Depletion

4. Chloride Depletion (Loop Diuretics)

5 Systemic Acidosis (Carbonic Anhydrase Inhibitors)

6 Potassium Retention (Aldosterone Antagonists)

Because these inhibitors are both potent and non-selective they can produce side effects by disturbing physiologic regulation. For example, except for the aldosterone antagonists, they all block reabsorption of potassium as well as sodium, which is why potassium supplements may be needed with these drugs.

These side effects can be very important and the use of diuretics demands knowledge of their risks as well as benefits.

Some Specific Problems

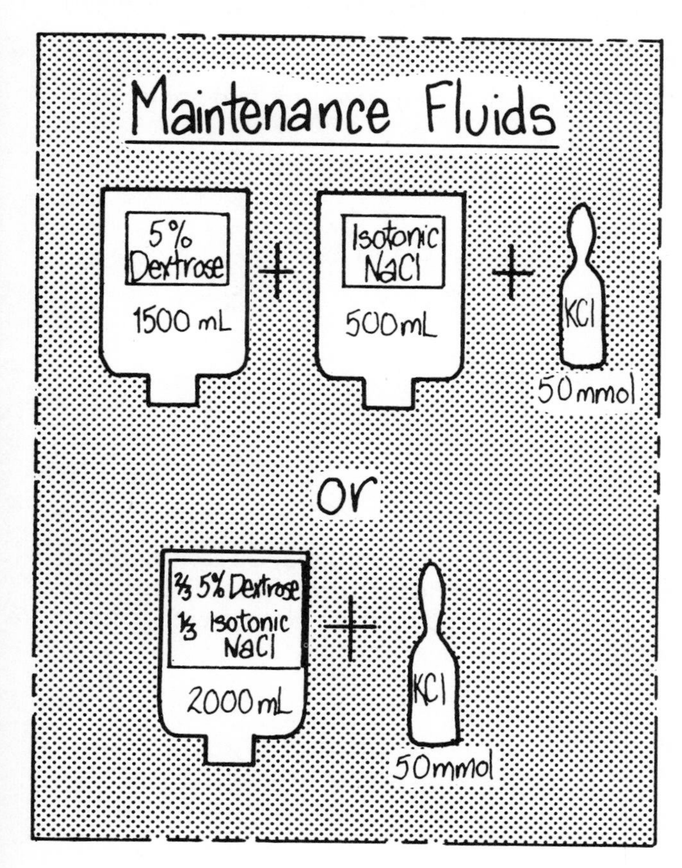

"Maintenance" fluid therapy for an average normal adult, with normal renal function, who cannot take orally for a few days can be based on the following general requirements.

Water	2000 millilitres per day
Sodium	75-100 millimoles per day
Potassium	50 millimoles per day

This can be provided by giving
1) 1500 millilitres of 5% dextrose and 500 millilitres of isotonic sodium chloride with added potassium
or
2) 2000 millilitres of "2/3 5% dextrose/1/3 isotonic sodium chloride" with added potassium.

With healthy kidneys the limits of error are wide, but with coincident kidney disease this regime must be modified.

Additional calories can be added by using 10% dextrose, but remember that "maintenance fluids" for a few days does not equate with "total parenteral nutrition".

E.C.F. Loss

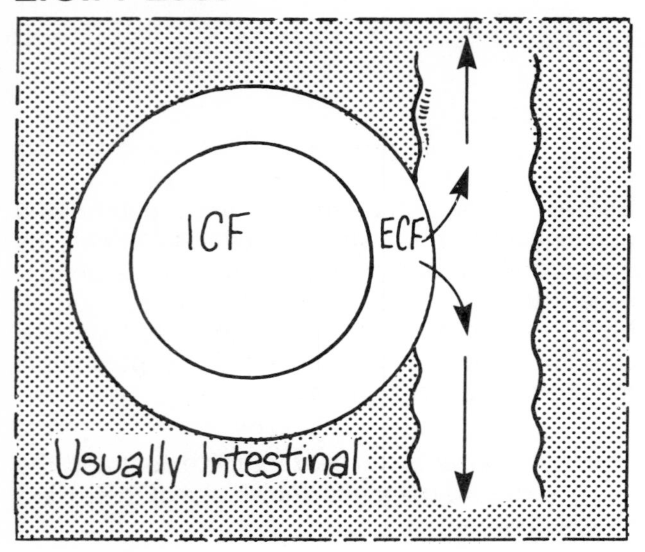

The common site for E.C.F. fluid losses is from the intestine. Such losses should be replaced by isotonic sodium chloride as a first step. Appropriate additions to this fluid will be dictated by the site of loss. With the exception of colonic diarrheal fluid, all intestinal losses are effectively isotonic with the E.C.F.

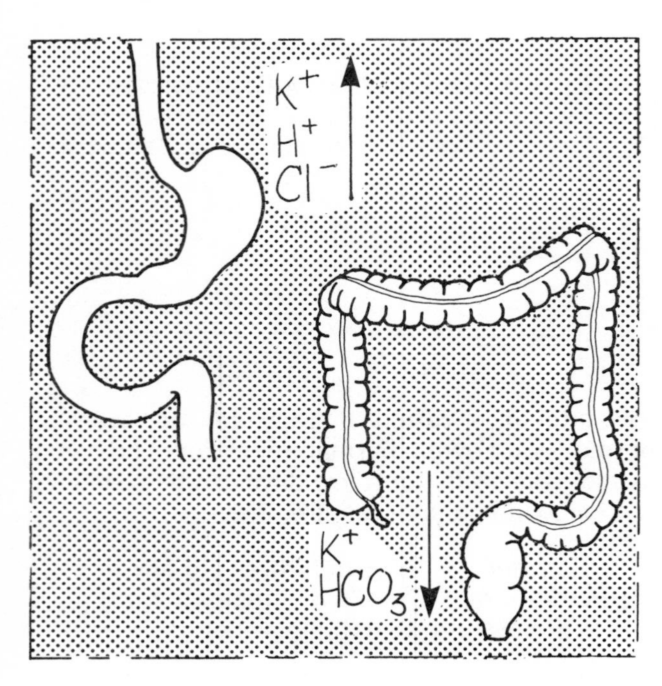

Vomiting will be associated with hypokalemia, hypochloremia and alkalosis as well as volume depletion.

Severe diarrhea will be associated with hypokalemia and sometimes hyperchloremic acidosis as well as volume depletion.

Colonic diarrhea will lead to hypotonic fluid losses.

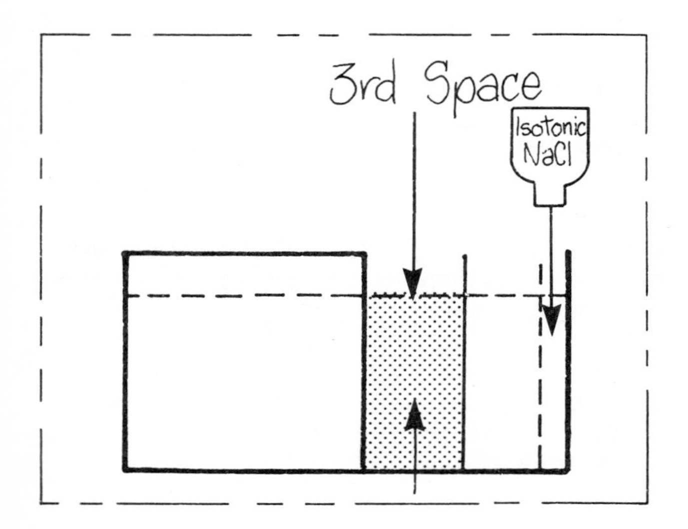

In general, a ''third space'' may be considered as being in communication with the E.C.F. and replacement therapy should also be with isotonic sodium chloride as a first step.

Burns

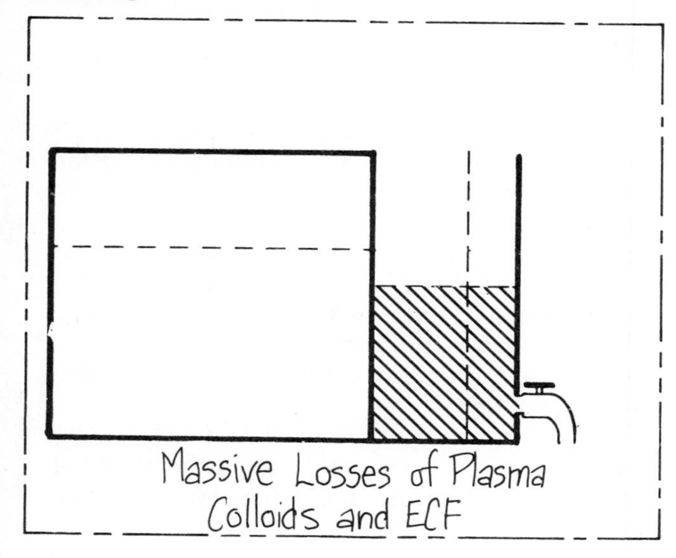

Burns are a special problem and involve massive losses of plasma colloids as well as E.C.F. Fluid replacement in such patients is the key to initial survival and the amounts of fluid required may be very large. Refer to detailed texts for methods of calculating fluid needs.

Bicarbonate Therapy

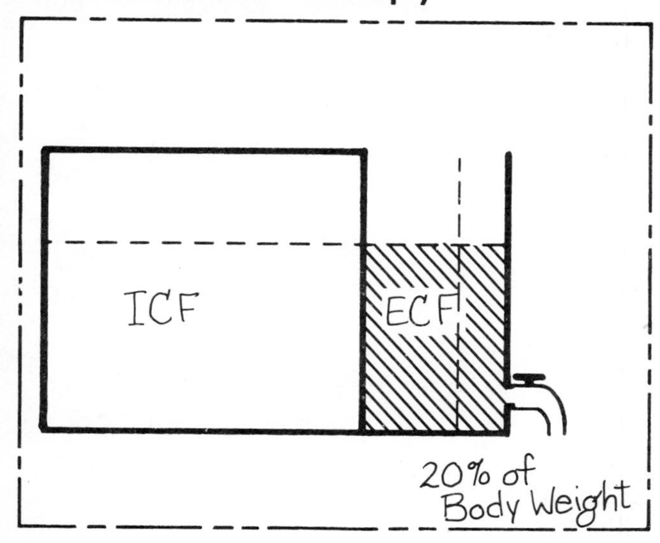

Sometimes, in a severe metabolic acidosis, bicarbonate is infused intravenously to correct extracellular pH. Deciding how much bicarbonate to give is an example of some of the problems inherent in replacement of one specific E.C.F. constituent.

Knowing that the total E.C.F. volume in an adult is 20% of body weight and knowing that the "normal" bicarbonate concentration in the E.C.F. is 25 millimoles per litre it is obviously a simple matter to calculate the deficit of bicarbonate in the E.C.F. in this way:

Deficit = (25 – measured plasma bicarbonate) x 20% of body weight

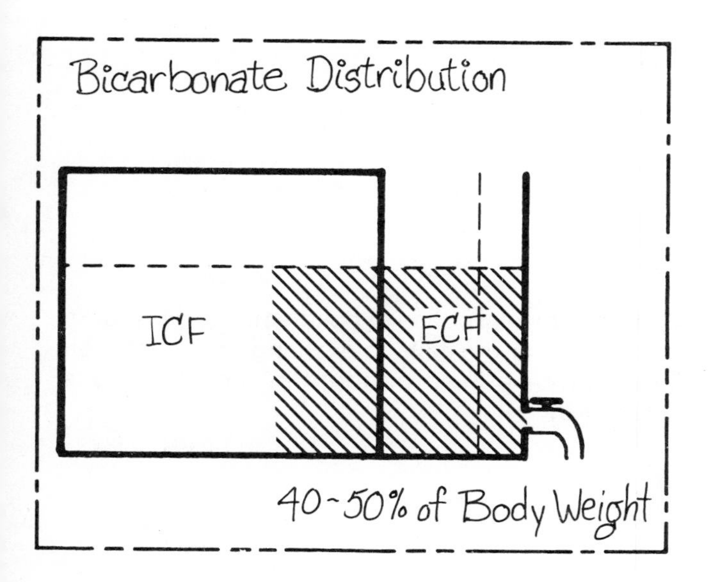

Unfortunately it is not that simple because

1) The true volume of distribution for bicarbonate exceeds the E.C.F. volume and probably approximates two-thirds of total body water (i.e. 40-50% of body weight) or even more

2) Rapid and complete replacement of a deficit can produce dangerous side effects. (See below)

An Example

Measured Plasma [HCO_3^-]	11 mmol/L
"Normal" Plasma [HCO_3^-]	25 mmol/L
Deficit	25 − 11 = 14 mmol/L
Half of Deficit	14 ÷ 2 = 7 mmol/L

Body Wt.	75 kg
ECF Volume	20% of 75 = 15 L

Amount of HCO_3^- Given Will be 15 X 7 = 105 mmol

So — it is usually considered practical wisdom to attempt to correct the measured E.C.F. bicarbonate concentration only half way to its "normal" value of 25 millimoles per litre and to assume a volume of distribution that approximates the E.C.F. volume. This is a *first step* in replacement and ensures that it will deliberately fall well short of complete correction. The amount given probably will be no more than half the actual deficit.

Having given this amount of bicarbonate a reassessment of clinical and laboratory data will determine the next step.

Severe degrees of metabolic acidosis may demand very large amounts of bicarbonate, but overenthusiastic treatment can have risks.

Risks of Bicarbonate Therapy

1) Too much, too quickly, may produce tetany or convulsions by lowering ionized calcium concentration. (See Chapter 10)

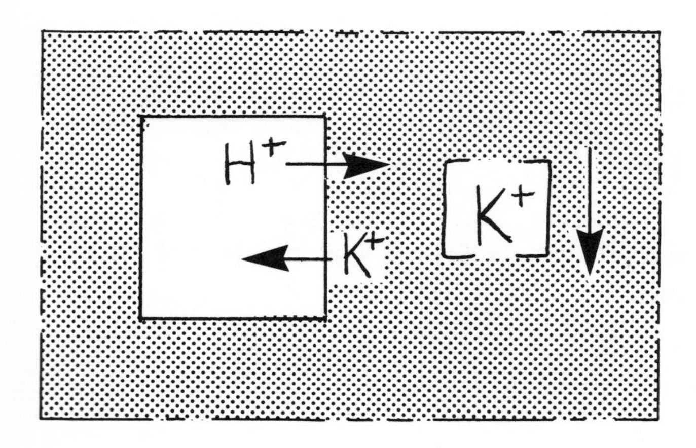

2) It may also produce hypokalemia as potassium ions return to the cells in exchange for hydrogen ions that have been buffered in the I.C.F. (See Chapter 9)

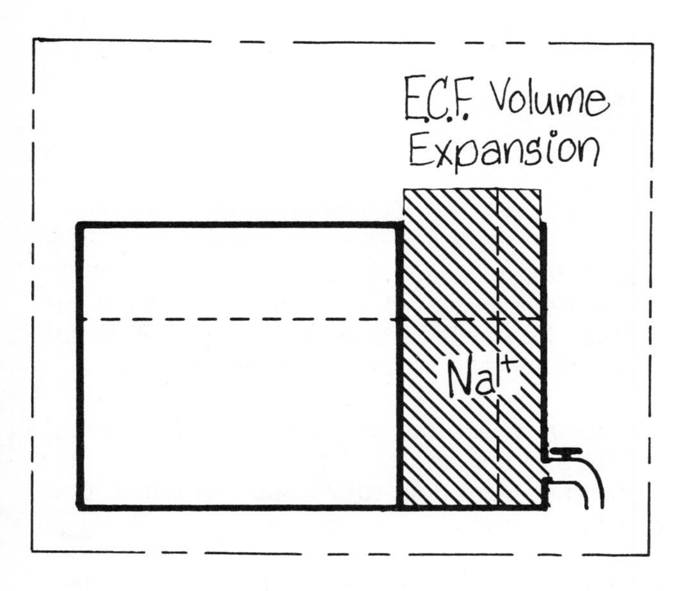

3) The bicarbonate must be given as a sodium salt and the sodium may lead to over-expansion of the E.C.F. volume. (See Chapter 6)

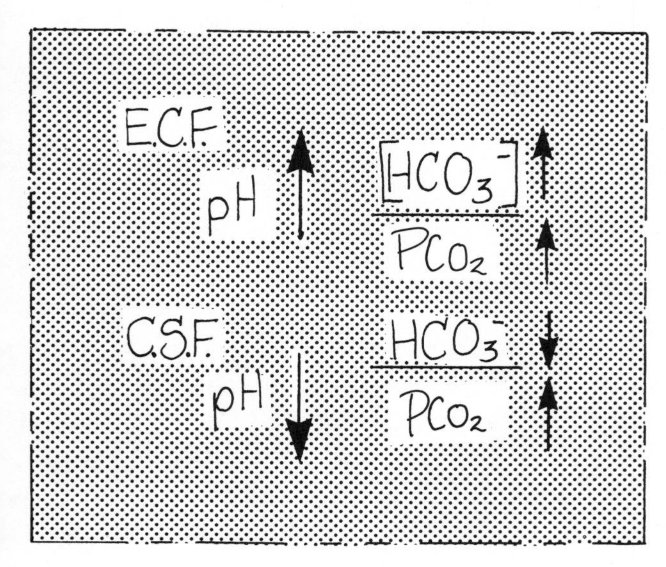

4) As bicarbonate concentrations rise, so the pH rises and the Pco_2 will also rise. But the Pco_2 rapidly equilibrates with the cerebrospinal fluid (C.S.F.) whilst the bicarbonate only *slowly* crosses the barrier between the E.C.F. in general and the C.S.F. Thus in the C.S.F. the Pco_2 rises but the bicarbonate remains relatively constant - as a result a rising pH in the E.C.F. may occur but a further *fall* in pH will occur in the C.S.F. leading to major disturbances of brain function such as convulsions.

Treating Hyperkalemia

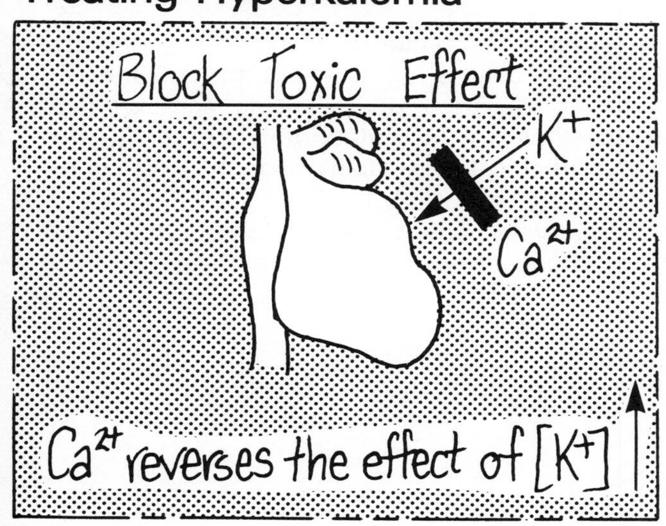

Because of the toxic effect of potassium on the heart, hyperkalemia may be an acute medical emergency. (See Chapter 9)
Management of hyperkalemia may involve three approaches; the choice of which to use depends upon the urgency of the clinical situation.

1) Block the toxic effect of potassium on the heart using the antagonistic effect of calcium. This does not alter the plasma potassium concentration but reverses (temporarily) its toxic effect on the heart. Intravenous calcium gluconate, or in an acute emergency calcium chloride, may be used.

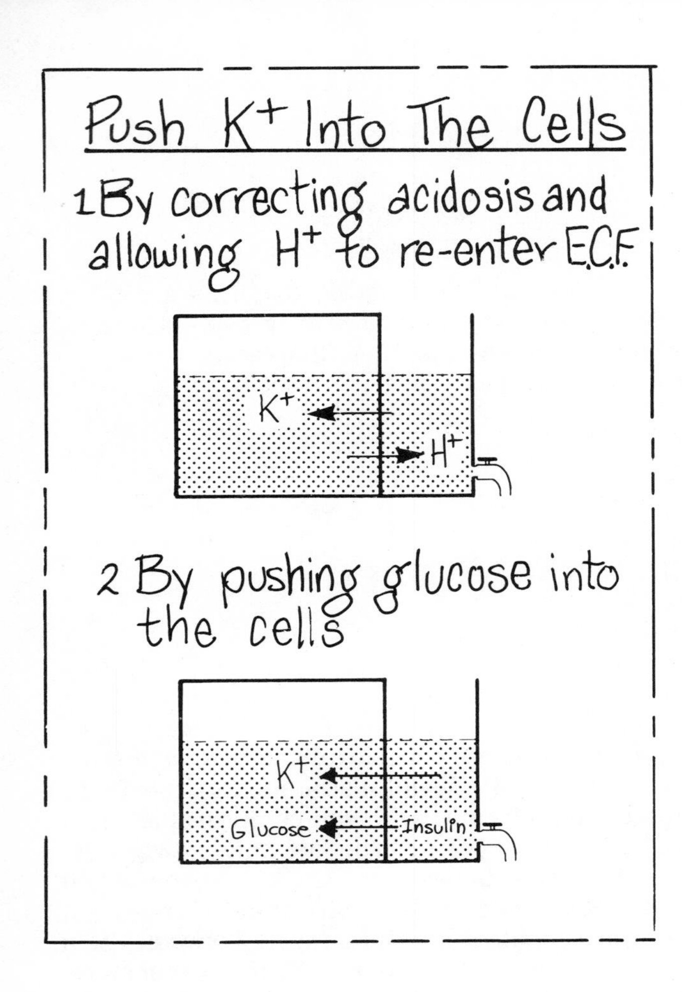

2) Push the potassium back into the I.C.F. This can be done by rapid correction of coincident acidosis, in which case hydrogen ions begin to leave the cells and potassium re-enters them.

The same effect can be achieved by infusing glucose and insulin. Insulin pushes glucose into the cell and results in co-transport of potassium with the glucose.

In diabetic ketoacidosis potassium will re-enter the cells rapidly as insulin is administered, an example of the same mechanism linking potassium and glucose transport.

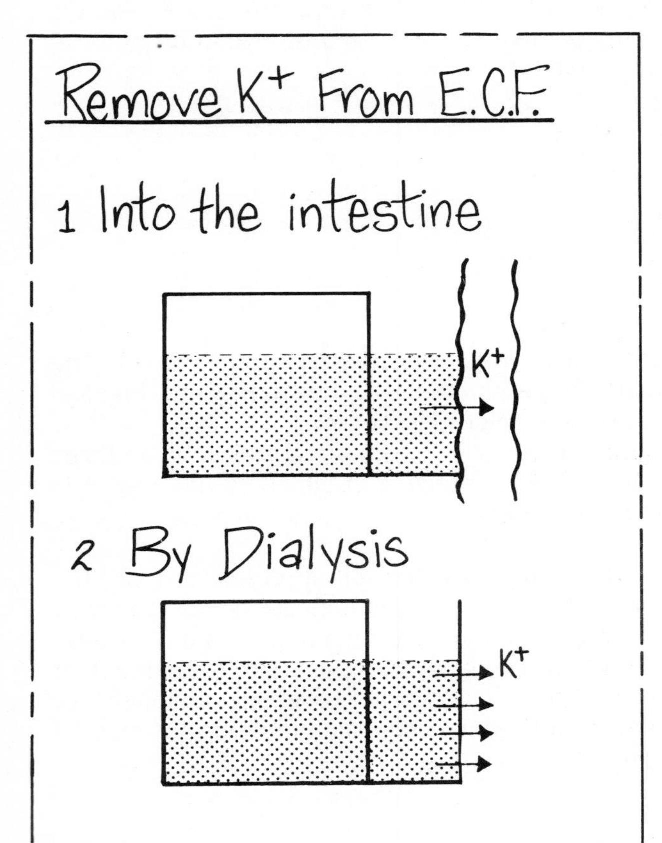

3) Remove potassium from the E.C.F. directly. Potassium can be removed into the intestine by inducing diarrhea or by using ion-exchange resins that will fix potassium in exchange for some other cation (usually sodium or calcium).

Alternatively artificial dialysis can be used either by peritoneal lavage or the use of an ''artificial kidney''.

Treating Hypokalemia

Hypokalemia is seldom an *acute,* life-threatening situation and is often associated with a shift of hydrogen ions as well as actual potassium loss. (See Chapter 9)

Treatment of hypokalemia includes
1) Correcting the cause of potassium loss.
2) Correcting the cause of alkalosis.
and then
3) Prescribing potassium supplements whenever possible by the oral route, since hypokalemia is seldom an emergency.

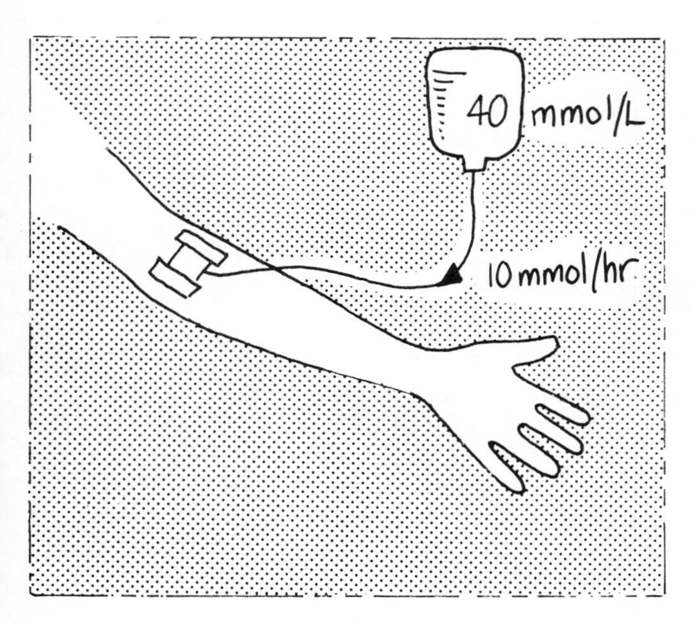

Intravenous Potassium - A Note of Caution
Wherever possible, the lower the concentration and the slower the rate of infusion, the safer becomes intravenous potassium.
Try to avoid concentrations above 40 millimoles per litre and infusion rates above 10 millimoles per hour.
Never exceed concentrations of 60 millimoles per litre and infusion rates above 40 millimoles per hour. These high levels require careful monitoring and experience, and should not be given for more than 2 or 3 hours.

References

General Reading

Brenner B.M. and Rector F.C., (1976). Eds. ''The Kidney'' W.B. Saunders & Company, New York.

Deetjen P., Boylan J.W. and Kramer K., (1975) ''Physiology of the Kidney and Body Water''. Springer Verlag, New York.

Ganong W.F., (1979) ''Review of Medical Physiology''. Lange Medical Publications, Los Altos, California.

Maxwell M.H. and Kleeman C.R., (1972). ''Clinical Disorders of Fluid and Electrolyte Metabolism''. McGraw Hill Book Company, New York.

Pitts R.F., (1968) ''Physiology of the Kidney and Body Fluids''. Year Book Publishers, Chicago.

''The S.I. for the Health Professions''. (1977). World Health Organisation, Geneva.

Schrier R.W., (1976) Ed. ''Renal and Electrolyte Disorders''. Little Brown and Company, Boston.

Smith H.W., (1959) ''From Fish to Philosopher''. Ciba Edition, with permission of Little Brown and Company, Boston.

Vander A.J., (1975). ''Renal Physiology''. McGraw Hill Book Company, New York.

Welt L.G., (1959) ''Clinical Disorders of Hydration and Acid Base Equilibrium.'' Little Brown and Company, Boston.

Sodium and Water

Brenner B.M. and Stein, J.H. (1978). Eds. ''Contemporary Issues in Nephrology; Sodium and Water Homeostasis''. Churchill-Livingstone, New York.

Cannon P.J., (1977). ''The Kidney in Heart Failure''. New England Journal of Medicine. 296, 26-32.

Kokko J.P. (1977). ''The Role of the Renal Concentration Mechanisms in the Regulation of Serum Sodium Concentration''. American Journal of Medicine. 62, 165-169.

Hydrogen and Potassium Ions

Emmett M. and Narins, R.G., (1977). ''Clinical Use of the Anion Gap''. Medicine 56, 38-54.

Kassirer J.P. (1974). ''Serious Acid-Base Disorders''. New England Journal of Medicine 291, 773-776.

Schwartz W.B. and Cohen, J.J. (1978). ''The Nature of the Renal Response to Chronic Disorders of Acid-Base Equilibrium''. American Journal of Medicine, 64, 417-428.

Tannen R.L., (1977), Ed. ''Symposium on Potassium Homeostasis''. Kidney International, 11, 389-515.

Calcium and Phosphate

Haussler M.R., and McCain B.S., (1977). ''Basic and Clinical Concepts Related to Vitamin D. Metabolism and Action'' New England Journal of Medicine 297, 974-983 and 1041-1050.

Lemann J., Adams N.D. and Gray R.W., (1979). ''Urinary Calcium Excretion in Human Beings''. New England Journal of Medicine, 301, 535-541.

Nordin B.E.C., (1976) Ed. ''Calcium, Phosphate and Magnesium Metabolism''. Churchill Livingstone. New York.